The Business of Poverty in Africa

This timely and thought-provoking book critically explores key theories, concepts and contemporary issues associated with the business of poverty through the lens of the travel philanthropy phenomenon and within the debates of sustainable development in Africa. Since the Band Aid era in 1984, and alongside the rise of the international aid and NGO sector, global travel and tourism has considerably expanded into the African continent and with it, face-to-face philanthropy too.

By drawing on the authors' extensive experience in the field of sustainable development and the African continent, and by using empirical evidence from case studies in Uganda, Ethiopia, and Zimbabwe, a systems approach is employed to explain the complexity of travel philanthropy as a growing exchange economy within tourism in Africa and the importance of considering the influence of multi-stakeholder perspectives. Through an interdisciplinary lens framed by a critical realism philosophy, this book provides a thorough investigation and new understanding of travel philanthropy as a system, challenging myopic views of gift-giving and gift-receiving practices in Africa.

Key system outputs reveal emerging moral dilemmas within gift-giving, leading to a call for adequate policy and practice interventions for the advancement of sustainable development. Incorporating observable and unobservable phenomena as matters of causality, a new conceptual model is presented, which redefines how impacts of travel philanthropy may be conceptualised and approached. By doing so the analysis offers significant insight into the real-world complexity of gift-giving and aid interventions through tourism, and how it differentiates from other forms of aid. The research presented in this book provides a solid basis to inform strategies for more effective and ethical pathways to gift-giving/receiving engagements, responsible tourism business management practices in Africa, and contributes to wider debates concerning international aid and donors' intervention implications in African contexts.

This book is a valuable resource for the tourism industry, policy makers, development professionals, students, researchers, academics in tourism, geography, business and management, economics, international development studies, anthropology, sociology, area studies, and applied ethics.

Amy Scarth (PhD, MA, BA Hons) is an applied social scientist and independent consultant specialising in socio-economic development impact and the sustainability of biodiversity in sub-Saharan Africa. Her expertise lies understanding real-world complex systems that private, non-profit and public sectors are embedded, the intersection of humans, nature and politics in Africa's protected areas, as well as global and local tourism strategies. Combining an advanced academic foundation with extensive practitioner experience, Dr Scarth also has a background in multistakeholder engagement amongst last-mile populations in fragile contexts of the world. Leadership roles across private industry, NGO, social enterprise, academia, and consultancy, have enabled thought leadership in diverse fields.

Marina Novelli (PhD, PGDip Panning, PgDipEd, BSc Hons Economics) is Director of the University of Nottingham's Sustainable Travel and Tourism Advanced Research Centre and Professor of Marketing and Tourism at the University of Nottingham Business School, UK. She is a globally renowned authority in the field of tourism for sustainable development, with an extensive track record of leading impactful research, consultancy and advisory services for international organisations such as the World Bank, the EU, various UN institutions, the Millennium Challenge Corporation, the Commonwealth Secretariat, the British Council and National Ministries. She is an Africa specialist having travelled and worked in 27 countries and over 50 destinations in the continent. She has publishing extensively on issues related to the continent and amongst her many recognitions, she is recipient of the 2022 Africa Impact Champion, the 2018 Africa Tourism Leadership Award and the 2018 Top 50 Awesome Scholars in Tourism.

Contemporary Geographies of Leisure, Tourism and Mobility
Series Editor: C. Michael Hall, *Professor at the Department of Management, College of Business and Economics, University of Canterbury, Christchurch, New Zealand*

The aim of this series is to explore and communicate the intersections and relationships between leisure, tourism and human mobility within the social sciences.

It will incorporate both traditional and new perspectives on leisure and tourism from contemporary geography, e.g. notions of identity, representation and culture, while also providing for perspectives from cognate areas such as anthropology, cultural studies, gastronomy and food studies, marketing, policy studies and political economy, regional and urban planning, and sociology, within the development of an integrated field of leisure and tourism studies.

Also, increasingly, tourism and leisure are regarded as steps in a continuum of human mobility. Inclusion of mobility in the series offers the prospect to examine the relationship between tourism and migration, the sojourner, educational travel, and second home and retirement travel phenomena.

The series comprises two strands:

Contemporary Geographies of Leisure, Tourism and Mobility aims to address the needs of students and academics, and the titles will be published simultaneously in hardback and paperback.

Routledge Studies in Contemporary Geographies of Leisure, Tourism and Mobility is a forum for innovative new research intended for research students and academics, and the titles will initially be available in hardback only. Titles include:

Travel Disruptions
Impacts, Responses and Resilience
Edited by Emmet McLoughlin, Domhnall Melly and James Hanrahan

The Business of Poverty in Africa
Inside the Travel Philanthropy System
Amy Scarth and Marina Novelli

For more information about this series, please visit: www.routledge.com/Contemporary-Geographies-of-Leisure-Tourism-and-Mobility/book-series/SE0522

Contemporary Geographies of [illegible] Tourism and Mobility

Series Editor: C. Michael Hall, [illegible] the Department of Management, College of Business and Economics, University of Canterbury, Christchurch, New Zealand

[illegible]

Place Disruptions [illegible]

[illegible]

The [illegible] of Poverty in [illegible]

[illegible]

For more information about this series, please visit: [illegible]

The Business of Poverty in Africa

Inside the Travel Philanthropy System

Amy Scarth and Marina Novelli

Routledge
Taylor & Francis Group
LONDON AND NEW YORK

First published 2025
by Routledge
4 Park Square, Milton Park, Abingdon, Oxon OX14 4RN

and by Routledge
605 Third Avenue, New York, NY 10158

Routledge is an imprint of the Taylor & Francis Group, an informa business

British Library Cataloguing-in-Publication Data
A catalogue record for this book is available from the British Library

ISBN: 978-1-032-80650-1 (hbk)
ISBN: 978-1-032-80647-1 (pbk)
ISBN: 978-1-003-49792-9 (ebk)

DOI: 10.4324/9781003497929

Typeset in Times New Roman
by Newgen Publishing UK

Contents

Figures

Tables

Foreword

By Judy Kepher-Gona (Sustainable Travel and Tourism Africa)

The sustainable tourism movement is dynamic. This dynamism is characterised by constant emergence of added terms, often led by academia and other ecosystem players, but rarely by the business community. The terms serve to capture trends and box them into manageable units. Travel philanthropy is one of a plethora of terms in sustainable tourism, which remains in turmoil. Policy makers, tourism ecosystem players, media, consultants, academics, and tourism operators use the term interchangeably with terms like charity, CSR, CSI, donations, and fundraising. The turmoil of term notwithstanding, travellers' philanthropy remains central to sustainability in tourism.

Over the last 20 years, working in conservation, tourism, and community development landscape in Kenya, I have interacted with this term in different fronts and encountered more than a dozen approaches to its application in sustainable tourism, with varied impact. My greatest lessons come from directly managing traveller's philanthropy, and assessing sustainability in tourism for local, regional, and global sustainability labels. I have learnt that there is structured and unstructured giving by travellers. Structured giving has promoted accountability and has potential to be effective, whereas unstructured giving is open to abuse by tourism businesses and individual beneficiaries. I also learnt that the purpose for which travellers give, the reasons for which they give, and expectations vary significantly. Often there are discrepancies between aspirations of those giving and needs of place. But this has not hindered giving since tourism businesses, responsible for midwifing the giving process, have flexibilities and can align gifts to meet expectation of a philanthropist. Sometimes they have influenced the choice of issues and beneficiary of gift.

Unlike other concepts in sustainable tourism that have been studied extensively and remain on the radar of professionals, travel philanthropy has escaped public scrutiny giving leeway to tourism operators to implement it in diverse ways. This freedom has been both a boon and a bane. In many instance tourism businesses have applied the concept as a fundraiser to encourage travellers to give for their 'do good' programmes. This corporate begging is deeply rooted in tourism and continues to evolve in form and approach. In one approach, businesses notify travellers of opportunities before travel and ask if they wish to pay a little more and contribute to a good course, which they can visit during their travel. In other approaches, businesses expose travellers to areas of need during their visit. Visitors are then offered banded options to which they can contribute. Other times guests have been asked to skip certain activities and the cost would go towards their contribution to a good course of their choice. Whereas all these approaches lead to giving by guests, the question arises as to whether this is fundraising or travellers' philanthropy. Is there a distinction between the two? And most importantly what is the impact on the ground?

My perception of travel philanthropy has been associated with voluntary acts of kindness by travellers in destinations, where trigger of giving differs significantly from individual to individual. Early in my career, I listened to the author of the book 'The Power of Half' at a traveller's philanthropy conference recount his family's philanthropic journey. Their story about awakening of an inner consciousness to share triggered by an encounter with a different reality. Their story illuminates the intuitive nature of philanthropy. Following their decision to 'halve their lives', they looked for good courses to donate to in Africa. In that same conference, the intrinsic desire in all humans to 'do good' and identify with those in need was discussed. This desire does not always manifest itself for all, but given the right conditions, it is likely that all humans will be philanthropic. If giving is guided, is that giving intrinsic?

Perceived from geopolitical perspective, travellers' philanthropy can be described as an acknowledgement of inequalities in opportunities and privileges between the Global North and Global South. This explains the occasional reference to travel philanthropy as voluntary transfer of wealth from the Global North to the Global South without barriers related to foreign policy. This discussion is of less interest to tourism businesses, but extremely important.

Beyond interpretation and application, another issue around travel philanthropy is the impact created through travellers' acts of giving. Being more a value driven giving as opposed to need driven, concerns have emerged around misplaced priorities that meet 'giver' expectations. A look at areas where travellers' giving in Africa is channelled shows that clean water, education, skills development in hospitality and tourism, healthcare, clean energy, and wildlife conservation lead. Coincidentally these are areas of conventional aid in Africa. In terms of nature of gifts, they vary from basics gifts like pencils to financing whole infrastructure investments, research, and long-term efforts to saving species, as well as securing land for wildlife conservation. Unique areas include human rights, microfinance, arts and culture, climate action, and public infrastructure. When it comes to longevity and frequency of giving, there are huge variations with many being one-off gifts. There are other ethical issues surrounding emerging approaches by tourism businesses and the role of tourism businesses in midwifing traveller giving.

How well are businesses prepared to receive philanthropic gifts? To what extent should businesses that receive gifts from guests be accountable to beneficiaries? Can travellers discern if there are right conditions for giving? Should tourism establishments actively invite travellers to give? Should travellers have expectations after giving? These are only few of the questions that one may need to reflect upon.

Travel philanthropy presents a unique opportunity for travellers to remain connected with the places they visit. Travellers automatically become ambassadors of these places which are useful for building destination image and even countering negative publicity. During Covid-19 lockdown, many tourism businesses in Africa, with travel philanthropy ambassadors, raised funds to support conservation and provide relief support to host communities. There are also threats inherent in this concept related to abuse by intermediaries. The areas of divergence in the practice of travel philanthropy need attention by all stakeholders to enhance understanding, practice, and management, in ways that add value to people of place, fulfil expectations of travellers, and contribute to elimination of prejudices in travel practices.

By delving into the complexity of the travel philanthropy phenomenon, this book provides a set of empirically grounded evidence responding to the need for critical reflections on its sustainability as a fast-evolving phenomenon, also associated with the growth of tourism in many parts of the African continent.

Acknowledgements

Our deepest thank you goes to each research participant from Uganda, Ethiopia, Zimbabwe, and many others from across the African continent that generously shared their time, experiences, knowledge, and perspectives. Their insights and wisdom have inspired new understandings of travel philanthropy and aid in Africa, paving the way for more critical understandings of what may (or not) constitute sustainable development. This book is the result of a journey of new discoveries, which have undoubtedly deepened our interest in continuing with our commitment to better understand this subject and facilitated new pathways of knowledge sharing that would not have been possible without their candid contributions.

We are very grateful to the Economic and Social Research Council (ESRC), the South Coast Doctoral Training Partnership, and the Wilderness Wildlife Trust for their crucial support in funding the PhD research that formed the foundation for this book. Their investment in research training and finance assistance made this work possible. Our sincere thank you goes also to Dr. Tim Laing and Dr. Vicky Johnson for the intellectual direction provided in earlier phases of the research that led to the work presented here.

Lastly, we are thankful to our friends and family for their unwavering support during the highs and lows of this challenging mission to unpack and explain something so complex, and the inspiration to embark on the next journey. We hope this work will inspire fresh perspectives and further research into the new concepts that have emerged.

1 Introduction

Despite Africa's numerous innovations and indispensable role in global supply chains – such as enabling the Global North's green transition and tech sector, or supporting the growth of worldwide plant-based diets – 'poverty' and narratives of need arguably remain one of Africa's largest exports, and increasingly a low-hanging fruit for setting up small businesses as well as philanthropic interventions. On the one hand, many have noted the significant role of the media in the 1980's in shaping the individual charitable gift-giving and philanthropy trends we see today and impacts of poverty-related narratives on Africa. Since the Band Aid era in 1984, global travel and tourism has considerably expanded into the African continent and with it, face-to-face philanthropy too. On the other hand, alongside the growth of a sizeable international aid sector in Africa, manifesting in various forms, face-to-face donation-receiving also expanded.

Whilst various forms of philanthropy have been practised for generations, it has become apparent that Global North individuals are increasingly shifting their compassion for *distant* suffering, towards alleviating it directly and achieving physical and personal encounters with beneficiary 'others'. Donors increasingly bypass armchair giving to make direct donations to their chosen beneficiaries, commonly legitimised by emotion – and often directly to young children, whether during their travels or upon returning home. This transactional relationship not only perpetuates the colonial gaze as widely discussed in the media right now, reinforcing narratives of Africa's woes and attracting possibly the wrong type of financial 'investment', but also invites a proper discussion about the relevance of the reverse gaze – from local communities in Africa towards the foreign donor, and this book addresses this.

The convergence of public and private interests, priorities, agendas, and beliefs about the role of tourism in sustainable development is in part reflected in the mounting number of initiatives, projects, activities, media articles, and academic publications on 'doing good' and/or 'being good' while travelling. When it comes to international travel and tourism to Global South destinations and Africa in particular, the commercial narrative is often interconnected with those prevailing narratives and imagery of the continent curated to drive international compassion for its woes more broadly. Individual values around the world continue to be influenced by this, driving an inclination to intervene in 'development' whether

DOI: 10.4324/9781003497929-1

through private sector organisations or as individuals, thus joining governments and non-governmental organisations (NGOs) in attempts to reduce 'poverty'.

In some destinations of the world, the growth of global travel and tourism is strongly correlated with the growth of travel philanthropy, a phenomenon worthy of an increasing attention to the subject and the specific focus of this book. The mobility of individuals worldwide, combined with the immersion of businesses into physical places or communities, can enable unique in-person or direct 'intervention' contexts, giving rise to specific forms of exchange relationships. Such trends, characterised by a lack of critical debate surrounding them, encouraged us to take a step back and unpack the concept of 'travel philanthropy' from different perspectives. We have thought more deeply about what this phenomenon comprises and presented an approach to better grasp the notion of impact within this niche form of 'tourism for development' and private philanthropy. Our investigation revealed some of the tensions between practices of 'doing good', 'creating positive change' and 'sustainable development', and the complexity of the system that characterises this.

The evolution of global tourism products are considered instrumental within this shift, particularly in Africa where experiences often incorporate 'the community' and corporate social responsibility (CSR)/Environmental, Social and Governance (ESG) blends with the core travel product, providing fertile grounds for travel philanthropy as a relatively new player within private philanthropy and the complex arena of international development, sustainability, and ethics. Beyond the success stories and admirable intentions, this book highlights the counterintuitive role of travel philanthropy in objectifying poverty and as an informal exchange economy system from which a form of 'business' for different stakeholders has emerged – hence the title of this book, and where traditional development paradigms, power relations, perspectives, models of giving and receiving, and the concept of impact, need to be revisited. This fundamental aspect of travel philanthropy has not been sufficiently addressed critically and empirically to date, whilst the system continues to produce outputs which demand attention.

Travel philanthropy is a social act framed by personal and political will and often ill-informed historical myths and misconceptions as well as nuanced agreements that allow, regulate, and control how people act while travelling across the globe for work and leisure. Inspired philosophically by critical realism and informed by systems theory to explain matters of causality, here the concept of travel philanthropy is challenged through the development of a conceptual model of causality of observable and unobservable phenomena.

Travel philanthropy is identified as an exchange economy, and by looking at the bigger picture of this fragmented aid landscape, beyond isolated interventions, it is interpreted as a collective impact generating system applicable across case studies. The research confronts the notion of passive aid recipients in Africa and presents the complexity of human behaviours and their causes/effects producing the system. It highlights for instance the significance of social/racial objectification amongst stakeholders, raising ethical dilemmas, as well as efficacy concerns and weaknesses in evaluation across this niche form of philanthropy, that is located within the development aid field. Here, the influence and complexity of giving–receiving pathways and the paradox of the 'free gift' in Africa are analysed.

Limited knowledge about the real complexity of gift-giving and aid interventions through travel and tourism, as well as its differentiation from other forms of aid, is addressed. The major void in understandings about travel philanthropy's meaning, significance, and impact, from an economic, social, cultural, political, and ethical point of view and, most importantly, from the perspective of intended 'beneficiaries', required deeper investigations. This book therefore presents transferable findings emerging from our research investigating what is occurring, how the travel philanthropy system functions, and why.

Building on our 40+ years of combined experiences including extensive periods of time spent on empirically grounded research, advisory work, and travel in the African continent, this book is intended to fill a gap in the literature by advancing a critical understanding of the business of poverty in travel philanthropy. By adopting a case study approach primarily centred around issues of corporate and traveller giving in specific communities of tourism destinations, and the local consumption of donations, this book challenges some of the myths and misconceptions associated with the travel philanthropy phenomenon more broadly.

Combining an overview of key theories, concepts, contemporary issues, and debates, and by investigating different African contexts, this book provides a new conceptual model defining travel philanthropy and its current 'impact' (see Chapter 11), with policy and practice implications for sustainable development and matters of causality. Through an in-depth exploration of 'travel philanthropy', arguably a multifaceted, ambiguous term that is theoretically weak, a broader theoretical framework is presented based on empirical evidence gathered specifically in Uganda, Ethiopia, and Zimbabwe and more broadly through critical observations emerging from extensive research and travel undertaken over a period of 20 years in other destinations in Africa, such as Botswana, Burundi, Egypt, Eswatini, Ghana, Malawi, Namibia, Nigeria, Rwanda, South Africa, Tanzania, The Gambia, Tunisia, and Zambia.

At present, while academic publication and mainstream literature in the arena of the 'history of tourism' in Africa (Cleveland 2021), 'tourism and development' in Africa (i.e., Dieke 2000; Novelli 2015; Adu-Ampong and Kimbu 2020; Novelli et al. 2021; Kepher-Gona and Atieno 2022), and 'travel philanthropy' in Africa (i.e., Honey 2011; Koot and Fletcher 2021; Novelli et al. 2015; Novelli 2015; Scarth and Novelli 2019, 2024) have grown, critical academic elaborations of these topics suffer from fragmentation of ideas and some confusion. Fragmented in the sense that the interdisciplinary nature of these topics means that it is still spread thinly through economics, geography, anthropology, business studies, development studies, and so on, often leading to a scattered discipline-centred interpretation of what development in this context is and its complexities, rather than a more holistic assessment. In the specifics of travel philanthropy as a growing phenomenon, extant literature lacks cohesive empirical evidence, at times also providing rather naïve elaborations of the subject. Only one book addresses the "rise of new benevolent trends in tourism" (Baptista, 2017), and while presenting some interesting anthropological elaborations, it remains limited to the Mozambique community tourism context. A critical perspective on the travel philanthropy phenomenon in Africa has not yet been brought together to date and our book intends to provide a cohesive elaboration of its conceptual definition, issues, and impact that are theoretically rich, empirically grounded, and transferable to other contexts.

The chapters that follow are intended to further readers' ability to question "travel and tourism for development in Africa" by casting light on the growing travel philanthropy phenomenon and the multifaceted relationships between those soliciting, giving, and receiving what is intended as a form of 'development assistance'.

Chapter 2 – The complex landscape of 'development' aid, private philanthropy and the 'gift' in Africa – sets the scene by offering a brief review of theoretical perspectives regarding the sustainable development debate and emerging approaches to 'ending' poverty. It looks at the role of the private sector and evolution of private philanthropy within the international development arena. Theoretical issues around the concept of the 'gift' in Africa are reviewed and questions concerning the good in giving introduced.

Chapter 3 – Impact evaluation and judging the 'good' in sustainable development systems – discusses the evolution of the impact evaluation debate in international development and sustainability, discussing how the sector is dealing with the complexity of poverty, today's interconnected world, and the generation of change. It draws on critical issues identified in the NGO and private sector, including tourism. The chapter highlights the growing relevance of complex systems theory in international development and the importance of interdisciplinary lenses.

Chapter 4 – Gift-giving and philanthropy through tourism: emerging theories and practices – offers a comprehensive review of extant literature, theories, and practices of gift-giving and philanthropy in tourism. The multifaceted nature of 'travel philanthropy' is discussed here and key conceptual themes identified, laying the foundation for critical knowledge gaps addressed in the subsequent chapters. The role of travel philanthropy in sustainable development is also introduced.

Chapter 5 – Redefining travel philanthropy: opening the black box – introduces a methodological framework that responds to the contemporary impact evaluation needs in development and addresses the complex nature of travel philanthropy as an open social system. By combining critical realism and systems theory, a fresh approach to developing this new conceptual model of travel philanthropy is explained, which helps to open the black boxes within an interdisciplinary system and to understand qualitative matters of causality. The chapter also introduces the empirical research methods that subsequent chapters will analyse, including what, why, and how case studies were selected in Uganda, Ethiopia, and Zimbabwe.

Chapter 6 – Actors, roles, and perspectives: shaping the system – identifies key travel philanthropy system actor groups/stakeholders (the people and organisations) and their dynamically shifting roles, compared across case studies. Here, systems perspectives are also analysed, identifying a taxonomy of five system functions or purposes, leading to the introduction of the business of poverty concept and its emerging significance.

By drawing on key interdisciplinary theories across business and management, economics, behavioural economics, social psychology, sociology, anthropology, and moral philosophy, Chapter 7 – Broadening the theoretical perspective: travel philanthropy as a social, economic and moral phenomenon – advances new understandings of travel philanthropy in relation to human behaviour. To understand the structure of travel philanthropy as a system, its underlying mechanisms and impacts, this chapter offers a helicopter view of broader issues emerging from the empirical research and frames how to zoom into social systems and human behaviour thinking.

Research conducted across Uganda, Ethiopia, and Zimbabwe led to the discovery of key underlying elements of the travel philanthropy system, with Chapter 8 – Travel philanthropy: an emergent exchange economy – introducing the elements identified across case studies which helped explain the system's emergent structure, including a common set of beliefs from which the system was found to be rooted. Additionally, a framework of principles and norms that enable exchanges in the travel philanthropy system is discussed, including for example informal rules influencing exchange relationships. Furthermore, it is argued that if core beliefs were absent, the travel philanthropy system would not exist and therefore recognising such underlying elements as an emergent exchange economy is essential to understanding the system presented in Chapters 9 and 10.

Chapter 9 – The travel philanthropy system and the business of poverty – introduces the travel philanthropy system. To simplify the complexity of this reality, a system map is presented, highlighting the key events and mechanisms that best explain the causal themes identified in our research. The analysis delves into different sections of the system map, where details are examined through a range of interdisciplinary theories. This chapter focusses on the local consumption of donations, including the business of poverty which is interconnected.

Chapter 10 – The travel philanthropy system: resource pathways – looks beyond the gift exchange, into the resource pathways and the interconnected questions about impact evaluation. This includes how resources are distributed and behavioural tendencies associated with the travel philanthropy gift. Here, we delve into how the informal nature of the travel philanthropy economy in terms of *roles* and *responsibilities*, the private nature of exchanges, and decision-making patterns, such as bounded responsibility, influence the types of giving and receiving in travel philanthropy, which produce three key resource pathways.

Chapter 11 – Redefining impact: the bigger picture, system outputs and the paradox of travel philanthropy – identifies what is produced by the system discussed in the previous chapters, the outputs. It brings together the findings from earlier Chapters 6–10 into a layered conceptual model which illustrates the underlying system elements, the system structure in the centre, and the more obvious system outputs found on the surface. Five system outputs are presented, which indicate the key impacts emerging from the travel philanthropy system in Africa. This chapter calls for further critical assessment of the intended and unintended consequences of giving–receiving exchanges in travel philanthropy, which is found to be a paradox, adding to the discourse on the 'unintended paradox of aid'.

Chapter 12 –Whose business is poverty? – summarises what we found behind scenes of travel philanthropy and by looking at the bigger picture beyond isolated interventions from the perspective of a collective impact generating system. An outline of how this advances knowledge beneficial to academia, policy, and practice is included. This chapter highlights aid recipients' considerable agency within the travel philanthropy system. In contrast to much of the international development literature and speak, where those receiving aid are generally portrayed as passive recipients of aid, our research highlighted that travel philanthropy recipients are in fact actively engaged in driving the system's existence and shaping its structure, contributing significantly (amongst other stakeholder groups) to what we call 'the business of poverty' in travel philanthropy. By reflecting on other African 'geographies of compassion', this chapter concludes that if one does not thoroughly understand the gift and decisions shaping giving–receiving pathways from multiple perspectives, any knowledge about system dynamics or meaning of impact will remain incomplete. Given the range of critical socio-economic issues linked to the business of poverty that emerged from our empirical research, this chapter further elaborates on the implications of the travel philanthropy system requiring a frame of reference beyond resource efficacy in a development context.

This book aligns with current academic and policy discourses around the UN Agenda 2030 for Sustainable Development and related Sustainable Development Goals (SDGs), informs sustainable and ethical pathways to gift-giving/receiving engagements, and accelerates responsible tourism business management practices in Africa. The research presented here also contributes to wider debates in social science concerning aid and donors' intervention implications in developing countries, and while the primary purpose of this research was not to measure the 'impact' of travel philanthropy but to understand it, it provides a much needed ethical and efficacy framework capable of capturing its complexity, paving the way for future research.

Furthermore, by providing a fresh interdisciplinary and multi-stakeholder lens on the travel philanthropy phenomenon, this book is a valuable resource for those investigating the role of tourism, CSR/ESG, or individual small-scale philanthropy in sustainable development, as well as those looking to advance impact evaluation in the social sciences or anti-poverty aid interventions more broadly. Impact understood as a system, its underlying elements and outputs, can be an invaluable approach to various forms of intervention in Africa, as local aid recipients often do not differentiate between development efforts, especially by 'outsiders'. It can be a useful term of reference for practitioners and academics as well as public, private, and third sector employees to better understand travel philanthropy in Africa specifically and in other regions of the world. In addition to students studying tourism, the book will prove useful to undergraduate and postgraduate students studying subjects such as development studies, international relations, geography, business and management, economics, anthropology, sociology, area studies, and applied ethics. Additionally, development agencies, consultants and international agencies, local government planning officials trying to understand tourism trends at a deeper level than simply marketing destinations will find this book useful as it will provide them with empirical evidence and the reflexive experience of the authors.

In conclusion, this book offers a wider perspective of theoretical frameworks and by using a case studies approach integrates theory with empirical evidence, making the narrative accessible, understandable, and relevant to both international development and tourism practitioners, academics and students of tertiary institutions willing to build upon the findings of our research and ask the difficult questions that hopefully will have been stirred through reading it.

2 The complex landscape of 'development' aid, private philanthropy and the 'gift' in Africa

This chapter sets the scene by offering a brief review of theoretical perspectives regarding the sustainable development debate and emerging approaches to 'ending' poverty. It also looks at the role of the private sector and evolution of private philanthropy within the international development arena. Theoretical issues around concepts of the 'gift' in Africa are reviewed, and questions concerning the good in giving introduced.

2.1 Development ideologies and narratives – 'the West' and 'Africa'

The international aid industry emerged after World War II, embedded partly in beliefs about improving the wellbeing and freedom of the world's poorest and most vulnerable people (Sen 1999). Historically, the African continent has especially been awash with international development funding and professionals endeavouring to create change. Its citizens have been living amongst ongoing interventions claiming needs and sustainable development goals for decades; yet in reality, many stakeholders whether local or global are in a quandary about 'international aid' and 'development'. Theories of 'development' influence people intervening in 'poverty eradication' (UNDP 2015). According to Sen (1999), for example, poverty is a matter of 'capability deprivation', the inability to acquire certain minimum capabilities rather than a matter of income, emphasising differentiation between means and ends. At the very basics, *development* is a term repeatedly criticised as a continuation of colonial discourse and imposition of Western ideologies, distinguishing between a modern *developed* world and a backward *undeveloped* world (Escobar 1995). Development policies endeavour to influence 'Africans' to think about themselves in new ways (Smith and Jeppesen 2017). *Modernisation* assumes Africa should follow in the developmental footsteps of 'the West' (Matunhu 2011). The hegemonic worldview presented in development discourse poses Western modernity as the model to be emulated in Africa, despite scores of examples where this simply fails (Francis et al. 2016).

Development theories have been critiqued for linear notions of modernisation, and causing economic vulnerability from dependency, yet continue to inform aid interventions (Ramalingam 2013). *Modernisation* theory locates the cause of

DOI: 10.4324/9781003497929-2

African underdevelopment in internal limitations associated with cultural practices, un-democratisation, and lack of industrialisation, and development is seen as an evolutionary process ending with a free-market economy under a liberal democratic political system (Rostow 1960; Francis et al. 2016). *Dependency* theory rejects some assumptions of modernisation theory, locating the cause of African underdevelopment in the way the continent is integrated within the global economic system, issues of capitalism, imperialism, neo-colonialism, and international political economy, arguing that the structure of the world economy expands wealthy countries to the detriment of the poor countries (Uche 1994). *Post-development* theory, or anti-development, whilst not rejecting change, is less of a companion to the traditional aid industry, advocating for alternative methods to bring about positive change, and more grassroots movements (Escobar 1992). *Sustainable development* is an aspirational model of growth that does not compromise future generations (Brundtland 1987), with Agenda 2030 for Sustainable Development and related UN Sustainable Development Goals (UNSDGs) posed to shape development across the world, although bearing heated questions about its effectiveness (Pinelli and Maiolini 2016). Reflecting on post-development theory, Cornwall and Brock (2005:1043) highlight this era of "one-size-fits-all development recipes spun into an apoliticised form that everyone can agree with". Although a substantial debate about the international aid industry exists, a call for an even broader and deeper critical investigation of the notion of 'development' efforts in Africa is still much needed.

The Development Economics' move away from terminologies such as *third world, developing economies*, and *less developed countries*, to the use of *the Global South* reflects the will to advance beyond paternalistic attitudes, somehow suggesting also that one is not to say Africa needs to catch up with 'developed countries' (Rudrum et al. 2016). Nevertheless, the polarisation between developed and developing contexts carries on regardless, just in different forms, with imagery replicating past colonial impressions representing people living in poverty and as objects of pity, reinforcing the problematic stereotyping of Africa (Jones 2017). Development as a concept and an industry is deeply contentious. Whilst Eurocentrism and freeing Africa from imperial narratives is not a new discussion (Collins 2017), there are now growing debates in respect to African agency in storytelling, within and outside the continent, revealing deep reverberations of history less talked about through concepts such as *coloniality* (Ndlovu-Gatsheni 2015). We are now seeing a greater shift from vocabulary (e.g., what 'development' means) to voice and viewpoint (e.g., who defines it and how viewpoints are formed).

Not to be confused with post-colonial theory, 'coloniality' concerns the enduring colonisation of power, knowledge and being, and according to Ndlovu-Gatsheni (2015), is part of the darker side of modernity and objectification of Africans. The obliteration of people's belief in their own capacities and achievements is found engulfed in a process of global dehumanisation through the 'thingification' of Africans (Ndlovu-Gatsheni 2015:490). Different from colonialism, and anticolonialism, coloniality is described as an invisible structure of power that emerged because of colonialism, producing, for example, Africans socialised into disliking the Africa that produced them, and liking Europe and the US, which often reject them. What Africa knows and likes about itself is argued to be profoundly influenced by the West. Impacts of coloniality are identified today in self-image, aspirations, psyche, language, aesthetics, and many more aspects of the social world, continually reproduced in schools, universities, churches, etc., and enabled by the international aid industry. Rodney (1972) remains one of few to strongly criticise the African education system to shape minds that feed bourgeois capitalist society rather than nurture minds meaningful for the majority of Africans. *Decoloniality*, Ndlovu-Gatsheni (2015) explains, aims to create space for ex-colonised peoples to re-tell the story for Africa and have the courage to judge Euro-American pretence, shifting the geography of knowledge by bringing identity into epistemology, towards the democratisation and de-westernisation of knowledge. This critical social theory is linked into development theory of a different nature that calls for 'post-racial pluriversality', as well as decolonisation of the mind, culture, and power (Ndlovu-Gatsheni 2015:493).

Many argue that the ways development interventions continue to be designed simply does not address the structural issues sustaining development challenges (Kalu 2017). There are also firm tones of resentment found within the current aid system criticising the dark side, rather than merely its misguided approach. Intellectual dishonesty, for example, is claimed as a 'bitter truth' in continuing to hide deliberate master plans to sustain Africa's poverty (Brenya and Adu-Gyam 2015:41). While the international aid debate is undoubtedly complex, it is clear that ideologies in African development continue to converge in both forthright and silent disputes.

Aid and the 'narratives of deservedness' claim to target 'the most vulnerable', yet rarely understand the meaning of poverty in specific contexts (Rudrum et al. 2016:1326). African people continue to be positioned as victims and subjects of development (Lahiri-Dutt 2017), with the African continent similarly depicted as a place of lack, absence, disorder, and somewhere in need (Routley 2016), bundled together under the tragic Global South crying out to be rescued (Howard and Burns 2015). The development aid narrative in Africa has become tremendously uniform, with customary buzzwords such as 'participation', 'ownership', 'empowerment', 'self-help', 'good governance', and 'sustainability' mechanised as a prerequisite for legitimacy, but in the opinion of some just words concealing donor power (Kamstra and Schulpen 2014). Scott-Villiers (2011:772) highlights 'invisible power' of knowledge and legitimacy hierarchies, and 'visible power' of organisational budgets. She refers to a 'theatre of political participation' within power struggles with elaborated rituals involving beneficiaries, brokers and benefactors, and strategic language used by local recipients of aid, for instance calling themselves 'oppressed people' or 'environmental custodians' to win external support. An emergent 'culture of aid' in Africa has been permeating both foreign and local participants (McNamara 2017) and, according to Chambers (2012), development aid is stuck in a paradigmatic syndrome, with feedback from inflexibility obstructing alternative paradigms.

2.2 Revised approaches to 'ending' poverty

International development is involved in inherently 'wicked problems' such as poverty, hunger, sustainability, and a global environmental crisis (Rittel and Weber 1973), with more recent debate recognising that these cannot be solved by linear planning because problems evolve as new 'solutions' are implemented. In merit to this, the UNDP (2015) has been promoting that the sustainable design of poverty programmes must recognise that poverty is both complex and multi-dimensional. As the aid industry is well and truly dealing with complexity (Lambe et al. 2020), the fundamental shift in thinking is that the traditional approach based on the reductionist view of development, policy, or any sort of social, economic, or environmental change intended in Africa is no longer acceptable. It is argued that smarter experts, more data, larger budgets, and pure economic theory is not the answer to managing the reality of consequences inherent in complex systems (Mueller 2020), with the aid industry requiring an entirely revised approach.

Currently, Africa abounds with international and locally managed 'projects', 'programmes', or 'interventions' grappling with the complexity of poverty. Interventions are broadly defined as "activities carried out by people at a particular place and time to achieve some kind of change" (Gates 2016). Most of these are administered by the non-profit sector or non-governmental organisations (NGOs/INGO). This is because, since the end of the Cold War, bilateral and multilateral donor agencies such as the World Bank and the European Union through the 'New Policy Agenda' gave more prominence to NGOs in poverty alleviation, viewing them as more efficient and cost-effective service providers than governments for channelling increasing amounts of money, especially to poor people (Edwards 1998). This availability of official funding led to an exponential growth of NGOs located in many countries, determining fragmentation and weak central oversight of the sector, and leading to doubts about their effectiveness as vehicles for sustainable poverty alleviation (Edwards and Hulme 1998). Joining NGOs in response to poor performance of local government institutions, growing aspirations for transparency, democracy, and participatory management, as well as citizen awareness about available resources, an escalation of community-driven efforts emerged in many African countries, in the form of community-based organisations (CBOs) (Tirivanhu 2020). Drawing a diverse set of warnings and praises, this array of non-governmental actors continues to be at the helm of many development interventions in Africa, ranging from sizeable global entities to micro grassroots organisations.

NGOs have been praised for their growing professionalism of development practice and potential to make development more equitable and effective, but heavily criticised for questionable legitimacy, abuse of power, amateurism, duplication, competition over resources, sub-standard work, lack of regulation, accountability, sustainability, and co-ordination with the public sector (Cavill and Sohail 2007). In this context, blame for failures of the international aid system has partly been directed to big donors for encouraging NGOs to maintain rather than challenge the status quo, pushing them into blueprint approaches reflecting Western paradigms and homogenising them (Kamstra and Schulpen 2014), so that they all resemble each other in a system of institutional mono-cropping (Evans 2004). Interventions often fail as designers do not understand the people they are meant to support, how they think, decide, and act (Lambe et al. 2020), with interventions not reflecting the cultural, social, and psychological dimensions of the poor, by being based on mere economic theories that 'tilt in favour' of the privileged in society (Singh and Chudasama 2020). They tend to miss a trick in terms of sustainability, where success is associated with keeping supplies of solutions coming, creating sheer dependency, and providing little or no institution building (Rudrum et al. 2016). For instance, many projects involve constructing buildings, which do little to address deeper structural problems necessary for change (McEwan et al. 2017), so when programmes end, the causes of poverty remain untouched, with any permanence of positive change becoming questionable (Taylor 2014).

Whilst some interventions generate no sustainable change, others development interventions create sustained change that is *negative*. In Ethiopia, for instance, much unregulated social assistance continues to work 'like giving alms to beggars', creating lasting recipients' dependency, providing fertile grounds for citizens to make "a lucrative business out of beggary", to the point that in Amharic there is the word 'tebakinet', meaning 'entrenched anticipation of aid', a culture of winning sympathy and using deception to tap aid resources in furthering one's own interests (Hailu 2015:829–832). Such kinds of preceding aid interventions prove challenging for newcomers with alternative development strategies in contexts where the 'dole out mentality' is rife and there is a strong sense that more outsiders are coming to 'improve' things (Mathie et al. 2017:58). One must be reminded that, although aid may have provided some considerable positive results such as reducing poverty, boosting life expectancy and reducing infant mortality (Arndt et al. 2015), charity, and/or aid can be 'wounding' (Hann 2017:986), with Kamstra and Schulpen (2014) calling for greater value-free research on NGOs.

2.3 Power in the 'aid chain'

How aid is distributed, controlled, and facilitated, and who benefits from it, are contentious moral issues (Sinervo 2015). Power relationships characterising the 'aid chain' have been written about extensively (Silk 2004; Morse and McNamara 2012; McNamara and Morse 2004). Silk (2004:245), highlighting that this is often discussed within 'a critique of post-modernity'. For instance, McNamara and Morse (2004) discussed the unequal power relations imposing donor perspectives and values, with rare examples of those at the end of the aid chain having real power (Silk 2004; Friberg 2015), aid recipient communities rarely disputing the goodness of aid, and the framing of their position as undeveloped, poor, ungoverned, and uneducated (Scott-Villiers 2011), all raising moral considerations. The 'romanticisation of the local' has also been debated, argued to emerge from the naivety of foreign donors, which can hide social agendas that appear at the local sites of power (Mohan and Stokke 2000). McNamara and Morse (2004) also argue that "local does not necessarily mean good", and evidently does not mean anything unless properly understood.

Local communities are often portrayed as weak and subservient, but it has been found that recipients do not behave like 'automatons'; they negotiate, try to capture advantages, resist control, and manipulate information going back up the aid chain (World Bank 1998 in Silk 2004:235). Actors along the chain, including aid brokers at the local level, are known to pursue different interests and cause power struggles (Song et al. 2013). Other aid brokers, such as global institutions of aid and development, also need to be differentiated from those who care (Western donors), and those that translate this caring into a caring-for the other sufferer (Korf 2007). Bebbington (2005) suggests that some people in these aid chains may well be committed to social change, but views on how social change should take shape may not be compatible without a common discourse. Discourse might centre on understanding issues of local power, or the existence of small-scale socially systemic corruption, or perhaps fierce competition amongst gift-receivers (Rudrum et al. 2016).

2.4 Questioning NGOs' effectiveness

Some say that the two functions of the 'development industry' are to produce an outcome – development – and to reproduce itself, thus keeping itself in business by justifying and legitimising development interventions (Ferguson 1994). This creates a bias when considering the effectiveness of NGOs in *assuming* that altruism and non-monetary preferences solely motivate them (Burger and Owens 2008). The funding market for NGOs is highly competitive and lucrative, which affects their behaviour. Burger and Owens (2008) therefore question if dishonesty (and confidentiality) have become symptoms of the sector in that many NGOs hide information from the public either to protect themselves in meeting unrealistic donor expectations, or to cover fraudulent, inappropriate, or inefficient efforts. Government regulations are traditionally far less stringent for the non-profit sector than the for-profit sector, partly due to the weak, informal, and arbitrary power of their consumers. NGOs often have silent beneficiaries, compared to the for-profit sector receiving consumers' feedback through their choices, and if they do not retain their customers, they close shop (Burger and Owens 2008). The issue with the non-profit sector is that mechanisms to motivate such care about their 'consumers' are not there. The result of this system is that many NGOs have an incentive to pay lip service to effectiveness and community involvement, and whether they have altruistic motivations or not, beneficiaries may tolerate white lies for more interest in ticking boxes for the donor (Burger and Owens 2008). Without this need for 'consumer' feedback, some NGOs assume an unquestionable right to continue operating social interventions without considering whether it is right for the recipient community (Berry 2014).

Furthermore, a scramble to prove legitimacy to funders causes isolated rather than collective impact (Patton 2019). When isolated impact dominates the NGO sector (Kania and Kramer 2011) so that a single organisation focusses only on its own mission, we can question how a lack of shared objectives and collective impact across NGOs limits sustainable progress (Patton 2019). A similar pattern has emerged amongst international donors too, typically prescribing a resource pathway or goal from a distanced silo, disconnected from what other donors in the funding landscape have prescribed from their silo. In the US, for instance, there are said to be so many donors gifting small amounts across so many charities that Null (2011) warned of inefficient allocations. There is a lack of research paying attention to the aggregate services provided by non-state actors, which by 2014, exceeded the forex earnings of Ethiopia's largest export commodity, coffee, yet are very unstructured in how they distribute these resources and remain uncoordinated (Hailu 2015).

Intervention effectiveness tends to be highly anecdotal rather than rigorously evaluated so it remains questionable what we really know (Davis et al. 2016). In the last 20 years, a greater awareness has emerged about the fact that development interventions have intended and unintended negative and positive consequences and that efficacy of development projects cannot actually be equated with meeting pre-defined measurable objectives (Picciotto 2019). As a consequence of this, calls for the international aid industry to "become less arrogant, more willing to learn from accidents, from failures, and from other people" (Green 2016:26) have emerged. Equally, aid projects can be instruments of 'development' if they are adaptable, managed in ways that broaden achievement goals, and if the purpose of monitoring and evaluation (M&E) mechanisms are for learning and accountability, rather than competition for funds to maintain their existence (Picciotto 2019).

What makes the landscape of aid and development in Africa more complex is the fact that it seems to remain predominantly guided by isolated and anecdotal knowledge, which adds to the complexity of what 'development' really means.

2.5 The role of the private sector in development

2.5.1 The evolution of private philanthropy in the public sphere

The private sector has gained agency in the international development arena in a number of ways, bringing even more complexity to debates surrounding sustainable development, philanthropy, the political economy of aid, and the ethics of gift-giving particularly in Africa. The evolution of private philanthropy and charity in low-income countries has been characterised by the growth of substantial private wealth shifted into independent initiatives detached from public sector approval and entering political grounds, such as the *Gates Foundation* (Partzsch 2017; Strachwitz 2015). The likes of *Band Aid and Save the Children* have also grown but found to be less prepared to tackle the problems' root causes and far more interested in focussing on fundraising efforts through representing the 'developing world' through a 'global spectacle' and 'consumer gratification' designed to alleviate the *symptoms* of poverty (Jones 2017:14). Equally, an increasing number of corporate brands, celebrities, and media navigating the right way to behave in the system have been consumer-led and publicity driven, often motivated by self-fulfilling priorities (i.e., profit or reputation), rather than solving problems they claim to care about (Talpalaru 2014).

Over the last 20 years, private foundations have gained significant influence in society (Fejerskov 2015; Strachwitz 2015). In the late 20th century, new foundations emerged – i.e., the Ford and Volkswagen Foundations – significant in size, supporting causes beyond those predefined by government, recruiting large, in-house teams bridging corporate interests with the foundation, thus representing a paradigm shift in public affairs (Strachwitz 2015). Foundations now have a unique role in the non-profit sector, and their resources and new approaches to international development have put them in the spotlight (Fejerskov 2015). For example, the Gates Foundation, which by 2006 became the biggest philanthropic foundation in the world with endowments of $35bn, is considered one of the 'superpowers' of philanthropy, comparing it to the US government in terms of international political tract (Fejerskov 2015:6).

The term *private philanthropist* generally describes persons who donate huge sums of their personal fortune to charity (Partzsch and Fuchs 2012). Large-scale private philanthropy has become deep-rooted as a practice of the global elite, and whilst praised for altruistic actions, some argue the real purpose is making disproportionate wealth more acceptable to the public (Maclean et al. 2015). Strachwitz (2015:832) also identify philanthropy as a means to legitimise 'misbehaviour'. Maclean et al. (2015) suggests private philanthropic giving is about the acquisition of philanthropic identities, narrative identities, or identity storytelling, narrated journeys of the philanthropist primarily a quest for meaning by individuals.

Entrepreneurial philanthropy and the popularity of the social enterprise has also grown in this space involving either 'ethical investments' by limited liability companies with financial and social/environmental returns, or impact investing (Maltby and Rutterford 2016), as well as non-for-profit models with social objectives funded through an entrepreneurial creation of wealth (Maclean et al. 2015). Corporate social responsibility (CSR) and the evolution of Environmental, Social and Governance (ESG) has become the name of the game for companies across most sectors, and with giving increasingly exhibited on websites and 'sustainability' reports, some argue to help legitimise capitalism (Jones 2017), by generally having a business case attached (Pinelli and Maiolini 2016). Having said this, the role of the private sector in aid and development is diverse and continues to evolve.

The 'marketised philanthropy' model is believed to have begun publicly in the 1980s with Band Aid and although Geldof was criticised by the NGO sector at the time, he arguably just mimicked successful models of emergency fundraising embedded in the NGO humanitarian sector (Jones 2017:4). What this caused, however, was the transformation of the charitable giving image, from an upper-class worthy, boring, and patronising activity to 'classless glamour, excitement and interest', and represented as individual philanthropy it "challenged the more selfish [UK] ideologies around Thatcherism to altruism and conscience" (Jones 2017:8), giving birth to 'individualist morality' on mass, bound up with lifestyle choices (2017:14).

There is less written about smaller scale private philanthropists and charity donors, but it is said that the rise of the competitive aid industry is now positioned within what Jones (2017:8) calls a "new era of slick, media-friendly celebrity humanitarianism" that uses the power of the globalising media to stimulate the consumption of ethical products and/or charitable causes. He points to a shift towards consumer-led philanthropy, CSR, and philanthrocapitalism as an ongoing depoliticisation of global welfare issues. Similarly, the power of online crowds and technology drives charity consumption of individuals 'endorsed with good conscience' (Yu 2017:2), and philanthropy has morphed into being small scale, individualistic, and random in nature.

From small to significant, charitable donations by private individuals across the world is now reported to be big business. In the UK, around £10bn is donated every year to charity (CAF 2019); Germany and France around EUR 4–5bn each (GDC 2019) by individuals; China around $21bn, and the US is said to be the world's most generous country with individuals, bequests, foundations, and corporations collectively giving $450bn to US charities alone in 2019 (Giving USA 2020). Large US foundations giving grants to Africa apparently doubled from 135 to 248 (private/community foundations) between 2002 and 2012, increasing their grant dollars 400% from $289 million to $1.5bn during this period, not including in-kind donations (Foundation Center 2015). The Bill and Melinda Gates Foundation alone contributed 70% of this total.

2.5.2 Corporate philanthropy – questioning the triple bottom line

'Corporate Philanthropy' has been defined as a company's support to a non-profit organisation or other charitable cause using company resources (Choi and Wang 2007). An uptake of CSR across the globe has evolved in recent years, with a wealth of literature covering concepts such as 'Strategic CSR' (Baron 2001; Burke and Logsdon 1996; Xueming 2005) and the notion of driving business competitiveness simultaneously as a result of 'doing good' and 'shared value' (Smith 1994). This moves towards creating products and services that purposefully create social and environmental gains (Porter and Kramer 2011; Valente and Crane 2010). Other forms include 'Altruistic CSR' (Lantos 2001) or 'Philanthropic Responsibility' (Carroll 1991), which intends no business gain, 'Cause Related Marketing', in which a percentage of sales revenue is contributed to an adopted cause (Kotler and Lee 2005), and Corporate Volunteering (Smith and Sypher 2010).

Literature on corporate philanthropy is growing and has become more complex (Pinelli and Maiolini 2016). Companies' goodwill is often assumed to drive CSR, with these positioning themselves as agents of development in countries or communities where they operate. There are, however, other factors emerging from the literature that may need further consideration, such as social licence requirements and trade for legitimisation (McEwan et al. 2017), moral and economic payback due to a community (Gardner 2015), favourable attention to the company (Fernandez and Hager 2014), and preventing 'groundswells of discontent' in a power equilibrium (McIntyre et al. 2015:307). McEwan et al. (2017), and Gardner (2015) argue that development interventions by extractive corporations tend to wilfully detach from moral and political economies in communities, building brand image for acceptance rather than longer-term economic potential, and neglecting local priorities over business ones, with little oversight of intervention efficiency.

The 1980s was also the turning point for this trend, with global media and corporations driving enthusiasm for corporate philanthropy programmes, social, and environmental responsibility more concretely integrated into business operations, not necessarily for altruistic reasons. Realisations that reputations could be enhanced through association with 'fashionable consumer philanthropy' (Jones 2017:18) and through exhibiting 'good works' in a range of ways (Shah et al. 2017) become part of the economic rationale for CSR's 'business case' and 'moral case' for sustainability discussed by Pinelli and Maiolini (2016). They debate different takes on *triple-bottom-line* thinking highlighting that owners of companies may regard 'sustainability' as a burden constraining the traditional 'single' bottom line, or as an external factor strengthening competitiveness and increasing profitability (Pinelli and Maiolini 2016). Companies are increasingly being perceived as "pluralistic socioeconomic systems that should become more inclusive of stakeholders' interests" (Ehnert 2014), whilst at the same time also understood to have limited resources and different responsibilities to governments in solving society's problems (Pinelli and Maiolini 2016). Furthermore, CSR has been criticised as a form of tokenism, greenwashing, insincerity, opportunism, pacification, imperialism, and neo-colonialism calling for a move towards developmental CSR whereby profit takes a less central position to development and sustainability (Jamali and Karam 2016).

2.5.3 Social enterprise models – philanthropy meets market principles

In the late 20th and early 21st centuries, solutions to social problems within a market framework were reignited from Victorian times, as so-called *ethical investments* (Maltby and Rutterford 2016). These became the focus of limited liability companies as models that proved restrictive and exclusive, discriminating between 'the deserving and the rest' (Maltby and Rutterford 2016:276). *Entrepreneurial philanthropy* grew and the entrepreneurial creation of wealth and philanthropic redistribution of that wealth were to serve specified social objectives (Acs and Phillips 2002). Similarly, *philanthropcapitalism* emerged as a phenomenon poised to stimulate "the deployment of market mechanisms on a massive scale to implement aid and development solutions, by wealthy private donors such as the Bill and Melinda Gates Foundation" (Jones 2017:20).

Social enterprises have been defined as "hybrids – operations that aim to generate a profit principally for reinvestment to serve their social purposes" (Maltby and Rutterford 2016:276). They have, however, been criticised for monetising social concerns by focussing on people's emotions and feelings. For instance, TOMS 'one for one' shoes is used as an example of social enterprise reflecting new neoliberalism principles and the result of 'conspicuous giving' or a call for public recognition by Talpalaru (2014:2), who also explain that "social enterprises explicitly connect their consumer products to forms of socially or environmentally conscious philanthropy". This is different from 'cause marketing' in which a company "conjures its audience as a participant in the very act of doing good" and donates to a particular charity. Talpalaru (2014) further argues that CSR is a fluid concept mocking the more recent growth in *social enterprises* for believing they transform the gimmicky or tax-write off reputations of many corporate CSR practices, and create a profitable business model instead.

2.5.4 The paradox of ethical consumption

Engaging in charitable action through consumption, including charities selling merchandise or running shops, has a history, although some authors suggest a paradox in this trend towards 'ethical consumption'. Jones (2017:14) critiques this so-called consumer philanthropy as a paradox enabling "donors to demonstrate their caring credentials... without fundamentally altering their own consumer behaviour or acknowledging the complicity of consumer capitalism in creating and perpetuating global injustice". Consumer philanthropy or ethical consumption is argued to be just another form of Western agency paying lip service to symptoms of poverty in the Global South by "altruistic action with 'feel good' gratifications" that legitimises the consumption of a product/service (Jones 2017:17) and 'myth of just capitalism' (Richey and Ponte 2011:137).

While donating to a humanitarian aid campaign gives donors the sense of having helped, buying a product from a social enterprise like TOMS shoes actually provides consumers with a *tangible* good, linked to giving and visible to others. This is what Talpalaru (2014:180) calls 'consumerism for a cause' or "buying recast as 'giving'"; the whole concept relies on affective connections between global aid, celebrity activism, and Western desire. The ethics of social enterprises that strategically connects their branding and products to 'underprivileged subjects' is questioned, as are enterprises that make giving through buying a "mandatory condition of a global citizenship" (Talpalaru 2014:182–184).

2.5.5 Private power or innovation in the public realm?

Debates regarding the privatisation and commercialisation of global politics have been gaining attention (Jones 2017). Private foundations are understood on the surface as entities benefitting the public, but they have come under scrutiny for lack of democratic governance enabling expressions of individual or corporate branding goals (Strachwitz 2015). Partzsch (2017) argues there is an insufficient debate on the significance of individual leadership, policy entrepreneurship, and social entrepreneurship in a discussion about big private philanthropy players – including Mark Zuckerberg (Facebook) and Bill Gates (Gates Foundation) – when linked to new and growing forms of governance and private authority in the global realm. Politics "became more commercialised, mediated and image-conscious" in the 1980s with a deeper penetration of the mass media into everyday lives determining the earlier discussed shift towards consumer-led charity and "more overt integration of philanthropy with business" (Jones 2017:15). Similar trends are more recently emerging in relation to the philanthropcapitalism phenomenon.

Private actors in global development are endorsed for bringing their business logic to inspire greater innovation and effectiveness compared to traditional donor organisations (Skarbek 2016), but this is also considered problematic in a 'development' context because simply transferring commercial logic can be inappropriate (Fejerskov 2015). Fernandez and Hager (2014) call for a more public outlook on foundation behaviours and outputs in regulatory terms because many private actors may lack accountability, spend money without real regard for community needs, and "act with privacy, as private actors" (2014:5). Similarly, for the corporate sector engaging in community development, McEwan et al. (2017) argue that whilst corporations are accountable to their shareholders, it is unclear who holds them accountable when they engage in community development because they are not expected to abide by the same principles of development effectiveness, such as those imposed by the OECD on donors. Fernandez and Hager (2014) argue that corporate foundations are the 'least redistributive' because their focus tends to be local and of a specific nature related to the interests and image of the parent company. Sustainability agendas of corporations are urged to be strategic, although individual private company goals for development are characterised by influential differences and interests, making the presence of private sector players in development a double-edged sword.

Schurman (2018) points out that 'new philanthropists' are alleged to engage more directly with grantees compared to older foundations and have greater capacity on the ground, developing a richer understanding of people's lives in specific places. However, whilst often hailed as positive in attempts to get the world out of poverty from the grassroots, superior to traditional bureaucratic and colonial ways, the private philanthropy sector is beginning to be criticised for exhibiting 'homogenising modes of practice' (Fejerskov 2015:13–14). Private institutional logic may contrast with discourses of international development, but despite initial distinctive characteristics and attempts to articulate how they remain 'different' to other donors, organisations such as the Gates Foundation end up adopting similar norms, principles, and organisational and programmatic traits to conventional donor organisations (Fejerskov 2015:2).

2.5.6 Two worlds collide? The NGOisation of private actors and vice versa

It is evident that the private sector is morphing into some sort of development actor, but it is also evident that the development sector is morphing into a set of organisations with commercial norms (Seu et al. 2015). Managerialist business logics have been steadily permeating the development field over the last decade as the notion of 'aid effectiveness' formally institutionalised by the Paris Declaration in 2005 combined with the broader results-orientated agenda of the private sector shares common ground (Fejerskov 2015). Seu et al. (2015) discuss the complex relationship between NGOs and the public. They argue that if the marketer model or transactional relationship with the public as current and potential donors is true, then NGOs that are supposed to be the 'moral entrepreneurs' are going to struggle when a significant "disillusionment, distrust and animosity towards humanitarian

NGOs" (Seu et al. 2015:223) exists in Global North locations such as the UK. This suggests potentially significant negative impacts for the non-profit charity sector and humanitarian aid in general. Fejerskov (2015), on the other hand, discusses how private foundations over time become more like NGOs due to external pressure, legitimate or successful model imitation, and normative pressure (see DiMaggio and Powell 1983, Fejerskov 2015).

This cross-pollination of behaviour regarding roles, norms, and objectives between the private sector in development and NGOs produces complex terrain. Arguments for and against appropriate roles, responsibilities, and capabilities for different types of development organisations are a hot topic, as public and private sector actors navigate this morally muddy, and potentially competitive landscape.

2.5.7 International development and the private sector – agent or tool?

The private sector interconnects with international development in a number of ways. Evidence suggests it is increasingly positioned as active *agents of development* (McEwan et al. 2017), or an *engine for development* (Mawdsley 2014), with donors also increasingly creating enabling environments to draw the private sector into development roles (McEwan et al. 2017). Di Bella and Grant (2013) present a framework for the different ways in which the private sector is currently being positioned within international development. They suggest a typology with three categories: (1) private sector *in* development, (2) private sector development, and (3) private sector engagements *for* development. Blowfield and Dolan (2014) also provide a framework contrasting the private sector as a development *tool* (contributing to 'immanent development' simply by operating in developing countries), with the private sector as a development *agent* (purposefully acting to bring benefits through 'intentional development'). CSR is often a term used for the private sector as a development tool, with some advocating "untapped potential to address social and economic development in the developing countries in which [corporations] operate" (Shah et al. 2017:2).

McEwan et al. (2017:7) flag a cautionary position that private sector-led development is "riddled with unresolved contradictions and blind spots", and argue the private sector is not a 'magic bullet' for community development, partly because it is found to exacerbate similar problems of states, NGOs, and donors. They examine the link with the wider debate around power relations, and question a lack of commitment and skills required to deliver effective development assistance when it is not their primary objective. They highlight mining and tourism because these tend to operate in geographically marginal areas, often where the state's presence is limited, and whilst some sophisticated interventions in communities are found, efforts are highly inconsistent.

2.5.8 International development and tourism – CSR beginnings

As an economic sector, tourism has increasingly claimed CSR merit across the world, as discussed in Chapter 4, and there are some issues identified about this trend in an intervention context, which requires further consideration.

A potentially urgent issue faced within the tourism industry is that CSR, for some players, aims to benefit business goals driven by decision-making enabled amongst those with below-par understanding of sustainable development, as opposed to the development 'causes' it claims to support. In terms of how tourism companies perceive CSR initiatives, Gligor-Cimpoieru et al. (2017:1642) highlight that CSR centres on relationship marketing and establishing trust with travel consumers, as well as attracting and retaining employees. Furthermore, they identify a confused understanding about the sustainable tourism concept amongst employees. Yet, they emphasise the importance of applying what *employees* think as part of a strategic approach to CSR, ignoring beneficiary perspectives, and the actual heterogeneity of values, beliefs, and competence beyond core industry skillset across the global tourism industry. We must ask *who* drives decisions about CSR interventions and motivations, especially in a workforce typically untrained or interested in development.

CSR activity tends to be driven by the underlying philosophy of an organisation, which is made up of its people. Even if companies aim to shift their strategies and philosophical values towards more meaningful, ethical, or legitimate CSR activities, it will not happen, according to Shah et al. (2017), overnight. A well-entrenched philosophy takes time because "corporate philosophy is not a product a consultant sells off a shelf" (2017:251). It is therefore important to realise that even if knowledge about meaningful CSR in tourism grows, to assume these principles are easily and automatically embraced across the industry is uninformed. Systemic capacities, interests, and ambitions may also need to be addressed.

In terms of CSR impacts within local communities in lower income countries, McEwan et al. (2017:4) use tourism and non-tourism case studies to caution against assumptions that such embedded corporations can succeed where governments, NGOs, and international development organisations have already faced 'intractable difficulties'. The characteristics they identified in CSR activities include the ways in which communities are defined, delineated, and constructed by corporations; the disconnect from local and national development planning policies and processes; top-down governance and lack of participatory processes. They also show a tendency towards highly conservative development visions that reflect self-interest and only "incremental improvement in 'safe' areas of intervention"; with transformational changes such as building local agency or bottom-up agenda setting lacking (2017:24). This reflects some of the earlier issues found in the sustainable development debate and raises concerns about the merit of some CSR claims by the tourism sector. Reflecting on ideas regarding the actual good in giving, rather than *that they give*, is therefore imperative.

2.6 Gift-giving and international development in Africa

2.6.1 The theory of the gift

The moral claims of development NGOs' interventions and the moral importance of gifts and donations has been ingrained in many people's thinking across the world

(Stirrat and Henkel 1997). The good in gift giving is a widely accepted narrative and, based on this idea, charity and NGO work has become even more widespread, especially in Africa. Much of the literature on gift giving continues to focus on the assumption that giving is good, ignoring potentially negative consequences (Sanghera 2016). Rather, gift giving is a complex system of exchanges, spanning beyond simplistic ideas of economics or pure altruism. We must therefore delve into these concepts when discussing international development and travel philanthropy, which are predominantly based around gift giving, and usually presented from the perspective of the 'altruistic' gift-giver.

Comte (1798–1857) coined the term altruism as a form of consequentialism in which an action was ethically right if it brought about good consequences to others, different from utilitarianism because the benefit is for others rather than the actor (Steiner 2017). Spencer (1879:232) also contributes to the debate on altruism, defining it as "all action that, in the normal course of things, benefits others, instead of benefitting self", but that all altruism starts with egoism in 'altruistic competition', thus contrasting Comte's pure altruism. Durkenheim (1897) also proposed a new approach, differentiating between interested and disinterested sacrifice.

Mauss's 1925 *Essai sur le don* (The Gift) significantly contributes to the altruism vs. egoism debates and remains influential in anthropology and economics. Combined with the work of Levi-Strauss (1987), Sahlins (1974), Parry (1986), and Bourdieu (1977), it leads to further understandings of the concepts of social relationships and gift-exchange. Whilst the common belief is that gifts are altruistic in nature, actual motivations fall on a continuum between altruism and egoism (Hyun et al. 2016), characterised by voluntary or obligatory acts of giving (Mauss 2016). The gift economy is defined not only by quantifiable values, but by invisible bonds and non-quantifiable components that economists arguably underestimate (Toffin 2015).

Mauss portrays the gift as a paradox, asking the following influential question: "What force is there in the thing that is given that makes the recipient return it?" (2016:33). He refers here to the symbolic nature of gifts as social phenomena, saying gifts create an obligation that lasts until the recipient can reciprocate in some way, but not by returning the actual gift. Gifts are known to involve expressions of affection (Belk 1979) and expectations of reciprocity (Mauss 2016; Malinowski 1922), and an absence of reciprocity shows modern society's lack of morality (Mauss 2016). Similar to Adam Smith's rules of market exchange, Mauss proposed explanations of the rules in gift-giving behaviour, involving the obligation to give, receive, and then give back, thus an alternative system of exchange in society with gift-relationships deeper than economic contracts (Mauss 2016; Steiner 2017; Boland and Griffin 2016). An aid industry broadly designed around supposedly one-way gifts thus has serious social implications.

Mauss explained that gifting is voluntary, but also obligatory because the giver must choose something to return, and with no way out of a gift relationship, one must give something even if it is an apology for being empty-handed (Boland and Griffin 2016). Mauss therefore argued that things given have power, the *Hau* of the gift, and to give is to show one's superiority and the unreciprocated gift makes the person who accepts it feel inferior. However, the sort of power is not wholly authoritarian because the actors can reciprocate in various ways (Boland and Griffin 2016). Mauss argued that when the debt of obligation has been discharged, the former creditor becomes the debtor, and this continues in a circular exchange with no way of it ending. Mauss proposed the relevance of the gift in the evolution of contract, from clan-to-clan exchange of gifts, to gift exchange and to contract. He argued that modern economic systems have emerged from similar institutions found in ancient economies. Levi-Strauss (1969) built upon Mauss's (1925/2016) work viewing gifts as political in nature, as means to security or power.

Debates over definitions of 'the gift' continue today. Derrida and Kamuf (1992), discussed in an analysis by Bornstein (2009), critiques Mauss's conception of the gift, arguing that a gift must not be bound, or it is no longer a gift, and as soon as there is an expectation of return, it is in the realm of debt and credit instead. From a multicultural perspective, particularly, there are further possible flaws in Mauss's theory. Parry (1986), Toffin (2015), and Sihlè (2015) provided examples of scenarios that question Mauss's obligation of reciprocity and suggest he does not pay enough attention to differences between societies, as we need to consider specific forms of complex economies, and different ideologies of the gift. A growing body of literature suggests that earlier theories leave us 'ill-equipped', as Sihlè (2015) argued that it's important to understand forms of giving in different societies, religions, and so on. Erie (2016), for example, also argued that, for Mauss, the gift assumes shared values between the giver and recipient, but multiple and competing systems of norms can affect the meaning of a gift, rather than being uniform in this way. It is thus important to note in the context of international development that different values are associated with gifts; it is not possible to generalise across contexts, so a single theory of the gift is actually unhelpful (Toffin 2015).

2.6.2 Moral obligations and gift exchange in Africa

The African worldview is said to divide humanity by social distance and that many Africans "perceive themselves as living within an imaginary circle" (Kragh 2016:59; Sahlins 1968). Gift giving is said to be a function that creates either social distance or closeness between giver and recipient (Cavanaugh et al. 2015). According to Kragh (2016), although those outside the extended family circle are treated as enemies, Africans enlist allies through giving and accepting favours. This is in line with Mauss's (2016) thinking about strengthening moral obligations and has great relevance in terms of opportunism and mistrust that characterise exchange of goods and services. Sahlins (1968) proposed an in-group/out-group continuum, which illustrates the connection between morals, social distance and reciprocity, with consequences for exchange. He suggested that a lack of reciprocation means ingratitude and increases social distance, and initial mistrust characterising relations between strangers can be transformed into trust and mutual loyalty through reciprocity, with self-interested transactions transformed into mutual concern for each other's wellbeing. Such thinking has been criticised by Parry (1986), who argued that this assumes an exchange between calculating individuals.

African society is argued to be qualitatively different from those operating according to principles of capitalism, neither preindustrial nor industrial, which tends to assume sharing and altruistic gift exchange to the 'preindustrial' and self-interested utilitarian to the 'industrial' in some sort of determined cultural superstructure (Kragh 2016). Despite expanding urban economies in Africa, Kragh (2016) argued how values, norms, and behaviour regarding cooperation and exchange are culturally distinct from the West. For instance, in Africa they highlight 'the big-man strategy' of giving 'extraordinary gifts' to create perpetual moral obligations of being generous and benevolent to enhance and maintain status and power, with acts of generosity being more calculated than these assumptive theories and significantly impacting principles of market capitalism of the 'West' (Kragh 2016:60). The concept of gift exchange in Africa requires deeper and more contextualized understanding than overlaying Mauss's theory of the gift onto this cultural setting.

2.6.3 'Doing development' and free gifts

Gifts can assert identity between the rich giver and poor receiver, reinforcing differences (Stirrat and Henkel 1997). Drawing on Mauss's theory of the gift, Stirrat and Henkel (1997) examined the relationship between Western NGOs and Southern counterparts, arguing that a seemingly free, altruistic gift often transforms into a heavily conditional object when it reaches the recipient, binding Northern donors and Southern receivers in symbolic forms of reciprocity and problematic partnerships, including recipients having to embody values of givers. Contrary to Mauss's (2016) theory, the NGO intervention is argued here to represent a pure gift in the sense that the receiver is indebted because there is a lack of reciprocity. The pure, free, or disinterested gift of a private donor, usually motivated by the recognition of difference argued by Stirrat and Henkel (1997), transitions into an interested, accountable, non-free transaction as it moves down the aid chain. Even if the original act of giving is symbolically the antithesis of the calculating relations of the modern world, by the end of its journey to the ultimate recipient, it becomes another resource to be manipulated and controlled by competing groups. Stirrat and Henkel (1997) recommend more participative, empowering and bottom-up approaches to development gifts, drawing on indigenous knowledge in place of top-down blueprint projects relying on 'expert' knowledge.

Conditional gifting can mean a gift is coercive and the giver has rights over the recipient (Boland and Griffin 2016; Rudrum et al. 2016; Zhan 2016). There has been a growing understanding in the different ways that gifts are offered and received, with debates about morality in the NGO context (Rudhrum et al. 2016; Gardner 2015; Sihlè 2015; Kragh 2016). According to Rudrum et al. (2016), in humanitarian aid there are sets of specialised producers with professional interests in encouraging gift giving and receiving. Swamy (2017) argues that humanitarian gifts are bound to structures and practices associated with the formal economy, as well as representing unequal exchange through giving without expectation (Swamy 2017). Bornstein (2009:643) argues that gift giving is 'riddled with ethical contradictions', but that both impulsive, spontaneous philanthropy, and instrumentally rational philanthropy need to exist for different purposes with appropriate structures to encourage them. She argues that, whilst impulse giving can create inequality, for example, more attention needs to be paid to how impulse philanthropy differs from welfare that involves rights and claims, especially in circumstances that 'warrant immediate, perhaps fleeting, attention'. In merit to Bornstein's (2009) discussion, she emphasises that, although giving to a beggar and giving to NGOs are both responses to human suffering, they are not placed in the same category. She recognises the historical progression from 'charity' to 'philanthropy' and positions the latter as a more formalised and rationalised form of giving than the former, which is generally characterised by impulsive giving and lacking in long term effects.

2.6.4 Transposing gift exchange – from the recipient perspective

It is evident across the literature that the power of the giver is usually the topic of debate in gift theory (Campergue 2015; Zhan 2016), including sociological impacts of unreciprocated gifts (Toffin 2015). A relatively recent study evidenced the lack of understanding of the gift from the perspective of the gift-receiver – "the other side of the gift" – rather than keep privileging the donor as the starting point of exchange by focusing on the anthropology of 'taking' and the ways transfers are accomplished through acts of soliciting (Retsikas 2016). Retsikas' (2016:5) aim was to reconsider the Other in this context and "recast the so-called gift-receiver from a *passive* into an *active* figure". He proposed that the gift might actually be the 'adjunct' of solicitation instead, often perceived by a needy receiver as their right and as pride for the wealthy payer, rather than a 'charitable gift given with condescendence'. Therefore, the gift-giver becomes a secondary function, and this 'taking' (which follows soliciting) is not the same as stealing, borrowing, or accepting, especially as the latter entails gratitude, indebtedness, and self-effacement. He argues that soliciting, as a practice that 'almost amounts to stealing', is unburdened with moral questions because it comes before the gift, as the *cause* of gift giving; therefore, it sidesteps the law and morality. Retsikas (2016) argues that the powers of *soliciting* and perspectives from acts of *taking* remain hidden in current Gift Theory, and that it needs attention because it is a space for new ethics.

2.6.5 Questioning the 'good' in giving

The travelling philanthropist leaves and the money or gift of some sort stays, with many donors thinking that their job is just to give the gift away. Travellers and NGOs do not necessarily take responsibility for the short- or long-term *impact* of gift giving. Emperor Mansa Musa (1280–1337) gave out so much gold on his travels in Cairo that he actually generated a decline in its value for a decade (Morgan 2018). If 'travel philanthropy' means gift giving by travellers, whether in a formal/managed or informal way, it has existed for very many years; however, good intentions do not always translate into positive impacts.

2.6.6 Efficacy or Myopia? The paradox of helping

In Western countries, helping the less fortunate is considered the right thing to do, but evidence demonstrates that aid to Africa can be "the disease of which it pretends to be the cure" (Moyo 2009:10). Nevertheless, scholars and practitioners' confidence in questioning the traditional construct of altruism or the ethics of giving is still a fairly emergent and cautious affair. Equally, there is small but increasing attention paid to possible negative consequences of giving, helping, altruism, beneficence, and the various constructs of 'doing good', challenging the common beliefs and assumptions that 'giving is good'. The arguments identified in the literature do not necessarily propose that "saving people's lives unnecessary or unethical" but suggest a need to more critically question the decisions behind giving and whether the impacts are negative or inferior to alternative ways of allocating that resource.

The power of the common narrative that altruism is "a virtuous action aimed at benefitting society", an 'ethical action', and that "the actor is operating from good intent" is significant (D'Souza and Adams 2014:186). Generosity and acts of charity are assumed universal, ethical social phenomenon, central to many religions, cultures, and philosophies (Von Bergen and Bressler 2017). The *moral obligation* of helping people in need is indeed argued by some philosophers (Kant 2011; Singer 1972) and well-meaning advocates of helping continue to dominate practice and academia. Mitschow (2000) argues that altruistic intentions become self-justifying, meaning that desire to assist the less fortunate relieves individuals of any responsibility of such action. People are regularly reminded of the moral imperative to assist those in need through charity promotions, beggars on the street, media figures advocating for more giving, and the respect altruists receive in society (Von Bergen and Bressler 2017).

Whilst questioning the good in giving remains taboo in many ways (Oakley 2013), and scientific inquiry remains weak due to a fear that publishing negative information will discourage people from providing more 'help' (Von Bergen and Bressler 2017), the benefits of 'doing good' are increasingly being challenged (D'Souza and Adams 2014; Von Bergen and Bressler 2017). As Von Bergen and Bressler (2017:114) argue, the aim is not to just ask if giving is 'good' or 'bad', but to "understand the conditions under which helping may negatively impact others", to challenge the 'blind trust in charitable causes', and to challenge beliefs that interventions should be judged on intention not results, to inform whether ignorance can be harmful, and to rethink strategies for helping. Regardless of intentions, the ethics of giving and the giver are beginning to be defined by impact (D'Souza and Adams 2014; Pummer 2016). With international development aid continuing and the growth of private philanthropic actors in the giving system of Africa, these are without a doubt critical questions that need to take centre stage.

Pummer (2016) also strives to contribute to the ethics of giving in theory and practice through challenging the common assumption that says if you *can* donate you can choose *where* to donate, and whilst arguing from a non-consequentialist perspective, he supports the idea of *effective altruism*. Recently gaining momentum, effective altruists use evidence and reason to determine the most effective ways to *do good*, such as seeking out the most effective charity rather than one close to their heart (Schmidt 2017). How the altruist decides effectiveness is not straightforward, however, with concepts such as confirmation bias or belief perseverance linked as people are found to be "blind and deaf to anything that disagrees" with their pre-conceived notions (Von Bergen and Bressler 2017:118). Effective altruists often confuse doing good with being good in virtue of consequentialism, when in actual fact, being a good person does not mean 'doing something well', as it requires informing oneself of ways to do good effectively and of ways to effectively avoid harm. Effective altruism may be about allocating resources to the most effective causes, but it does not tend to focus on the consequences, the broader impacts of giving.

The harm that giving can cause can be unobserved and systemic. Regardless of whether those who receive help may really need it, feel appreciative and grateful to those who aid them, receiving help can also have negative consequences, such as dependency, inferiority, lowered self-esteem, indebtedness, etc. (Von Bergen and Bressler 2017; Hailu 2015). Some broader negative impacts of aid are known, yet the narrative of good intentions still tends to prevail.

Concurrently, *unenlightened altruism* – framing unintended consequences of well-intentioned but ill-conceived giving with negative impacts on society – pushed interpretations of altruism beyond the feel-good factor, and stimulated a debate addressing well-entrenched assumptions, positioning altruism as a far more complex social phenomena than widely assumed. The key argument is that *if* the consequences of an altruistic act are not well understood by the actor, and they proceed 'within an intellectual vacuum', this act under an altruistic banner runs a high risk of being unethical, as well causing devalued efficacy, in which case it might be ethical *not* to act at all (D'Souza and Adams 2014:185).

Disregard for the real problem, ignorance, and myopia can result in counter-productivity, antithetical to the original intent, and, worse, create 'deleterious' or 'pernicious' effects on society and economic sustainability (D'Souza and Adams 2014:186–189). D'Souza and Adams (2014:189) argue that "an enlightened altruist would attack the root causes of suffering rather than simply alleviate the suffering of the victims" and apply evidence and reason to maximise social good, while unenlightened altruists do not use much reason or evidence before acting.

Furthermore, by reflecting on the work of Partzsch and Fuchs (2012) and Goldfarb (2011), philanthropists cannot be simplistically defined as do-gooders, ignoring conflicts of interests and values. Similarly, private sector corporate actors can be argued to be 'laden with power and meaning' and paternalistic relationships with recipients (Banks et al. 2016:254). Goldfarb (2011) provides a distinction between philanthropy and misanthropy, the latter referring to some non-profit organisations intending well but leaving the public worse off than before, or directly or indirectly conflicting with the aims of other non-profits, with the sum of one effort cancelling out the others. She underscores errors we can make when determining the value or legitimacy of a cause, and how we can be prone to ignoring the truth. She also argues to know what is needed is not easy, as it requires "the courage to acknowledge one's ignorance", but also because causes to problems change, ethical judgements evolve, and feelings, opinions, and intentions are not homogenous within communities (Goldfarb 2011:730).

2.6.7 Travel philanthropy – questioning the good in giving

The underlying theme within the travel philanthropy literature is centred on 'luring more donations from travellers' and the assumption that 'giving is good'. The actual impacts of transitory gift giving are rarely debated, as discussed earlier in this chapter. Therefore, in line with Liket and Simaens (2015), one needs to ask how much do we *really* know about the consequences of philanthropy? Pollock (2016:1) argues that:

> *Much is made of the human right to travel and turn up on another's doorstep unannounced or without adequate preparation for the consequences, yet very little is said of the right of a destination community to say either 'no thanks' or to try to shape the kind of visitation that ensues.*

In the development context, Scott-Villiers (2011:772) describes the "strategic pose in the presence of the powerful" in which the beneficiary can seem ungrateful and unwise to suggest the problematic nature of aid. This also emphasises the issue of *who* defines the good in giving.

Within the travel philanthropy discourse, there is extremely limited evidence about broader implications of gift giving in an ethical, political, economic, social, and wider context, and its role in sustainable development. Travel philanthropists have been accused of lacking time to understand a problem or solution in detail, pushed to think fast within the timescale of their trip, and as a result they behave instinctively and emotionally as opposed to strategically, slowly, logically, and deliberatively (Brest 2015; Abernethy 2011). As part of these implications, Macy (2011:177) is one of few to discuss the potential issue of 'unwanted philanthropy' and the widely referenced 'paint syndrome', meaning 'doing something charitable' that is actually not needed. To the same end, while Crouch (2011) discusses Bali's orphanage scams as a lucrative business, often legitimised under religious banners, Thomas (2014) identifies travel philanthropy as an unexpected resource that, without empowerment of local authorities in its management, produces income inequality between tourist areas. Similarly, Korf (2007) mentions post-tsunami aid fuelling social jealousy, and Honey (2011) refers to ad hoc giving causing children to skip school to seek handouts, in some cases encouraged by parents to beg from tourists, and leading to harmful practices like the gathering of sweets within locations with inadequate dental care and malnutrition (Honey 2011). In such contexts, Silk's (2004:236) 'unintended paradox of aid', sustaining the root cause of the problem, is further validated here, and, although there are isolated cases of positive impacts (Novelli 2015; Honey 2011), arguably these do not stretch to broader impacts or to what extent they reflect 'enlightened altruism'.

What is unclear is if the concept of travel philanthropy is concerned with real-world impacts of giving, or consuming the experience of giving. Literature refers to travel philanthropy as a means to capture "the satisfaction that comes from making a personal contribution – and seeing people and nature benefit as a result" (Western 2011:13), or motivations tending to be focused on enriching the donor's lives (Bertini et al. 2014) as opposed to 'targeted' beneficiaries. Consumption branded as ethical is often considered to be morally good without question (Burgold and Rolfes 2013), justifying the behaviours of *new moral tourists* seeking a personal understanding of global problems through the pursuit of meaningful experiences, no longer appearing as 'civilisation-weary voyeurs'. Rather, they are positioned as just (Butcher 2003), or "as a kind of aid worker with moral integrity whose presence in a slum or township is morally integrated", drawing upon 'moral geography' which suggests philanthropists tend to project their criteria on to distant places, not necessarily matching local perspectives of a poly-contextual world (Burgold and Rolfes 2013:172).

Within corporate philanthropy, Smith and Sypher (2010) echo Burgold and Rolfes's (2013) point by emphasising personal or socially desirable values of leaders reflected in philanthropic activities of an organisation to legitimise their importance. Charitable donations also often come with conditions that the recipient is expected to meet (Abernethy 2011), which in Africa may be an exchange for reinforcing the image of poverty in an improved tourist experience. 'Symbolic domination' is arguably present consistently in tourism-related gift giving (Korf 2007), with the handing of books to school children, or the dancing of an indigenous group followed by the drop of a note into the donation basket being only a few symbolic elicitations of the debateable good of giving in travel philanthropy.

It is therefore not a surprise if there are doubts about whether travel philanthropy just reinforces the mistakes of unenlightened altruism and replicates traits of ineffective development aid, or if it contributes to sustainable and ethical development in Africa. Thus, this topic requires further, in-depth investigation, particularly in merit to how this can be best evaluated in Africa. The next chapter explores the evolution of the evaluation debate in the sustainable development context.

3 Impact evaluation and judging the 'good' in sustainable development systems

This chapter discusses the evolution of the impact evaluation debate in international development and sustainability and how the sector is dealing with the complexity of poverty, today's interconnected world, and the generation of change. It draws on critical issues identified in the non-governmental organisation (NGO) and private sector, including tourism. It highlights the progression in evaluation approaches and identifies a changing but nascent narrative attempting to address moral and efficacy issues of anti-poverty aid and its impact. The chapter discusses the growing relevance of complex systems theory in international development and the growing importance of interdisciplinary lenses.

3.1 Evaluation, impact and complex systems

3.1.1 The evolution of the evaluation debate in international development

The United Nations defines evaluation as:

> *The level of achievement of both expected and unexpected results by examining the results chain, processes, contextual factors and causality using appropriate criteria such as relevance, effectiveness, efficiency, impact and sustainability... an evaluation should provide credible, useful evidence-based information that enables the timely incorporation of its findings, recommendations and lessons into the decision-making processes of organisations and stakeholders.*
>
> (UNFPA 2020:7)

One cannot *assume* that 'giving is good', informed by or as part of 'sustainable development', and since philanthropy has the potential to generate both positive and negative impacts, travel philanthropy needs to be further understood as a complex and multi-dimensional phenomenon (UN 2020a). Furthermore, it is increasingly realised that development interventions aimed at addressing poverty challenges and creating impact in Africa need to understand the complex, adaptive systems in which they intervene (Douthwaite and Hoffecker 2017; Banson et al. 2015; Ramalingam 2013). This has stemmed partly from a desire or need by

DOI: 10.4324/9781003497929-3

those intervening to avoid further 'failure' (Mueller 2020) and address ethical dilemmas associated with interventions in sub-systems that reverberate around other sub-systems, which may produce counter-intuitive and system-wide effects (Jackson 2019).

In the practice of international development, evaluation is considered a critical 'tool' for knowing what works and what does not for stakeholders (York and Bamberger 2020). The evaluation debate has evolved over the last two decades, politically as well as technically (Webster et al. 2018). Impact evaluation and understanding the nature of social problems is critical to make things better, but many donors simply pay lip service to the political pressure of the results agenda and funding outcomes rather than exhibit a genuine desire for sustainable change that deals with complexity (Davis et al. 2016; Gates 2016; Mowles 2013; Arensman et al. 2017). Evaluation can be *ex ante* – i.e., the approach to strategy and development – or *ex post* – i.e., the approach to monitoring and evaluation (M&E) (Arensman et al. 2017; Richards 2019). Over the last three decades, the transformation of evaluation practices was first influenced by stakeholder approaches, then result-based M&E, and now systems thinking and complexity science have become the latest trends (Gates 2016).

3.1.1.1 Disinterest in the truth? The issue of giving aid regardless of results

The development field has been plagued with horror stories of negative social and environmental impacts of misguided projects, playing "into the hand of aid pessimists", and in more recent years, corporate social responsibility (CSR) of the private sector is said to stimulate similar debates (Picciotto 2019:8). The NGO sector has been in a crisis of accountability and transparency, remaining reluctant, unwilling, or unable to provide information, which has damaged its credibility (Burger and Owens 2008). What NGOs say they are about can be mismatched with how they sell their missions and allocate resources (Kirk 2012; Wisor 2012). This awakening of the aid industry's reputation has brought tougher measures for project selection, evaluation, and management, but only for some. Much of the extant evaluation practices show limited ambition and gather information about achievements rather than causal mechanisms underlying intended and unintended change (Wessel 2018). Even when producing more harm than good, many aid agencies continue to give aid independent of results (Skarbek 2016).

With larger donor funds channelled through NGOs, pressures for multiple accountabilities in some of the third sector have risen, but have also driven organisations to avoid reporting anything where possible, and led to dysfunctional effects, including less flexibility (Edwards and Hulme 1998; Goddard 2020). However, an absence of accountability can cause a greater probability for illegitimate actions, ineffective interventions, or terrible harm (Edwards and Hulme 1998). Berry (2014) raises flags about silences in the literature regarding some NGO humanitarian endeavours that have no regulatory body, relying on self-evaluation and individual judgement, whitewashing robust analyses of actual impacts and instead measuring what matters to the NGO staff. They argue that some foreign interventions can do more harm than good, yet the discontinuation of an intervention is not an option incorporated into evaluation frameworks, and so such silences overlook potential thresholds where harm outweighs the good.

3.1.1.2 Transparency and accountability in the hidden world of NGOs

Transparency is argued to be a requirement for good governance, successful feedback systems, and reliable information for effective regulation and monitoring of NGOs by beneficiaries, donors, and governments. Due to the high degree of private information and hidden actions in the 'NGO-beneficiary-donor nexus', transparency in the NGO sector is a key issue and 'fraught with information asymmetries' (Burger and Owens 2008:3). Burger and Owens (2008) highlighted that specific challenges to transparency are distance, which makes it prohibitively expensive for donors to visit to observe outcomes; reliance on second-hand information usually by the NGO itself; donors not being familiar with local circumstances and culture to detect fraud; and beneficiaries' lack of incentive to scrutinise NGO outcomes. They warned of presuming that decision-makers in non-profits are all honest, altruistic, and non-economically motivated employees, rather that opportunities for deception in the NGO sector are rife due to lack of transparency, weak monitoring and regulation, a lure of tax exemption, and often intangible and subjective provision of goods and services. For this reason, Tirivanhu (2020:2) suggested "an approach towards building accountability that relies on civic engagement" for community-based organisations (CBOs) and views the poor as central to M&E. He indicated positive outcomes from social accountability as improved governance, increased effectiveness, and empowerment, despite acknowledging the complexity of these. He also cautioned against a naive view of NGOs and information that is reported in a climate that lacks independent M&E.

Accountability, intended as "the duty to provide an account of the actions for which one is held responsible" (Gray et al. 1997:334 in Cordery and Baskerville 2011), is also important. Accountability is not just about governance; it is essential for improving the performance of development interventions (Sarker and Rahman 2015). Charities are criticised for privileging accountability to their most salient stakeholders (upward accountability), but, as suggested by Cordery and Baskerville (2011), a detailed stakeholder analysis based on systems theory may need to take place for a more ethical and inclusive perspective of accountability, including downward accountability to beneficiaries. Goddard's (2020) work in Tanzania highlighted that NGOs primarily use accounting intended as a political instrument or a device to improve effectiveness and efficiency (Cavil and Sohail 2007) and to enhance legitimacy status, although a slight shift towards more holistic accountability, considering both upward and downward accountability, seemed to be emerging. The private sector involvement in development opens new debates on top of the NGO world because, for example, "most of what happens inside the [Gates] foundation remains shrouded in mystery" and can be biased towards commercial models they are already familiar with (Schurman 2018:181).

3.1.1.3 Private sector philanthropy, tourism, and measuring public impact

Private sector involvement in giving requires understanding impact from a private and public orientation because the sectors' characteristic focus on support for others rather than widespread systematic change (Fernandez and Hager 2014) suggests private sector interest in sophisticated evaluation approaches found in the broader international development sector may be tenuous. Some models have emerged to evaluate decisions in the charity and philanthropy sectors, such as investment models (also referred to as the financialisation of the charity sector), that focus more on individual return than need (Maltby and Rutterford 2016). Such cost-effective models using econometrics or technocratic indicators to measure aid do not view all potential beneficiaries as equally deserving but rather ask how efficient they are as beneficiaries from the donor perspective, tending to incentivise isolated projects with short-term outlooks (Cochrane and Thornton 2016). Cochrane and Thornton (2016) argue that how companies such as Charity Navigator rank charities reflects broader trends in development in that the professionalisation of the sector continues to promote technocratic and calculating approaches, maintaining the dominance of the neoliberal agenda. On top of this, they warn that cost-effectiveness metrics may miss important components, which result in unintended outcomes negatively affecting beneficiaries and the development sector.

The tourism industry's development as a mechanism for private sector sustainable development, poverty alleviation, and environmental conservation has been criticised for its 'lofty goals', often failing after development donor funds are no longer available (Tyrrell et al. 2012:280). For instance, the Overseas Development Institute (ODI) has been amongst the first organisations contributing to gain a better understanding of the impacts of tourist development in low-income households at the destination level and recommended four evaluation tools: livelihood analysis, enterprise analysis, local economic mapping, and pro-poor value chain analysis (Mitchell 2012). Tourism impact assessments, however, have often ignored a range of features beyond quantitative income variables (Thomas 2014). Mlynek (2017) argued that giving must come with dignity, and in tourism, whilst there may be value in viewing what life is like in other countries, one must be mindful not to fall into the poverty tourism trap, using poor people like any other tourist site or wild animal attraction, which is a relevant consideration for the travel philanthropy domain. It is evident that there is a need to look beyond standalone economic impacts to fully understand the real socio-economic world in which the practice occurs. Equally, 'unobserved tourism' as suggested by Cantis et al. (2015) does not exist politically if it is not measured and reported, meaning it is less planned for, with travel philanthropy possibly falling into this trap. However, how to 'measure' impact is a thorny debate.

3.1.1.4 Towards the art and science of 'development'?

A key debate in international development is whether the sector can progress from evaluation as "institutional rituals" (Hailu 2015:818) to evaluation as meaningful and reflexive learning practices in a complex world (Green 2016). Management of development projects has for decades been concerned with social change solutions to advance development, and project evaluation has been the focus of much discussion since the 1960s (Richards 2019). Development interventions in Africa persist under the banner of creating changed realities for the poor, yet processes of social change, how change happens, and why change does not happen are still not well-understood (Green 2016).

Policy fails when complex problems are addressed using linear reductionist approaches, having expectations that past data will accurately predict the future (Mueller 2020), and many poverty alleviation interventions have failed to deliver results (Sarker and Rahman 2015). Evaluators may not be blind to the impacts of complexity and the influence of contextual elements (Weiss 1997). However, progressions in approaches have had to go against the grain of non-systemic and mechanistic ways of thinking that are far more familiar with focussing on isolated programme effectiveness rather than supporting change in dynamic social systems 'fraught with complexity' (Moore et al. 2019:75). Interventions in low-income countries are also found to present complex environments to carry out evaluations (Pawson and Tilley 1997, Emerson 2020, Webster et al. 2018), especially due to complex causal relations difficult to establish, invisible achievements, information access difficulties, interventions often taking place 'behind closed doors', and research resources often being limited (Wessel 2018:401).

The contemporary evaluation literature is filled with prominent and conflicting arguments. Fear and lack of willingness amongst NGOs to see failure as a necessary learning rather than a lapse in reputation is an issue (Kirsch 2014; Green 2016). Intended and unintended consequences must be actively searched for (Wessel 2018). Contextual influences must be considered beyond our own intuitions/assumptions (Reinholz and Andrews 2020). Evaluation frameworks 'enshrined into results-based contracts' and 'superficial' visual representations do not open the important black box or 'missing-middle' of causality, and thus more in-depth, critical, and flexible theories of change are now higher on the agenda (Vogel 2012:19). Moreover, most evaluations fail to capture system-wide effects and complexity (Mutale et al. 2015). Real people in the real world need to be able to use evaluation findings, which present opportunities as well as difficulties. Thus, there has been a shift in embracing what might feel like 'disappointments', to use these learnings as "catalysts for rethinking, redirection and improvement" (Patton 2019:34). Therefore, adaptive iterative management informed by a living theory of change (ToC) is argued to better reflect the demands of today's complexity (Richards 2019), drive superior evidence-based decision-making to advance our efforts in international development (UN 2020b), and avoid irreversible damage.

3.1.2 Re-evaluating 'evaluation'

The ethics of giving traditionally focussed on whether and how much to give to organisations helping people in poverty, but in more recent years discussions have focussed on where to give effectively, predominantly focussing on accountability of doing the 'most good per dollar donated' (Pummer 2016; Emerson 2020). Cost–benefit analysis to guide resource allocation dominated international development projects' evaluation for many years, serving as "an emblem of rationality and professionalism", and is still integral to the evaluation toolkit, although increasingly discredited (Picciotto 2019:7). Wessel (2018) argues that much of monitoring protocols demanding numerical accounting of outcomes convey little meaning of achievements. Picciotto (2019) argues that to improve impact judgements, distinctions need to be made between 'doing the right things' (relevance) and 'doing things right' (efficacy and efficiency). All three need to be evaluated because a project does not qualify as success if goals are actually wrong (relevance), if goals are right but not realised (efficacy), or if scarce resources are misallocated despite the best intentions (efficiency).

There may be a shift within the development sector towards understanding the impact of interventions better, but evaluating the merit, worth, and significance of a social intervention is a value judgement. Therefore, how and what values are selected as the basis of these criteria is a critical issue in international aid and development (Gates 2016; Emerson 2020). Value judgements about interventions are both unstable and temporal and vary depending on the purpose or perspective one takes. Thus, evaluation needs to step it up to account for this instability and conditionality (Gates 2016).

3.1.2.1 Moral philosophy and judging the 'good'

Outside of international development, moral philosophers directly and indirectly contribute to the evaluation debate. What is clear from debates regarding the ethics of 'doing good', ideas of *consequentialism* or *kantianism*, is that context is obviously the crux of the matter, and so broad-brush evaluations of the good in giving can reflect on these philosophical perspectives. Ethical theories discuss whether we are obligated to bring about the best available consequences or if we are obliged to make self-sacrifices to prevent the suffering around the world (McElwee 2010). Ethics concerns how our conduct affects ourselves and how it affects others (Gomberg 1986), and according to Singer (1972), we have a duty towards victims of emergencies anywhere in the world. In virtue consequentialism, good persons have specific character traits that have good consequences, i.e., virtues (Schmidt 2017). Kant's deontology is the idea that there are moral principles that we should not give up every time we see they might lead to undesirable consequences, so this ethical perspective locates value of action in its maxim or principle, not in its foreseen or actual causal impacts (Williams 2019). For virtue ethicists, such as Hume, the moral quality of a person rather than their action takes centre stage (Schmidt 2017). Consequentialism says that the righteousness of an action depends on the outcome or consequence of the action, rather than the intent (D'Souza and Adams 2014). Singer (1972) argued that we ought to give to the point that our gift creates for us a cost equal to the benefit it would bestow on the victims, but Gomberg (1986) suggested consequentialist moral philosophies ask too much of us.

Applied to international development, Wisor (2012) argued that consequentialism should not guide NGOs, yet development agencies are moving in this direction. In NGO resource distribution, the question of consequences is critical, whether in terms of benefits or harm. Wisor (2012) argued that whilst consequences must play an important role, there are other moral reasons that should guide NGO resource allocation, such as equity, rather than just value for money or a strict cost-benefit analysis. If consequentialism informs spending, it will end up being allocated to a limited number of cost-efficient locations or causes. This approach is likely to create implausible and insensitive interventions, treat beneficiaries as mere units of consumption and production, and NGOs will grab the lowest hanging fruit doing the easy things, like building schools, rather than creating systemic change more difficult to measure. He suggests non-consequentialist theories such as deontology or pluralist positions are less trouble-some. Also, according to consequentialism, an act is wrong if it promotes a worse outcome than another available act would have done. Pummer (2016) similarly argued that although consequentialism may say to give all your money to charity, and whilst promoting the moral good is agreed to be relevant, there may be multiple moral factors that are relevant that also need to be weighed up, using common sense morality. On the other hand, not being guided by consequentialism and leaning on Hume or Kant instead effectively takes the responsibility of giving impact away from the donor.

Traditional ethical theories undoubtedly helped in many ways to structure and contribute new perspectives on evaluation theories in international development. However, Williams (2019) argued that we have yet to create a morality that is adequate to today's complex world. The reasons are that many of the moral philosophers focus their arguments on the 'individual', whereas in contemporary times, we increasingly face collective consequences, including harm, locally and globally, so how does the moral philosopher tackle that? Williams also asserted that "retrospective responsibility [does] not serve us well" (2019:459), when indeed we need to be both forward and backward looking.

3.1.2.2 Progression of evaluation approaches in international development

One of the first theory-driven approaches to international development programmes was the linear logical framework model, still used by many governments and NGOs today (W.K. Kellogg Foundation 2004; Reinholz and Andrews 2020; Floate 2019; Richards 2019). Criticisms shifted approaches to 'Theory of Change' with deeper cause–effect cycles presented in road maps, blueprints, or theory for action (Vogel 2012; Floate et al. 2019; Douthwaite and Hoffecker 2017). Refinements drew more attention to programme assumptions of how something works and why (context–mechanism–outcome), thus unpacking the ToC through a realist approach (Weiss 1972; Pawson and Tilley 1997; Chen 1990; Vogel 2012; Douthwaite and Hoffecker 2017; Wessel 2018; Webster et al. 2018; South et al. 2019; Richards 2019; Reinholz and Andrews 2020). The most recent trend is reflexive M&E, real-time learning or organising frameworks through developmental evaluation shaped around making use as a design feature of the intervention (not just knowledge ex post), embedded into the ongoing management, with rapid feedback throughout the life of an intervention (Patton 2011, 2016, 2019; Gates 2016; Bailie et al. 2020; Reinholz and Andrews 2020). The use of 'big data' has also now entered the debate, with the international development sector being criticised for being slow in integrating IT into programme evaluation and better solving global challenges (York and Bamberger 2020).

'Developmental evaluation' may change the way the international development sector operates because it may create a learning environment involving constant theory refinement rather than the pre-determined targets that have dominated the discourse thus far (Davis et al. 2016; Richards 2019). Distinct from formative (refine a strategy), summative (judge merit, value, and worth), and accountability-orientated evaluations (efficiency), developmental evaluation advocated for what in the current literature means ongoing iterative adaptation in complex situations (UNFPA 2020) and improving rather than proving, embracing non-linearity, feedback, emergence, relationships, boundaries, and adaptation (Webster et al. 2018; Bailie et al. 2020). Understanding systems and complexity thinking has therefore gained traction. Practices of developmental evaluation have rapidly spread across philanthropy, social innovation, and international development, but the approach has become messy with various interpretations, so questionable in terms of application in the current time (UNFPA 2020).

Ideas from systems thinking and complexity science (STCS) nevertheless increasingly challenge and transform the way international aid and development is conceptualised, practiced, and evaluated (Gates 2016; Williams and Hummelbrunner 2011; Ramalingam 2013; Mowles 2014; Douthwaite and Hoffecker 2017; Mueller 2020). Gates (2016) stressed that social problems cannot be assumed to be separable, solutions cannot be studied independently from other problems, and because social problems are not straightforward, linear, and rational processes, we need to stop trying to 'treat' problems with the 'right' solution. According to her, the sector requires ongoing political–moral inquiry, debate, and scientific inquiry beyond the inadequacy of only demonstrating ineffectiveness and negative consequences without challenging underlying logic and disregarding for dynamic complexity, such as feedback. A systems lens is argued to be different from a 'what works' lens by embracing messiness and complexity (Gibbs et al. 2020; Gates 2016; Green 2016). It is the patterns of causation said to be the fundamental source of complexity – recursive causal connections of positive, and negative feedback loops (Homer-Dixon et al. 2013). Evaluations through a complexity lens therefore include non-linearity of change and the emergent nature of outcomes (Wessel 2018).

The key arguments driving these changes to evaluation models were to advance from the unrealistic, predictable linear pathway, the flawed focus on rigid intended outcomes, not processes, single stakeholder views, inflexibility, superficial reports and ignoring the reality of complexity. The evolution, advocacy, and criticisms of contemporary evaluation processes are still ongoing, critiqued particularly for remaining a formulaic template often not properly understood or applied (Floate et al. 2019; Vogel 2012), as well as remaining blind to outcomes that are not anticipated or intended (Arensman et al. 2017; Vogel 2012; Douthwaite and Hoffecker 2017). However, the more recent evaluation models drawing on complexity science, realist evaluation, and reflexive M&E aim to create an empirically-based and complexity-aware change theory for inclusive and sustainable development (Douthwaite and Hoffecker 2017). From systems thinking, Senge's (1990) self-reinforcing processes have become part of the ToC complexity models, informing higher-level causal assumptions (Douthwaite and Hoffecker 2017).

3.1.2.3 The bigger picture – individual action and the collective level

Despite this growing awareness about the need to face up to complexity, a reductionist view remains – i.e., that a particular intervention can be assessed using a single outcome, even though it is clear responses are always system-wide (Mutale et al. 2015). A call for a shift from programme-level ToC towards the development of middle-range theories of change is present in the literature, to transcend the individual program level and have the potential to apply to a wider set of interventions. The aim is to produce sustainable development impact through creating abstract analytic frameworks for accumulating learning, which can abstract across cases to identify common mechanisms (Pawson 2013; Douthwaite and Hoffecker 2017). This approach demands viewing the 'context', which requires thinking in systems to understand how change happens and what needs to change because programmes operate within webs of informal and formal systems; formal may be 'seen by someone', but informal systems are argued to be very real too (Moore et al. 2019:77). For instance, 'community context' must not be viewed as a 'static backdrop to an intervention', but rather a dynamic feature of a wider system (South et al. 2019:11). Contexts are defined as the settings in which activities occur (Gates 2016), but boundaries are blurred for many development interventions (Mutale et al. 2015).

Within this debate has emerged the argument that project outcomes tend to be explained through individual responsibility rather than systemic effects, which is inaccurate when interventions are rarely bounded (Jackson 2019; South et al. 2019). Collective impact evaluation is therefore included amongst innovations in developmental evaluation in complex dynamic situations, focusing on levers of change that operate at different levels of a system (South et al. 2019; Patton 2019). As Reinholz and Andrews (2020) point out, it can be hard for individual actors to recognise how more distant parts of the system influence change efforts. Having a central oversight of the NGO sector is, however, found problematic because it is not only made up of large, well-funded NGOs but also a 'patchwork quilt' of interventions of varying size and quality (Edwards and Hulme 1998:10). Patton (2019) argued for learning processes designed around a collaborative approach to evaluation. There are indications of advancements in evaluation, and in Africa, according to Bailie et al. (2020:2), in agricultural research and health, where 'innovation platforms' are starting to bring diverse stakeholders together from different parts of the system to learn, share ideas, and solve problems. Efforts are growing to increase the quality of evidence and improve the ability to transfer findings to other contexts in international development (DFID 2019).

3.1.2.4 The changing but nascent narrative about systems thinking and complexity science in international development

Conversations regarding STCS, including methods, methodologies, and conceptual and theoretical issues, have been emerging since the 1980s, so they are not entirely new (Ulrich 1988), but broader and deeper discussions in international development are more recent. Douthwaite and Hoffecker (2017) argued that this familiar impact pathway narrative locked in over the last few decades does not support organisations to harness beneficial system dynamics nor reflect the difficulty in forecasting impacts ex-ante or the production of unexpected impacts; thus, real intervention impacts remain invisible. It also remains challenging for those embracing complexity-awareness to communicate such impacts to traditional donors, who still expect results that are based on the adequacy of their new 'solution', the number of 'beneficiaries', and specific livelihood impacts of that solution on those 'beneficiaries'. Evaluating complex and systems change interventions by considering emergent processes and outcomes has only more recently become the narrative in international development and is arguably still nascent (Douthwaite and Hoffecker 2017).

There has been progress in complex systems thinking approaches and some evidence of the difference this makes in evaluation; however, more recent studies have begun to question how complexity is actually understood and used by policy makers and other evaluators or development professionals, which is part of the challenge (Barbrook-Johnson et al. 2020). This theoretical advancement may very well sound like the answer to formerly ineffective development evaluation, but it also requires appropriate data and reaching the right people in a way they can understand to make changes and have the power to make decisions (Webster et al. 2018). Barbrook-Johnson et al.'s (2020) research in the UK indicated that surprisingly few evaluators are embracing complexity – issues highlighted included that the value proposition needs to change and framed as an opportunity not a cost, and that the skills and tools to communicate complexity up the hierarchies of the aid chain can be a barrier.

Complex challenges to be addressed in international development nevertheless demand this knowledge and these new approaches to grapple with the deeply systemic nature of real-world issues. The importance of embracing systems and complexity clearly warrants the battle against the norms, bureaucracy, and overall structure of the aid sector. To embrace systems and complexity, it has been found necessary to 'zoom in' to understand the influences on human decisions and behaviour, 'emergence' in systems, and consequences of interventions and also 'zoom out' to understand the broader socio-economic context in which they operate (UNFPA 2020). Evaluation in international development demands a deeper understanding of systems and complexity thinking, which is the focus of the next section.

3.2 Complex systems and international development

3.2.1 Living in an interconnected world: systems thinking and zooming out through an interdisciplinary lens

Climate change, the financial crisis, aid dependence, and the Covid-19 global pandemic are just a few of the more obvious examples of the interconnected world in which we live, associated with macro-systemic decision-making and action consequences that can reach even the remotest corners of this planet. In response to global challenges, there are other micro-level interconnected development interventions, like flood control efforts that may lead to more severe floods (Gates 2016) or Ethiopian irrigation and hydro-power in one area becoming detrimental to livelihoods of fishermen in another (Ratner 2020). Other negative impacts include those from long-term welfare programmes generating a lack of resilience, and efforts to raise self-esteem creating narcissistic behaviour (Von Bergen and Bressler 2017). Positive, unintended consequences can also be identified (Green 2016). However, despite a growing awareness about such unintended consequences and the complexity of addressing specific problems, there remains a scarcity of evidence regarding the application of systems thinking to evaluate complex interventions in low-income settings (Mutale et al. 2015) and we remain 'complexity-blind' (Douthwaite and Hoffecker 2017:89).

Self-organising social systems or systems of spontaneous order are the result of interconnected human actions rather than purposeful design (Meadows 2008). Dynamics in systems include emergence and feedback processes, with agents adapting and co-evolving in processes of self-organisation (Mingers 2014). They organise based on feedback from their environment rather than a central authority (Smith 1759; Polanyi 1944; Hayek 1967; Meadows et al. 1972; Senge 1990; Mingers 2014; Jackson 2019). Created and controlled by no one, many systems in the world demand to be understood by grasping such emergent social phenomena, insofar as they arise from individuals, yet possess novel features irreducible to the properties of individuals (Bhaskar 1989; Lewis 2000). Social entities that emerge do at the same time co-occur with individuals, so we therefore find that some explanations can be reductive to their parts but also possess irreducible causal powers not identical with the causal powers of their parts when they are isolated – i.e., powers individual human agents have when separated from the whole. The change in one system component can affect another, thus promoting self-organisation and emergence at different levels (Banson et al. 2015).

In complex systems, non-linearity is a distinctive characteristic, which means that there is no proportionality between cause and effect, the sum is greater than the parts, and we find what is called 'emergence'. Small changes can therefore lead to large, unanticipated impacts throughout the system, passing thresholds or tipping points (Gladwell 2000). The key characteristic of complex systems is that their behaviour is difficult to understand; they are unpredictable and characterised by their adaptive and dynamic nature, feedback loops, multiple scales, thresholds for change, high and low areas of stability, and open or ill-defined boundaries that span areas of expertise and responsibility (Barbrook-Johnson et al. 2020). It is impossible to understand these influential dynamics if we are only zoomed in, so problems need to be framed by seeing the "whole forest rather than a particular tree", looking beyond the details to the context of relationships in which they are embedded (Sherwood 2002). In development interventions, complexity refers to when short- and long-term consequences of a decision result in entirely different outcomes and/or when intervening in one component results in unexpected consequences on another component (Mutale et al. 2015).

Systems thinking requires an interdisciplinary framework to see the root causes of challenges because 'problems' are part of an overall system. Thinking in systems aids conceptualisation and action towards integrating socio-economic dimensions and ecological, political, and technological aspects of sustainability (Banson et al. 2015). Real-world cases illustrating phenomena, such as imitation, fads, and habits, partially invalidate traditional economic theory. Rather than market forces, actors affect other actors, and adapt to what others do; thus, culture, beliefs, institutions, norms and technologies change in response to changes in the other (Mueller 2020). Change in complex systems may occur in slow processes – i.e., demographic shifts – and sudden unforeseeable jumps, such as rises in oil prices (Green 2016).

Whilst such systems may be dynamic and unpredictable, they produce recognisable patterns (Douthwaite and Hoffecker 2017). In complex systems, actors often interact in unstructured ways that lead to the emergence of cross-scale patterns that influence interactions amongst actors (Lambe et al. 2020). The unpredictable nature of systems does not lend itself to tight control and management, but by understanding their interconnections and dynamics, systems can be more successfully intervened in probing certain activities, and the best we can do is think in terms of probabilities (Douthwaite and Hoffecker 2017; Mueller 2020). Mueller (2020:312) believes systems can be 'harnessed, influenced, constrained, tweaked, hedged' and, to some degree, understood. Whilst there is order in self-organising systems, due to the sheer number of relationships and feedback loops amongst their elements, human systems are without a doubt complex (Green 2016). Whilst we cannot predict the future with precision, such approaches are found to help governments, aid professionals, policy makers, and others influencing the system anticipate longer term and broader systematic consequences of decisions and actions and to help avoid unintended consequences resulting from the silo mentality (Banson et al. 2015).

STCS are sometimes two distinct traditions within the systems and complexity fields, with different theorists, concepts, and methods, but also realised to incorporate ideas that span both (see Gates 2016 for analysis). Complexity has received increasing attention, particularly in the evaluation literature since the early 2000s (Gates 2016). Gates (2016) introduces the term STCS to combine the two, focusing on viewing systems as ontological phenomena, such as healthcare systems, rather than viewing systems as social and epistemological constructs that people use to understand complex situations. Critical systems thinking (Flood 2010; Jackson 2019; Mingers 2014; Ulrich 1988) uses social theory to reveal underlying assumptions in systems thinking and a shift from ontology to epistemology (Jackson 2019).

Extant literature calls for developments in middle-range theories that describe how interventions operating in contexts characterised by complexity produce system change (Pawson 2013). There is also a growing body of literature in international development advocating for the use of systems thinking. Fields include agriculture (Douthwaite and Hoffecker 2017; Banson et al. 2015), public policy (Seddon 2008), finance (Davis et al. 2016), health (Jackson and Sambo 2019; South et al. 2019), community-based research (South et al. 2019), advocacy (Wessel 2018), and organisational management (Senge 1990).

3.2.2 Beyond bounded problems in Africa's development

Breaking systems into their separate elements and applying siloed quick fixes has shown underwhelming success after decades of development efforts. Creating boundaries around 'problems' in international development to make them seem solvable is ineffective, and the fragmented landscape of programmes that form as a result needs to be replaced with a systems approach (Davis et al. 2016). Complex natural, social, and economic systems demand new ways of thinking beyond the linear approach to solving problems, especially in addressing 'wicked problems' and subjectively experienced problems such as poverty (Rittel and Webber 1973). South et al. (2019) echoed this argument from a public health perspective, stating that the dominant research paradigm continues to focus on defined individual-level interventions rather than being 'context-centred', which would explore levers for change that operate at different levels of the system and the contribution of different actors.

This realisation partly stems from growing evidence that fixing one problem can shift the problem somewhere else, whether to another component in the system, across disciplines, or across time, creating bigger problems in the future (Sherwood 2002). As Jackson (2019:148) says, "the real-world is not divided up into academic disciplines". Studying and solving complex issues requires interdisciplinarity (Lowe et al. 2013), systems thinking (Checkland 1981; Senge 1990; Meadows 2008; Seddon 2008), and an application of theory and methods to understand how systems can respond differently to interventions depending on local conditions and broader environmental factors (Ramalingam 2013; Barbrook-Johnson et al. 2020). Elements of a system must be identified, but this involves broadening our perspectives towards 'apparently unrelated factors' and understanding how structures, agency, and the broader context interact to create change (Green 2016:10). To strengthen any system, we first need to understand it (Jackson 2019).

Only relatively recently has this more holistic perspective been embraced by development practitioners, and the drive to focus on root causes and interconnectedness amongst system components is increasingly being prioritised. Although in 2020 this approach had been embedded in business, health, conflict management, natural science, education, population, and various other fields, it arguably remains recognised and adopted by a very limited proportion of all those intervening in Africa's 'development'. It is nevertheless evident across global development literature that systems thinking is an approach that has become a burgeoning debate and a positive step forward. Whilst identified by Banson et al. (2015:676) as a 'new concept for sustainable management' in agriculture, one of Africa's most important sectors, he argues that it is "yet to be exploited by many researchers, managers and policy makers in Africa". It seems very obvious to some in the 'development' sector that a systems approach is critical for sustainable development in Africa, but for others it remains a mystery, and linear approaches continue to be adopted.

3.2.3 Understanding complexity in development

Mental models traditionally used in development tend to be linear plans from the perspective of the donor, so if A then B, but this has led to repeated consequences, which include failure, frustration, and missed opportunities (Green 2016). When evaluating non-linear processes involving multiple stakeholders, we must address different interests and demands amongst these actors (Arensman et al. 2017; Banson et al. 2015). In systems thinking, the concept of understanding mental models of diverse stakeholders involved in systems through gathering different systems perspectives has become a key part of this approach (Senge 1990), which, when combined, are used to visualise a system. One is cautioned by Lambe et al. (2020), however, that whilst it is important to study actors within a system and the system itself, assuming that people's interests, expectations, and experiences are the same homogenises social complexity and trivialises experience-based behaviour, cultural context, and meaning.

Complex systems theory is not new; it emerged from general systems theory, cybernetics, chaos theory, systems dynamics, and other schools of thought, and the origins reach way back to Aristotle (Jackson 2019), but for international development interventions, complexity is a more recent discussion (Wessel 2018). Sustainability challenges are complex and constantly changing, and anyone attempting to positively contribute towards the sustainable development goals (SDGs) has no choice but to be adaptive. Systems thinking supports the management of complexity more effectively, and looking for system traits drawn from complexity science seems an increasingly obvious direction for anyone involved in sustainable development in Africa. Policy making and evaluation settings are putting more attention to complexity, but a continued ambiguity around the concept hampers its usefulness in practical terms (Gates 2016). As systems are realised to be constantly changing, we are urged to become more 'reflectivist' (Ramalingam 2013) and dance with the systems (Meadows 2008). Using Green's (2016:20) analogy, we must shift from being architects and engineers to 'ecosystem gardeners'. The way research questions are framed and evaluated requires a paradigm shift (Mutale et al. 2015), and STCS are invaluable for evaluations with any purpose (Gates 2016).

3.2.4 From ideological blindness to the truth of emergent outcomes

Understanding intervention outcomes as emergent rather than planned, looking at interactions during the process of implementation, is now being advocated (Arensman et al. 2017). We have heard many times that "yesterday's solutions become today's problems" (Sterman 2006:507) and that "good intentions may not always translate into good outcomes" (Von Bergen and Bressler 2017:114). There are also various references to unwelcome effects from bureaucratisation, tunnel vision and measure fixation (Zwart 2015). Giddens' (1993:765 in Zwart 2015) unintended consequences are consequences that result from behaviour initiated for other purposes.

Emergent outcomes in complex systems can be intended or unintended. In international development, we often hear about short- and long-term consequences (Mutale et al. 2015). In social science, unintended consequences, especially unwelcome side effects of policy, have been the topic of debate (Zwart 2015). Because unintended outcomes are often mistaken for ineffective management, their identification in international development still carries the stigma of failure rather than an evolving, strategic approach. Growing evidence across the literature argues for a shift to focus on all outcomes, intended and unintended, or else we risk critical emergent outcomes being overlooked (Arensman et al. 2017). However, there is also evidence that there remains great confusion around the concept of consequences in a development and a complex systems context.

Zwart (2015) expanded the consequences debate, arguing that to assume unintended consequences are always unanticipated is naïve; rather, expected negative effects are more likely to be traded off against intended consequences or gains. He also pointed out that theories such as bounded rationality offer explanations about why there is a failure to anticipate unwelcome consequences and systemic effects of aggregate outcomes caused by numerous unorganised individuals. However, he argued that some theorists are prepared to accept this as the rule rather than the exception, so unintended yet anticipated effects, he argues, receive less weight than they should. This issue, he argues, is entangled with justification and responsibility, the 'double effect' amongst moral philosophers, unintended harm carefully weighed on a moral scale. He argues that if unintended effects are anticipated, they are a different phenomenon as they follow from choice.

Many outcomes found in social living systems demand understanding human decision-making, so evaluation professionals advocating for grasping non-linear change in a complex world, and so on, really cannot achieve this without reflecting on interdisciplinary knowledge, which can inform what is behind relevant decisions of those in systems. For instance, beliefs, principles, and norms have been flagged as a critical element in understanding development better, although they are recognised as difficult to articulate (Reinholz and Andrews 2020) and constantly changing (Mueller 2020).

Travel philanthropy is arguably much more complex than travellers and tourism businesses doing what they consider good in tourism destinations. Therefore, our understanding of travel philanthropy and sustainable development requires broadening and deepening, using an interdisciplinary lens and a systems approach. The interrelationships within travel philanthropy as a phenomenon largely remain unexplored, leading to the need for an interdisciplinary theoretical framework that looks beyond international development and gift-giving thinking. The next chapter 'zooms in' and addresses this perspective, thus delving into travel philanthropy as a system.

4 Gift-giving and philanthropy through tourism

Emerging theories and practices

This chapter offers a comprehensive review of extant literature, theories, and practices of gift-giving and philanthropy through tourism. The multifaceted nature of 'travel philanthropy' is discussed here and key conceptual themes identified, laying the foundation for critical knowledge gaps addressed in the subsequent chapters. The role of travel philanthropy in sustainable development is also introduced.

4.1 Changing landscapes of philanthropy

The prevailing narrative and image of the African continent curated to "whip up international compassion for Africa's woes" (Ayittey 2015:345) has influenced individual values around the world. This has driven a growing inclination to intervene in 'development', a concept often misunderstood by organisations and individuals attempting to change the world to one they would like it to be (McDonald and Scaife 2011; Hüncke and Koot 2012; Harrison 2013).

International Official Development Assistance (ODA) aid chains, as well as some larger charities (i.e., Oxfam), are being widely critiqued for weak connections with local realities and financial wastage rating up to 30% (Pettit and Beresford 2009). As a result of these criticisms, private donors, including so-called philanthropists from higher income countries, have looked for alternatives to supporting institutionalised foreign aid (Korf 2007) and achieve greater geographical proximity to overseas causes they believe in (Harrow and Jung 2016).

During the last two decades, private philanthropy for international development has expanded to comprise a larger number of relatively small donations (Atkinson 2009) in addition to traditional foreign aid. There is, however, no reliable data source available to estimate the aggregated public and private financial aid value received by African recipients. Atkinson (2009) argued that public policy should pay more attention to private willingness to make charitable transfers overseas.

Philanthropy has been differentiated from 'charity' (Bertini et al. 2014) and hailed for its potential to be more impactful and sustainable than government foreign aid, as it is characterised by innovation, less bureaucracy, and fewer economic spending constraints common to aid budgets. Bertini et al. (2014) argued that, with

DOI: 10.4324/9781003497929-4

aid budgets under pressure, global philanthropy is becoming more important, with philanthropists often praised for their new approaches. They also highlight the evolution of new approaches and mechanisms within philanthropy, such as 'venture philanthropy', as well as renewed interest in old ideas, such as collaborative giving. As a consequence, a growing critique around the lack of governance and hidden political agendas of private philanthropists, such as high net-worth individual donors or those driving the so-called philanthrocapitalism, has emerged (see McGoey and Thiel 2018).

Meanwhile, Global North citizens have shifted their compassion for distant suffering towards a desire to directly act across that space to alleviate it (Korf 2007) and achieve physical and personal encounters with beneficiaries or 'distant others', whilst creating in-lifetime legacies (Silk 2004). The evolution of global tourism, of which growth has been well documented (UNWTO 2024; WTTC 2024), is considered instrumental and increasingly embedded within this shift. Within this overall growth, an increasing number of African countries have been making their way onto the popular destination list of travelling philanthropists to transform Africa's woes. While the African continent may remain dominated by tour operator-led wildlife safari products, in the last ten years we have witnessed a major growth in travel products shaped around philanthropy for 'transformative tourism' (Valdez and Magio 2016), reinforced by the more recent call for better tourism (UN 2020a).

The influence of tourism on generating financial donations to destinations has existed for decades (Ardoin et al. 2016), although meaningful quantifications of the phenomena are currently not available. Over the last decade, evidence exists that financial donations have been amplified, partly due to the travel industry's response to a growing travelling population segment wanting to donate or 'give back' (Phocuswright 2015), increasing corporate social responsibility (CSR) efforts (Goodwin et al. 2009, Novelli 2015), and individual philanthropists travelling more to Africa (Honey 2011). Phocuswright (2015) stated that the future of travel philanthropy was bright and expects a third of travellers to increase their travel-related giving in the following two years. This amplification is likely to continue with the call for solidarity (UN 2020a) that has emerged from the pandemic arguably providing further grounds for tourism's growing role within the realm of philanthropy.

There are well-established assumptions that giving is good, which undoubtedly inspires ambitions of private philanthropists, including travellers and tourism businesses, to generate more donations (Phocuswright 2015; Honey 2011). However, there is limited evidence about the real complexity of gift-giving through tourism, as well as its differentiation from other forms of aid. There is a major void in our understanding of its meaning, significance, and impact in an economic, as well as social, cultural, environmental, political, institutional, and ethical context – and most importantly from the perspective of intended 'beneficiaries'. This research explored the concept of *travel philanthropy*, arguably a multifaceted, ambiguous term that is theoretically weak.

4.2 Travel philanthropy: a multifaceted term, conceptual elaborations and meanings

A range of authors within tourism studies have touched on 'philanthropy' in 'tourism' *per se*, such as improving pro-poor approaches (Ashley and Haysom 2006), with labels associated with 'philanthropy' used interchangeably within the tourism industry, academic discourse, and narratives. Some examples include Altruistic Travel (Weaver and Jin 2016, Jose 2013); Adventure Philanthropy (Michelson 2013); Compassionate Tourism (Nankervis 2009); Meaningful Tourism (Coghlan and Filo 2013); Charity Tourism (Coghlan and Filo 2013); Philanthropic Journeys (Fernandez 2010); Donor, Private and Urgent Service Travel (Elevate Destinations 2012); and Social Tours (Burgold and Rolfes 2013). A handful of organisations, such as the Centre for Responsible Travel (CREST) and Sustainable Travel International (STI), have published guidelines and handbooks for 'travellers philanthropy' (Global Giving 2014; Sustainable Travel International 2008; Honey 2011), mainly to assemble descriptive case studies and practitioner accounts. In 2008 and 2011, CREST international symposium focused on travel-related philanthropy echoing this topic, in an attempt to generate interest in the subject. It is during the 2008 conference, that a group of experts comprising of Prof. Marina Novelli (co-author of this book) and practitioners from across the world became acquainted with what appeared a generally very superficially explored subject in need of critical assessment.

While 'travel philanthropy' and 'travellers philanthropy' are relatively new terms (see Honey 2011; Seltzer 2004; Novelli 2015; Novelli et al. 2015; Bekkers and Weipking 2011; Ardoin et al. 2016), also used by public, private, and third sector organisations, Seltzer (2004) is cited as the one who coined the term 'travellers philanthropy' in 2001 to encompass "3 T's of support – time, talent, and treasure – to host communities" (also see Honey 2011). Travellers philanthropy has since remained a term seemingly agreed upon, and the subject of expanded and morally charged debates. It does, however, seem that Seltzer's (2004) definition of 'travellers philanthropy' may not be sufficiently explanatory of the wider meanings associated with the term, leading to vague interpretations and inconsistencies in application. Sources questioning the actual definition, language, or meaning are sparse (Novelli et al. 2015, Scarth and Novelli 2019), yet a range of broad-brush issues and advice surrounding the presumed concept prevail. The paradoxical meaning and application of the term and concept thus remain a challenge faced by all stakeholders dealing with such growing and complex phenomenon.

The term 'travel philanthropy' (Novelli et al. 2015) is better for the purpose of this study because it is more inclusive for defining the concept. The use of the 'travel' and 'travellers' terms in the specific philanthropy context is not addressed in the extant literature and appears to refer indiscriminately to all 'tourism' stakeholders or as an act by 'travellers' within the 'tourism' phenomenon. Such inaccuracy echoes other assumptions equating 'tourist' with 'foreigner' (Mitchell 2012), despite clear definitions of what tourism is (UN 2010). If we want to ignore domestic tourism and diaspora markets by only associating the term with 'foreign' travellers, it needs to be more explicit. Furthermore, if the term means anyone travelling to places outside their normal environment for any philanthropic-related action, merely intended as philanthropic donations of time, talent or financial resources/treasure (Seltzer 2004), this will exclude the involvement of other key stakeholders, such as businesses, communities, and other actors necessary to understand the concept under investigation.

Similarly, the term 'philanthropy' is also a contested term and concept argued to be sitting in a 'loose field of theorising' (Liket and Simaens 2015:286) across a wide spectrum of debates (Harrow and Jung 2016). A substantial body of knowledge is available in the social sciences domain on philanthropy *per se*, and research appears across different disciplines and their fields of study, including for example marketing, economics, social psychology, biological psychology, neurology and brain sciences, sociology, political science, anthropology, biology, and evolutionary psychology (Bekkers and Wiepking 2011). In tourism studies, we must reflect on this ancient phenomenon and meaning to understand its significance within the tourism and sustainability domains specifically, to inform any further associated discourses.

The most conventional contemporary definition of philanthropy is "private initiatives, for public good, focussing on quality of life" (Bertini et al. 2014:29). According to Goldfarb (2011), much of the non-profit management literature equates philanthropy with money. Reflecting on Salamon and Anheier's (1992:130) definition of philanthropy as "the giving of gifts of time or valuables (money, securities, property) for public purposes" and Payton's (1988) definition as a "voluntary action for the public good", Goldfarb (2011) raises the important question of what constitutes a 'public good', which also opens up a much wider interdisciplinary debate.

The deeper one goes into the literature, the less philanthropy is understood to be an outward act of charity, but rather one that emerges from inner ideals, or the disposition of human nature that motivates outward acts of benevolence (Sulek 2010). Academic work on the subject of philanthropy comes largely from outside the tourism field and context focussing mainly on the characteristics of those that engage in philanthropy, such as how intentions are influenced (Bekkers and Weipking 2011). Work also looks at reasons and motives for giving (Nickel and Eikenberry 2009; Bekkers and Weipking 2011; Etang et al. 2012; Goldfarb 2011; Mickiewicz et al. 2016; McDonald and Scaife 2011; Harbaugh 1998). More recently, these concerns have also become the predominant area of investigation in tourism studies (Weaver and Jin 2016, Ardoin et al. 2016). Nevertheless, despite definitional disputes, Goldfarb (2011) discusses the increased attention that philanthropy is receiving, particularly as non-profits face more financial constraints these days, thus being an increasingly important phenomenon to grasp as it is included in different agendas (i.e., international development, CSR, environmental conservation). Sulek (2010:204) provides a detailed account of the historical underpinnings of the term 'philanthropy', arguing the importance of clear definitions to direct the discourse.

There is still very little knowledge available about how philanthropy may manifest differently in the travel and tourism context (Ardoin et al. 2016). Concepts such as the 'warm-glow' have emerged with related literature touching upon the power of donor emotions in gift-giving and how this may impact levels of concern regarding the 'public good' vs. personal benefits derived from the act of giving (Atkinson 2009; Null 2011; Harbaugh 1998). Weaver and Jin (2016:661) emphasise that the focus usually returns to "fleeting tourism experiences that focus on the attainment of selfish stakeholders' goals", and that compassion has generally been neglected across the tourism literature, certainly lacking empirical investigation. Novelli et al. (2015) are the only authors who scrutinised our understanding of philanthropy and related debates over time within the tourism field of study, positioning their discussion around traditional, modern, and post-modern philanthropic approaches, and connecting the emergence of travel philanthropy within the African context (Novelli 2015). However, despite the constructive progress made in this argument towards a wider consideration of the phenomenon, there is a lack of clarity as to what philanthropy specifically means and/or encompasses within the actual context of travel philanthropy.

A deep dilemma about the 'do's and don'ts' of travel philanthropy therefore requires establishing consistency in what 'travel philanthropy' actually incorporates, fully acknowledging the diversity within it, and raising questions about the meaning and value of current discussions, towards a clearer direction in the discourse. The review of the term therefore urges some caution, particularly as it informs the research process and findings later discussed, which indeed challenge current fuzzy meanings and practices.

However, travel is thought to stimulate philanthropy amongst consumers (Ardoin et al. 2016), leading tourists to realise that they can make a difference and motivate them to put something back (Goodwin et al. 2009). Honey (2011:3) proposed a more comprehensive vision that travel philanthropy is not just about 'collecting loose change for charities' but rather helps tourism businesses become involved as 'good citizens' in the destinations where they work. In her perspective, tourism assists local projects with a 'hand up', not a 'hand out', by promoting social empowerment, education, and entrepreneurship in an effort to foster longer term sustainable conservation and development of destinations. It also enriches the travel experience for visitors through "meaningful, culturally sensitive, and productive interactions with people in host communities" (Honey 2011:3).

CREST, a key international advocate for 'travellers philanthropy' eager to raise awareness of the concept within the media, governments, and development agencies, highlighted the need to deepen, enrich, and embolden the concept (Honey, 2011). However, despite this continued lack of conceptual understanding, travel philanthropy appears to have become more widespread and professional in the ten years that preceded the work of CREST. Many of the travel philanthropy efforts are thought to proceed without substantive theoretical understanding of the entire phenomenon and remain anecdotal or conceptual (Ardoin et al. 2016). Without clear definitions and perhaps standards, there is a risk of this becoming another 'greenwashing' term, like 'ecotourism', which is erroneously used, loosely applied, and poorly understood, with little environmental or socio-economic significance, and in some cases negative ecological and social implications (Biggs 2008). It is therefore critical to investigate current definitions to clarify specifically what the term encompasses or perhaps should encompass, before examining some of its deeper characteristics.

Spalding (2011) partly questions the definition of travel philanthropy and queries what the concept encompasses, whether it refers to giving back to the host community, collecting donations, enhancing the traveller experience, voluntourism, or actually effective strategic philanthropy. Novelli et al. (2015) have pioneered academic, critical elaborations of travel philanthropy in the context of values and practices, placing it within the global development arena. Whilst recognising the potential benefits, they question the ethics and impacts of travel philanthropy and its place in sustainable development, with further elaborations of these provided in Scarth and Novelli (2019).

Inconsistent perspectives currently frame 'travel philanthropy' as a concept and the need for a deeper look into its associated or inferred meanings is needed to have a better understanding. In *Table 4.1* we attempted a rigorous analysis of the interchangeable use of the term present in the literature and identified a set of emergent themes/categories, enabling a better mapping of the travel philanthropy bigger picture – the whole (Senge 1990). These include the following: (1) The Input (to/from tourism), (2) The Vehicle (using tourism), (3) The Medium (through tourism), (4) The Intervention (by tourism), (5) The Output/Impact (from tourism), and (6) A form of sustainable tourism or a movement.

A. THE INPUT – to/from tourism

This is the philanthropic resource generally associated with money and its source, often for a specific cause in a tourism destination. This study identified two categories under this theme: (1) donation by a traveller and (2) corporate fundraising for CSR – voluntary donation or involuntary fee.

1 ***Donation by a traveller*** – Travel philanthropy is defined as the donation by a traveller on-trip or post-trip (Talladay 2011), as a (voluntary) contribution in addition to the cost of the trip already purchased (Maathai 2011), and an amount that generates "a new stream of cash, goods, and volunteer services beyond what is generated through the normal tourism business" (Honey 2011:207; Bertini et al. 2014).

   From this perspective, it would seem that 'travel philanthropy' is only the actual individual donation by a travel consumer. How, where, why and to whom that donation is allocated appear a separate issue. It is also unclear if travellers' donating direct to 'beneficiaries' independently of a tourism business are included or excluded, if they have travelled to the location of the beneficiary or if this is an incidental event.

2 ***Corporate fundraising and CSR donations – voluntary and involuntary*** – The scope of travel philanthropy is broadened by considering not only the above type of 'concrete contributions' (Honey 2011:3) by the 'traveller', but also by

Table 4.1 Travel/travellers philanthropy: use of term by other authors

Travellers philanthropy as …	*Including*	*Example authors*
A. Input – to/from tourism *Philanthropic resource and source – who and what*	• Traveller donation • Corporate fundraising for CSR – voluntary donation or involuntary fee	Talladay (2011), Maathai (2011), Honey (2011), Bertini et al. (2014), Crouch (2011), McLachlan and Binns (2014)
B. Vehicle – using tourism *Using travel and tourism to give – why and how*	• Demand-led (consumer trends) • Supply-led (product development and innovation)	Jamieson et al. (2004), Crouch (2011), Ashley et al. (2007), Ham (2011), Burgold and Rolfes (2013), Phocuswright (2015), Sustainable Travel International (2008), Spalding (2011), Hüncke and Koot (2012), Burns and Barrie (2005), Mostafanezhad (2012)
C. Medium – through tourism *Aid chain to beneficiary – distribution*	• Tourism company as channel/agency for donations • Unmanaged/direct traveller donations	Honey (2011), Sustainable Travel International (2008), Valdez and Magio (2016), Bertini et al. (2014), Spalding (2011), Lindkvist (2011)
D. Intervention – by tourism *Allocated resources – to what*	• Formal programme (implemented by tourism company) • Partnership (community-led intervention with tourism company) • Informal intervention (by traveller or tourism company)	Honey (2011), Thomas (2014), Phocuswright (2015), Bill (2011), Ham (2011), Bertini et al. (2014), Novelli (2015)
E. Output/Impact – from tourism *Results and consequences*	• Creating sustainable destinations	Spalding (2011), Ardoin et al. (2016), Brightsmith et al. (2008), Novelli et al. (2015), Valdez and Magio (2016)
F. Form of [Sustainable] Tourism Development OR Movement *Tourism as a force for good*	• Responsible tourism segment • Pro-poor benefits • Ecotourism expectations being met • Transformative travel and tourism • Capacity building	Pollock (2016), Spalding (2011), Ardoin et al. (2016), Barnes and Eagles (2004), Weaver and Jin (2016) Western (2011), Honey (2011), Novelli (2015).

the tourism business operator, as donations of 'time, talent and treasure' (Bertini et al. 2014; Honey 2011; Crouch 2011). The scope also includes contributions from a company and its guests to communities in tourism destinations through a variety of fundraising activities (Crouch 2011).

What this shows is that the concept is also associated with *voluntary* corporate philanthropic donations and charity and/or CSR, as charitable funds raised by a tourism business from any source and/or a proportion of their profit, is used as philanthropic contributions to a specific cause too (Honey 2011; Goodwin et al. 2009). This somewhat blurs the boundaries between private philanthropy by travellers and corporate philanthropy of businesses evident in many other industries (i.e., mining and pharmaceutical).

Examples of this practice include global tour operator Thomas Cook's allocating over $7.7 million in travel-related donations reported between 2005 and 2009 (Goodwin et al. 2009). Moreover, 39 other companies have raised approximately $249 million this way (Goodwin et al. 2009; Ardoin et al. 2016), evidencing that corporate philanthropy in the tourism industry is an established and potentially growing endeavour. With evidence showing that tourism businesses are voluntarily participating in CSR practices, it is unclear if this aspect of [tourism-related/based] philanthropy is indeed part of 'travel philanthropy', or if tourism industry CSR is a separate concept altogether. Honey's (2011:3) suggestion that travel philanthropy is not just about 'collecting loose change for charities', but rather helping tourism businesses become 'good citizens' in the destinations where they work, again takes the focus from the traveller onto the tourism business.

Whilst some authors (see Honey 2011) praise tourism for its fundraising successes, other authors argue that CSR lags behind in tourism compared to other sectors (Ashley et al. 2007). They say it should become mandatory (involuntary) for tourism operators to donate in the destinations they operate or implement a kind of tax to consumers to implement this (Blanco et al. 2012). For instance, in her study on Tanzania, Wolff (2011) reports on consensus amongst tour operators on the suggestion that the government should create a formal requirement for all operators to provide financial support to communities and conservation projects through a proportion of their corporate earnings. Similarly, while McLachlan and Binns (2014) support this view in their study on Zambia, some operators around the world openly add a fee to the cost of the holiday per traveller (e.g., Lets Go Travel Kenya 2011, i.e., for four+ days trip a $100 charge per person). This sits outside the voluntary domain and leads to questions about whether we can define such funds as travel philanthropy, CSR or effective pricing strategy to pay a kind of rent for operating a business in a public or community space.

Many tourism operators tend to be small and perhaps not keen to sacrifice their typically small profit margins, which raises questions about travel philanthropy becoming a clever means for the industry to maintain their social license in return for public cultural and natural assets under the banner of good works funded by others (McIntyre et al. 2015). 'Donations' in general, whether travel related or not, 'obligatory' or 'voluntary', can be associated with different moral motivations (Silk 2004) adding to the confusion caused by an earlier statement that travel philanthropy is "beyond what is generated through the normal tourism business" (Honey 2011:3).

The potential impact of giving is clearly enormous as the tourism industry jumps on the opportunity for additional funds and projects to become tangible, meaning "poverty and deprivation now have real faces" (Crouch 2011). According to the UNEP (UNWTO 2014), just $5 of each $100 spent by a tourist from a developed country travelling in a developing country remains in the local economy, which is far from ideal. This does not include philanthropic donations. Based on anecdotal evidence, millions of dollars arguably go into tourism host communities all over the world through donations, grants, and voluntary services, all of which are predominantly untraceable and unrecorded through any consistent method.

Ashley et al. (2007) identified greater potential in adapted business practices, which provides increased benefits to local economies and development impact. They offered evidence from Spier Estate in South Africa indicating that a small increase in local procurement injects more money into the local economy than what would be possible through philanthropy. This derives from the assumption that while the local economy has the capacity to support several linkages, the business would be able to maintain them. Linkages to a local economy have been fairly well researched (UNWTO 2013). However, limited evidence is available to question why traditional philanthropic programmes are so high on the agenda of tourism companies. Many have yet to even consider developing 'inclusive business models' or "the integration of development concerns into their core business models" (Ashley et al. 2007:23), thus "going beyond philanthropy to explicitly use their value chains to create and expand local economic opportunities" (Ashley et al. 2007:16).

While Pollock (2016:3) urged the sector to "stop using CSR and philanthropic practices to justify ignoring some of the structural and systemic flaws of the current system", Ashley et al. (2005) suggested going beyond making donations and investing in changed business models for mutual tourism business and local benefits. They highlighted that "community economic development is viewed as a philanthropic cause rather than a core business risk or source of opportunity" and that most tourism businesses were still at the philanthropic end of the spectrum (Ashley et al. 2005:14). Moreover, Ashley and Haysom (2006:268) reinforced the critical question of whether tourism businesses form potential linkages into local economies from their actual business practices, thus helping drive a pro-poor approach for the sector. They suggested the need to distinguish philanthropic donations from adapted business practice by indicating if it "is integrated into business practice or is a stand-alone activity". Arguably, over time net increases in local income can outweigh philanthropic spend. However, without measurement, even on an aggregate basis, how can we differentiate this?

For some, it is confusing whether 'travel philanthropy' is in reality just another term for/form of CSR. Organisational entities involved in 'travel philanthropy' are diverse. Some are fully commercial and some are non-profit. Some follow sustainable principles, and others may not. Kourula and Halme (2008) suggested that when companies develop new products or services as a solution to an environmental or social problem that creates a revenue stream, this is a kind of CSR that aims to fulfil a win–win condition, contrary to 'philanthropy'. Kourula and Halme (2008) distinguished three types of CSR: (1) *Philanthropy* (emphasis on charity, sponsorships, and employee voluntarism); (2) *Corporate Responsibility Integration* (emphasis on conducting existing business operations more responsibly); (3) *Corporate Responsibility Innovation* (emphasis on developing new business models for solving social and environmental problems including new products inspiring planned or impulsive giving, addressing social or environmental problems).

A handful of reports are optimistic about tourism companies' financial contributions in destinations from a 'CSR perspective'. Examples include McLachlan and Binns' (2014) research in Livingstone (Zambia), looking at 'proactive', 'active', and 'reactive' behaviours of tourism operators in terms of CSR, and Burns and Barrie's (2005) analysis of CCAfrica (now &Beyond) and The Africa Foundation, positioning these as an excellent example of CSR, albeit questioned for promoting colonial style behaviours.

A wealth of discussions covers the incentive for tourism companies to invest in CSR due to the nature of the business. As private tourism businesses are often based upon public assets (e.g., nature and culture), as previously indicated, evidence suggests a strong rationale for CSR in tourism in terms of maintaining a social licence locally (McIntyre et al. 2015). CSR has also been linked with pro-poor tourism (PPT) (Ashley and Haysom 2006), raising the importance of differentiating philanthropy as CSR, a 'donations approach' (or external action), from an 'alternative approach' of integrated business practices (or internal change), which is necessary for local pro-poor impact to be durable and purposeful, questioning *how* rather than why philanthropy takes place. In addition, CSR in tourism is said to have been driven by transparency of the 'production' process to consumers (Ashley et al. 2007) as well as meeting what are increasingly considered standard customer expectations, of which they [consumers] themselves are unwilling to pay extra for (WTTC 2015).

B. THE VEHICLE – using tourism

A significant proportion of the literature on travel philanthropy focusses on the role of tourism businesses in maximising travellers' giving, as well as creating platforms for philanthropic interventions. Are travellers the low-hanging fruit of philanthropic financial resources? Is it the role of the tourism industry to 'pick them'? If businesses do not respond to growing travellers' demand for 'giving back', will they lose their competitive edge? Within this conceptual category and in response to these questions related to using tourism as a vehicle for donations, two themes emerged: *demand-led* – i.e., travellers, and *supply-led* – i.e., tourism operators.

1 ***Demand-led*** – Travellers often actively look for ways to make a difference in the world, waiting to be shown or asked (Crouch 2011), and seek to improve the conditions of the destinations they visit, with Jamieson et al. (2004:9) highlighting that "the tourist must be provided with opportunities to directly participate in the poverty reduction process". Market research has demonstrated that consumers increasingly consider social impact when designing their trips (Ashley et al. 2007); personal satisfaction usually drives travel philanthropists who believe that a place is better off because they visited it (Ham 2011).

An investigation of whether travel philanthropy includes or is being driven by the demand for the 'consumption of giving/philanthropy', aside from donor motivations previously highlighted, suggests that marketable 'products' are sought by the industry to hit on travellers' desires or soft-spot for giving and 'lure' them to give or 'do-good'. As a result, some companies have realised the potential of this philanthropic revenue stream with marketing strategies. These include designing experiential products around caring for a place, as "a chance to serve their [donors/travellers] own interests" and personal satisfaction or 'intrinsic motivations' (Ham 2011:142), as well as serving 'conscientious travellers' (Burgold and Rolfes 2013) and their belief in being part of the 'poverty reduction process'.

Furthermore, companies are frequently urged to 'increase the flow of donations' (Bill 2011:210) and to integrate the communication and marketing of travel philanthropy programmes into their overall operations. Phocuswright (2015) advises addressing the growing demand of this segment of travellers and recommend creating CSR initiatives; educating and informing travellers about giving programmes; arranging site visits; and facilitating interactions with local communities. Thus, the tourism industry or suppliers responding to the demands of their customers act as a vehicle or platform for giving, often when they return home.

2 ***Supply-led*** – Travel philanthropy spans the market and is associated with a diverse range of travel and product types, which fall into the realm of giving as the central purpose of the tourist experience or giving as the by-product of the tourist experience (Honey 2011). As a competitive strategy serving the increasing demand for 'philanthropic travel', this has generated competition within the industry to appeal to contemporary consumer demands – thus become a strategy "to strengthen their foothold in this market" (Sustainable Travel International 2008:8).

The word 'lure' is common in the literature on travel philanthropy, whether demand-led or supply-led. Ham (2011) argues that the 'best wins' are likely to be when an appeal is focussed on the caring of place, both nature and people, leading to immediate effect on travellers' emotions (see Exquisite Safaris 2010) and projects' visits creating a "virtuous cycle of deeper [travellers'] commitment that leads to more donations" because 'seeing is believing' (Bertini et al. 2014:19).

The utilisation of 'productive interactions' mentioned above may suggest a market-driven approach to philanthropy (see Spalding 2011). This alludes to the manufacturing of relationships between a business and/or host destinations, based on learned travellers' behaviours, needs and motivations discovered when host communities (usually those considered poor) are physically encountered on trip and from which the philanthropic impulses emerged (Goodwin 2011; Burgold and Rolfes 2013). In order to achieve the so-called best wins and to 'lure' tourist donations, travel philanthropy is widely discussed in the context of travel product development, innovation, and consumption, framing tourism as a vehicle for philanthropy.

It is evident that some travel products are indirectly associated with giving but designed in a way that travellers are tempted to give. An example of this is the South African 'social tour' (Burgold and Rolfes 2013), in which giving is often spurred by impulses generated by being immersed in 'poverty', yet not the core focus of the tour. There are debates that tourism operators capitalise on the vulnerability of the destinations in which they operate (Dawson et al. 2011), and that they often promote a 'dual mandate' by drawing people to an authentic homogeneous group as a commercial experience – i.e., visiting indigenous people, whilst also supposedly promoting real development goals (Hüncke and Koot 2012). Burns and Barrie (2005:469) describe these travel products as "Much needed social and economic development [turned] into a 'holiday experience'". Burgold and Rolfes (2013:171) stated that small-scale tour operators, non-governmental organization (NGO) aid projects, as well as luxury travel companies (i.e., Abercrombie & Kent, Wilderness Safaris, and Base Camp Safaris) increasingly organise tourists' visits to development projects, centred around the notion of 'ethical consumption', and/or under social, community or volunteer tourism labels and "non-intrusive and sustainable ways to experience a country". This reflects what Honey (2011) referred to as a way to deepen traveller understanding about inequalities.

The charity travel market based on what is commonly known as the 'challenge product', including charitable organisations turning to tourism as a platform and facilitating alternative fundraising events centred around philanthropic adventure and sport tourism, has existed for some time (Coghlan and Filo 2013). These blend "the voluntary act of fund-raising with the more hedonic pleasures of a packaged adventure tour" (Lyons and Wearing 2008:15). Scant attention addresses whether travel philanthropy is separate from this blend of travel and fundraising by a charity, with research remaining limited to motive behind the participation of consumers.

Studies tackling the 'time' and 'talent' aspects of manufactured volunteer programmes within the travel philanthropy domain (Fernandez 2010) have evidenced, like Seltzer (2004), these as two of the three segments or the T's of travel philanthropy, although Honey (2011:208) seems to believe volunteering is separated from 'philanthropy' by stating that "some travel companies have philanthropy or voluntourism as their central mission". Although not central to the purpose of this specific study, it is worth noting that, to date, volunteering is arguably the most researched and debated philanthropy-related travel product, with Wearing and McGehee (2013) providing a detailed review of volunteer tourism literature, most of which has appeared in the last decade.

While the literature generally refers to homogeneous schemes, products, activities or effects, there is limited research clearly mapping a travel philanthropy typology, the fragmentation of the tourism sector and related implications of giving within tourism are normally brushed over or simplistically investigated, leaving a recognisable gap in the knowledge that demands further investigations into the travel philanthropy umbrella concept.

Using tourists to generate donations is a scarcely discussed issue. Butcher (2003) talks about the moralisation of tourism and highlights that alternative forms of tourism, such as ecotours or voluntourism, have one thing in common: tourists understanding themselves as the moral alternative to traditional mass tourism. These consumers follow the belief that their travel consumption contributes to the cultural and/or natural preservation of a tourism destination. Influences on tourist behaviour may include 'the pristine and exotic other' (Hüncke and Koot 2012:671), and idealistic notions that "we visit and seek to locally support sustainable change by proactive investments in long-term professional partnership projects" (Lindkvist 2011:136). These influences support the assumptions about the homogeneity of tourists, and tourism organisations across the board, posing romantic visions of today's traveller and industry behind the growth of giving.

Shifting values, increased wealth, technology, and contemporary needs are highlighted as just some factors shaping how, why, and where people give, and travel, which have contributed to the 'evolving product of its time' (Bertini et al. 2014:18). In addition to the products mapped in Section 2.2, some philanthropic-related tourism experiences also sway towards a philanthropic by-product of a wildlife safari, such as visits to orphanages or local charities, as part of a wider nature-based or adventure trip, rather than the core purpose (Novelli 2015). The encouragement for and assumed legitimacy of the predominantly commercial tourism industry to engage in this potentially complex arena continue (Honey 2011), with indications that certain destination characteristics (e.g., poverty and/or natural assets under threat) can affect the level of travel philanthropy or perhaps influence the type.

C. THE MEDIUM – through tourism

Beyond using tourism as a vehicle to generate donations, extant literature considers travel philanthropy as a medium through which tourism industry players act as channels for philanthropic donations into the destination where they operate, or as agencies distributing donations to 'projects'.

1 ***Tourism company as channel/agency for donations*** – Travel philanthropy is also defined as a 'giving programme' of donations from the company and/or those solicited from guests being allocated to *specific* community and/or conservation projects (Honey 2011:208). This is not about managing the 'project'

but only channelling funds into a project. Travel is widely viewed as a means to advance philanthropy (Sustainable Travel International 2008), but also 'as a powerful medium' positioning itself as a legitimate solution and funding for social justice, economic development, biodiversity conservation, environmental protection, and cultural preservation (Valdez and Magio 2016:12). The travel philanthropy discourse has positioned travel companies as the channels of compassion (Lambert and Lester 2004) responsible to 'marshal' positive community development and conservation through this 'form of charity' (Western 2011:13). Whether we should question the legitimacy of tourism operators' advice in remote sub-Saharan African communities for private donations opens up a broad debate (see also Novelli 2015).

2 ***Unmanaged/direct traveller donations*** – Few isolated narratives about 'small amounts' of philanthropic giving by tourists – including the informal, sometimes impulsive, giving of travellers, as opposed to only formal programmes (see below in 'intervention'), or donations channelled via tourism businesses as above – accuse this form of distribution of boosting "the power of corrupt leaders and creating a 'hand-out mentality' where residents expect gifts" (Western 2011:16). The notion that tourists will try and give back appears to drive the need for more strategic action (Bertini et al. 2014), based on the perception that "donations and volunteerism here and there at a travel destination do not create a reliable and sustainable form of support" (Krantz's interview response in Bertini et al. 2014:52). While Spalding (2011) suggests that there has been a shift in the practices of travel philanthropy, moving away from impulse charity by travellers, there are no critical studies on impulsive/informal travel philanthropy substantiating this.

Abernethy (2011) discussed the ethical dilemma that tourists often face around *who* to give the gift to, given the limited time spent in a destination. Whilst not taking into account emotion and individual encounters, he outlines three options for leaving the donation: a travel supplier or intermediary – e.g., hotel, tour operator; a village member with recognised status – e.g., chief; a local organisation with close ties to the village – e.g., village council or NGO. He further argues that it is the responsibility and moral obligation of the tour operator to identify legitimate local projects and reduce spontaneity, thus gaining legitimacy and sustainability. So, while Zijlma (2016) recommended that travellers always donate via an operator and hand it to the director, Bertini et al. (2014) emphasised the importance of someone trusted, and Fernandez (2010) suggested that it is the responsibility of the service provider [tourism operator] to guide clients [traveller] towards purposeful observations of the impact of the beneficiary organisation. However, as the management of tourist accommodation is commonly foreign and white, tourists, also predominantly foreign and white, tend to believe that the management are legitimate advisors, trusting their professionality in running their business and abilities in managing philanthropic donation wisely (Abernethy 2011).

D. THE INTERVENTION – by tourism

It is apparent that travel philanthropy does not end at fundraising and donating, or channelling funds to other stakeholders' projects as a 'medium', but extends to direct *interventions* (typically in the destination a company operates or a traveller visits) and therefore involves power to make decisions about the allocation of financial or other in-kind resources. 'Intervention' as a category does in some ways crossover with 'medium' discussed above but it requires separation. Travel philanthropy is understood by some as a 'medium' in terms of being an intermediary of a donation (so in the 'aid chain' of that donation) and the 'intervention' refers to the actual allocation of that donation, usually the end 'project' or 'beneficiary'. Three conceptual categories have emerged from the literature: (1) formal programme; (2) partnership; (3) informal intervention.

1 ***Formal programme (implemented by a tourism company)*** – Travel philanthropy has been compared to a 'progressive development strategy', a "cost-effective and innovative means of delivering social and conservation services" (Honey 2011:207–208) encompassing the mentioned 'time, talent and treasure' and "flowing from the travel industry and travellers into host communities" (Honey 2011:3). Travel philanthropy is commonly defined as a formal 'programme', an *intervention* established and managed by the tourism operator, including small-scale local programmes run by individual lodges to global programmes and destination-wide funds (Honey 2011). In some cases, a formal philanthropy programme has also been hailed as a company asset that can 'lure' new customers and increase customer retention (Bill 2011:138), bond customers to the destination, prepare them to act philanthropically (Ham 2011), and, through facilitated higher interactions with the community, generate a greater level of affinity and engagement over a longer time (Phocuswright 2015). Therefore, charitable support mechanisms highlight formal programmes, often as a benefit for tourism businesses and their guests (Bertini et al. 2014). While Sustainable Travel International (2008:11) refer to "leveraging business assets and contributing...", Bertini et al. (2014:52) suggest using "travel as a mechanism for stewardship of the Earth and its people".

  In some cases, such programmes are considered channels for funds only. In others, they are a form of charity offering a 'new type of assistance' that is 'supporting and empowering local communities' and encompassing everything from core travel and tourism operational value chains, such as jobs and skills, to directly claimed "improvements in healthcare, education, local livelihoods, and environmental stewardship" (Honey 2011:3). In this context, the line is blurred between charitable funds and actual core business activities benefitting local communities (Ashley and Haysom 2006).

  Ashley et al. (2007), however, argue that tourism companies are not development agencies and rarely have the choice or desire to prioritise development over profit, and Song et al. (2013) point out that tourism does not exist primarily to benefit the poor, mirroring earlier thinking that "tourism is a business not a charity" (Krippendorf 1987:137). An increasing number of tourism operators, acting as channels of compassion or intermediaries for donations, have nevertheless begun to formalise their 'travel philanthropy' channels. Some companies are creating hybrid models and becoming formal charities, thus leaping into the development aid and community development arena. Due to the complex ecosystem and heterogeneity of tourism, operators have been warned not to uniformly implement travel philanthropy initiatives or programmes into travel businesses (Sustainable Travel International 2008). Although some believe that the partnership of charities and tour operators can be symbiotic (Turner et al. 2001), this is not necessarily what travel philanthropy appears to be; rather we are actually witnessing a blurring between tourism operators and development agency/charity approaches, thus formal 'programmes' being operated from *within* the tourism business, rather than as external charities.

2 ***Partnership (community-led intervention with tourism company)*** – Travel philanthropy often assumes connotations of a 'partnership' between community projects, tourism business and travellers (Spalding 2011). At times this partnership becomes even more powerful and effective than prominent development aid NGOs, as they appear to have greater ability in providing direct support at the local level through grassroots organisations 'on the front lines' rather than aid that goes to 'large international groups' (Honey 2011:9). Lindkvist (2011:124) highlights also that one of "the absolutely unique aspects of travellers' philanthropy [is its] potential for long term planning", arguing that tourism businesses are likely to be around for longer than traditional NGOs, although acknowledging that tourism can be volatile, countering the sustainability of partnerships. Notwithstanding Honey (2011) claims that this is a move from charity and handout philanthropy models to sustainable development projects; there is no evidence to support this.

3 ***Informal intervention (by traveller or tourism company)*** – If formal programmes or channels do not lure travellers, do they still give gifts when they travel to Africa, and is this part of travel philanthropy? Do tourism companies without formal programmes still give? A few studies have been made towards challenges due to personal tourist contributions (e.g., Western 2011), and it may be said that "impulse charity by travellers and/or contributions by travel operators or hotels to meet what they perceive as local needs and priorities" is outdated (Spalding 2011:19). However, the literature highlights that the travel philanthropy debate includes informal interventions, sometimes impulsive, as opposed to only formal programmes or partnerships. Calls for more strategic action that harnesses goodwill indicate the importance of considering this form of intervention in travel philanthropy (Bertini et al. 2014), which remains under-researched.

In Africa, particularly, larger companies tend to have a philanthropic arm – i.e., Wilderness Safaris, Abercrombie & Kent and G Adventures in Wilderness Wildlife Trust, Abercrombie & Kent Philanthropy (AKP), and Planeterra Foundation, respectively. Although Lindkvist (2011) argued that tourism businesses are more legitimate than [non-tourism] NGOs for implementation and assessment of local needs and required deliverables because they are based in the destination and are not affected by time constraints; this remains rather assumptive and with little evidence that tourism businesses have created ethical and/or positive impact. Time and presence may not be the most important ingredients, with traditional aid chain interventions more tightly linked to the achievement of results (Lucea 2010). In particular, travel businesses may operate in an arena outside of their expertise and in a way that may be far less accountable and transparent, as the donors go home and have no way to monitor progress besides what they may be told.

In terms of partnerships, there has been a huge interest in participatory development in recent decades, but the topic remains highly contested and the literature inconclusive. In tourism, a number of different approaches have emerged off the back of participative theories such as community-based planning (Murphy 1988), incremental planning (Getz 1987), integrative planning (Inskeep 1991), and collaborative planning (Jamal and Getz 1995). However, theory and practice can be very different. Authors in development literature still question if development is really a vision in which "one side can choose to give while the other may have little choice if they want to survive" (McNamara and Morse 2004:253–254). Whether these are relevant theories for travel philanthropy is open of debate.

Honey (2011:29) suggested that some operators in CREST have changed their strategies from 'donor-driven model' to a 'needs focussed development model'. Moreover, Novelli et al.'s (2015) criticism about travel philanthropy potentially mirroring the mistakes of traditional aid models may reinforce Honey's (2011:216) argument that, whilst travel companies play a role, it is the 'community organisations' that need to be the drivers in 'philanthropy-based development initiatives'. To what extent this is the case and a move beyond idealistic notions of communities and local organisations' participation beyond top-down and charity models is limited, evidently demanding further empirical enquiry.

Bertini et al. (2014) claimed that philanthropists are the agency getting to the root of problems reflecting the new 'world citizens' and 'ethical' consumers of the public North (Korf 2007), for whom giving back is ingrained in their value system (Phocuswright 2015). Media campaigns have developed a passion for the plight of those 'at a distance' (Silk 2004), and in theory people are more educated about sustainable development than a few years ago (Tomas et al. 2017).

Whilst viewing institutionalised foreign aid, for example, the World Bank, as driven by Western self-interest, philanthropists are less inclined to identify similar patterns in their own private donations; rather, they perceive it as 'pure' altruistic generosity towards suffering others (Korf 2007). NGOs as the 'sources of development alternatives' have continued to be highly criticised too (Silk 2004), and although some questions about how travel philanthropy really compares to known failing models were raised by Novelli et al. (2015), further research may be needed on whether these new world citizens are legitimate decision-makers in Africa's development.

E. THE OUTPUT/IMPACT – from tourism

A noticeable shift has emerged conceptualising travel philanthropy beyond the *input* (donation), plus a channel of compassion (intermediary, agency or *medium* for donations), to the concept also encompassing the actual *output* and *impact* of interventions in a tourism destination. This somewhat echoes previous calls for clarity over the sustainability of 'eco-tourism', 'pro-poor tourism' and 'responsible tourism' models.

1 ***Creating sustainable destinations – the results and consequences*** – Some have framed understandings of travel philanthropy around the meaningful conversion of donated resources directly benefitting target communities (Spalding 2011, Ardoin et al. 2016) as the solution to achieve sustainability in destinations (Bertini et al. 2014, Brightsmith et al. 2008). For example, they might contribute improvements to the "environmental and social conditions in the destinations they serve" (Sustainable Travel International 2008:11), or further "the well-being of local communities and conservation in travel destinations" (Bertini et al. 2014:51). Others have suggested that travel philanthropy solves the need to strengthen local support for tourism activities and provides a welcoming environment for tourists by building greater happiness through tackling welfare needs (Valdez and Magio 2016).

   In an attempt to enable local beneficiaries to move from the role of mere observers in the development process, Novelli (2016) suggested that the co-construction of knowledge and capacity building through peer-to-peer and trade-plus-aid approaches facilitate outputs and impacts based on the needs of a destination. Looking at Wilderness Safaris, for instance, she highlights how their travel philanthropy activities aim to directly affect the communities they operate in through targeted, structured projects based on long-term plans that empower people, aligning with local community needs, their culture, and the natural environment. This could either focus on individual (i.e., scholarships) or group beneficiaries (i.e., infrastructures), as long as they are sustainable (Novelli 2015). However, Abernethy (2011) called for a shift towards 'ethical traveller philanthropy' and for a checklist to help improve consequences, which is matched by sparse research. Novelli et al. (2015) and Novelli (2015) investigated the real-world impact of travel philanthropy's cash or in-kind donations to destinations, communities, and/or individuals, making any conclusive remarks on its impact, accountability, and/or transparency empirically unfounded.

Thomas (2014) sustained that programmes generated through travel philanthropy are not well organised and generally have limited pro-poor impacts but also fails to fully assess actual 'impact' of donations in his Malian study, rather only listing items donated, such as ambulances. Equally, Wolff (2011) highlighted the role of travel philanthropy in safari-based tourism, and whilst offering insightful indications of the sum of in-destination financial and in-kind donations, only merely attempts to analyse the input rather than focusing on results – tangible or intangible, positive or negative. From a more cautious perspective, Western (2011) acknowledged that philanthropy is important in destinations but equally warned of the dependency risks of tackling destination sustainability with donor money and the challenges of managing travel philanthropy impact. Overall, although acknowledging exceptions related to the volunteering segment of travel philanthropy (e.g., Fernandez 2010, Mostafanezhad 2013, Wearing and McGhee 2013, Griffiths 2015, Turner et al. 2001), the impact of other forms of travel philanthropy remains theoretically addressed rather than empirically investigated.

Jamieson et al. (2004) highlighted the complexity of the tourism value chain, involving many sectors and bringing about both favourable and unfavourable change to economies and societies. They suggest that we need auditable quantitative methods to assess the impacts of development activities beyond financial interventions. While economic benefits may be evident, possible negative implications for social or cultural assets may be ignored (Biggs 2008). Travel philanthropy has become part of this tourism value chain, and whilst not measured as a distinct segment, or at all, it evidently has both quantitative and qualitative effects on a destination. In their advancement of the PPT value chain model, Jamieson et al. (2004) are probably the closest to considering the role of travel philanthropy as a qualitative output of the industry. They recommend reporting charitable giving cash values by tourism enterprises, tour operators, or tourists, and while this might inform a more comprehensive evaluation model, the key issue to address is the lack of measurement of the actual impacts of such contributions, as opposed to the spread of the economic value to the poor and non-poor in the community.

For instance, there is a need for critical understanding about the actual benefit (and unwanted change) of such financial or in-kind injections into destinations that addresses, for example, the welfare gain or non-monetary measures of poverty and development, rather than just its absolute value (Biggs 2008). Whilst the value is important to measure, it is crucial that we do not simply focus on the economic 'benefits' alone in any measurement development framework, because, as Biggs (2008) pointed out, where ecotourism, for example, might benefit a community economically, it may well result in damage to social and cultural life.

While there is a growing body of literature of value chain approaches to assess the economic impact of tourism (for example, Mitchell 2012; Simpson 2009), yet there is still a serious lack of understanding of the value chain beyond economics considering the holistic net benefit to the poor or the 'local community'. In fact, Simpson (2009) suggested that we cannot measure all aspects quantitatively, such as behavioural change, power relations and exploitations, to name just a few, and that a qualitative assessment would be more useful.

While the structural and systemic flaws of the tourism economy have been denounced on multiple occasions, the industry needs to stop ignoring the lack of mutually beneficial linkages between their operations and local communities, and hiding behind philanthropic practices (Pollock 2016). Possible ways forward include doing business differently and going beyond making donations (Ashley et al. 2005). The role of tourism businesses in the development of 'poor' communities is a well-traversed subject, with literature also focusing on advice for tourism enterprises on how to become more sustainable NGOs and/or charitable organisations themselves (see Crouch 2011). However, differentiating between a corporation as an 'actor in development' (e.g., creating jobs and investing in infrastructure) and a business as a 'development actor' taking responsibility for the outcomes of its 'intentional' programmes is important (Blowfield and Dolan 2014). Banks et al. (2016) explained that companies are both part of immanent development landscapes, which would be the localisation of their value chain, for example, as well as intentional development interventions, perhaps the philanthropic activities on top of core business. The differentiation between what is a philanthropic practice and what should sit under standard business operations when operating in generally vulnerable locations adds to the complexity of determining the impact of travel philanthropy.

F. FORM OF [SUSTAINABLE] TOURISM DEVELOPMENT OR MOVEMENT

Travel philanthropy as a concept is believed to have emerged from the responsible and sustainable tourism movements (Pollock 2016), responding to calls for "integrating tourism company and visitor support for local communities into the core definition of responsible travel" (Bertini et al. 2014:51).

While Spalding (2011:19) explained travel philanthropy as the "next development in responsible tourism [...] as ecotourism businesses and their guests responded, out of compassion or guilt, to needs they saw in the host communities", other authors highlighted that philanthropic motivations have been strongly found in the context of ecotourism and nature-based travel (Ardoin et al. 2016; Barnes and Eagles 2004). Concurrently, Western (2011:13) positioned travel philanthropy as an approach to tourism or a niche within it and highlighted that "the personal contribution distinguishes travellers' philanthropy from ecotourism". Calls for a distinction between prosocial or pro-environmental behaviour and compassion (Weaver and Jin 2016; Sgalitzer et al. 2016; Ardoin et al. 2016) emerged within responsible, sustainable and ecotourism domains. At the same time, geographies of compassion (Mostafanezhad 2012) became evident as a consequence of growing compelling motivations "to act in ways that alleviate suffering and distress" (Weaver and Jin 2016:665), or part of CSR and responsible travel practices (Honey 2011).

Sound evidence of responsible and/or sustainable characteristics specifically related to travel philanthropy are unreported, with few examples of it historically linked to grants awarded to ecotourism related projects (Biggs 2008) using social-entrepreneurship business models or as innovative sources of funding. Continuing with such blurred definitions of travel philanthropy poses a risk of replicating the shortfalls of many ecotourism-related interventions. Many have become just a meaningless 'jagged wooden signboard' (Jose 2013:2) or a mere sorry cry from (eco)tourism operators who are struggling to operate substantial local impact from fees alone (Western 2011), contrary to what was expected at the outset of the movement.

The assumption that giving through travel is responsible and the call for "a 'hand up' not a 'hand out'" (Honey 2011:3) remain generally unsupported by solid evidence proving the actual impact of philanthropic activity in host destinations. Given the complex and fragmented nature of tourism characterised by organisations of very different sizes and missions, as well as travellers and residents with myriad values and cultures, there cannot be any solid claims that travel philanthropy is a form of sustainable or responsible tourism contributing to sustainable development. This is due to the lack of a holistic understanding of the impacts of travel philanthropy – e.g., economic, social, cultural, political, environmental – and its interactions with the social system in which it operates.

While the six themes/categories discussed may provide some foundations to better understand the concept of travel philanthropy, the growing phenomenon associated with it is far more complex than one could ever envisage.

4.3 Travel philanthropy in the sustainable development debate in Africa

Tourism in Africa increasingly intertwines with philanthropy or gift-giving. Again, monetary and in-kind resources flow to, from, or through the tourism industry directly into African communities in the form of 'philanthropy' or 'development', interventions managed by commercial tourism operators or not. This thesis requires an understanding of evidence concerning travel philanthropy's role in 'sustainable development', and knowledge of its qualitative knock-on effects within the communities in Africa where gift-giving practices occur. Emerging as a relatively new player in this segment of private philanthropy for international development, the tourism sector is entering an extremely complex arena in sub-Saharan Africa – 'development', 'aid', socio-economic impacts, and the complexity that surrounds this.

Discourse of tourism's role in 'sustainable development' is not new. Tourism entered the sustainable development debate in the 1970s in response to the realisation of mass tourism impacts. In the 1980s, forms of 'alternative tourism', including terms such as ecotourism, were increasingly adopted, and since the 1990s, 'sustainable tourism' hit the agenda. In 2015, there was a significant shift when tourism came under the global radar through the UN Sustainable Development Goals (SDGs) and positioned less as a threat to sustainable development, but as a potential tool for sustainable development, particularly for less developed countries. Whilst the sector is both dynamic and complex, studies have quite widely covered the potential positive impacts of the sector, as well as also cautioning negative implications too (Tyrrell et al. (2012) for a list of relevant authors). Tourism often entails a stable population (resident communities in the destination) meeting a variable population (tourists), and so this can enable new cross-cultural relationships, emotions, and knowledge (Bimonte and Punzo 2016), thus opening a range of potential impacts to the natural, socio-cultural, and economic environment. Today's shift in compassion for distant suffering to physical encounters with 'beneficiaries' in Africa is located within similar ambitions and debates (Korf 2007).

Tourism as an efficient tool to alleviate poverty does not always receive unanimous support; yet, there is considerable evidence to prove the contribution of tourism to the economic growth of less developed countries (e.g., Thomas 2014) and the tourism value chain concept has gained importance (WTTC 2015; UNWTO 2013). PPT, highlighted by Jamieson et al. (2004:3), is "an overall approach designed to unlock opportunities for the poor", which also gained momentum over recent years. However, in reality for travel philanthropy, although charitable giving may be simple to carry out, alleviating poverty is much trickier and more complex (Ashley et al. 2007). We cannot assume the geographies of generosity influenced by travel, fuelled by compassion and emotions, and rich in moral motivations (Silk 2004) are positive for the 'addressee' (Korf 2007:370). A range of research has demonstrated that giving gifts can create the superior self (those without suffering, or the donor) and the passive receiving other (those identified to be suffering and in need). The concept of Northern actors or travellers as carers who are active and generous and of Southern actors as cared for, passive, and grateful in this 'Resource Transfer Paradigm' (Silk 2004:230) raises moral debates. Silk (2004) emphasises that such North–South encounters lack the effort of listening to the poor, rather choosing to give, and what to give.

Rather than arguing which public good is the best allocation of donations (a theme in the 2011 CREST handbook, for example), questions surrounding actual impacts within the informed sustainable development debate are critically weak. We also cannot decide what constitutes this generic 'public good', because, as Goldfarb (2011) emphasises, it is complex and multifaceted: Who should decide what that is? Moreover, we would need to question if giving good in the first place. There is a concerning lack of knowledge regarding the overall complexity associated with travel philanthropy and impacts on a local level – and, there is most importantly no evidence to show evaluations that incorporate, let alone prioritise, the perspectives of local people that reside in the destinations in question and whom the donations are potentially affecting. This issue is an urgent gap in interdisciplinary knowledge.

As Carlson (2012) emphasised, no one wants to think their generosity hurts people, but we have to learn from the debates within the international development and aid literature, which have raised complex issues around power relations, dependency, legitimacy, sustainability, ethics, and other implications of gift-giving and consider the actual right of the recipient community to shape what comes next. The paradoxes for development in the context of travel philanthropy have been touched upon but not extensively (Burns and Barrie 2005). We need to understand the specific characteristics of travel philanthropy and how these might play out in terms of the sustainable development paradigms. There is no solid evidence in the tourism field of study addressing such issues empirically, especially as we remain in a stage of lacking inclusive definitions and conceptual frameworks to base meaningful analysis upon.

Given the involvement of multiple stakeholders interacting, to fully understand the meaning, significance and impacts of gift-giving and philanthropy through tourism in Africa requires a system level understanding (Senge 1990; Meadows 2008), requiring us to turn to international and sustainable development knowledge as well as systems and complexity theory.

This chapter framed the bigger picture of travel philanthropy by illustrating what the phenomenon is understood to incorporate by different stakeholders. The findings identified travel philanthropy's role in the African development landscape, which arguably faces critical questions of 'impact'. The next chapter will explain our methodological approach to advance the understanding of travel philanthropy from extant literature by reflecting on the discussion in Chapter 3 about impact evaluation within the complex and evolving landscape of sustainable development.

5 Redefining travel philanthropy

Opening the black box

This chapter introduces a methodological framework that responds to the contemporary impact evaluation needs and addresses the complex nature of travel philanthropy as an open social system. In the words of Bhaskar and Callinicos (2003:27) "[w]e're all bound up together. We sink or swim together. Never before has global interconnectedness become such an empirically identifiable fact" so thinking in systems is fundamental, especially when even a belief or behaviour in one part of the world can seriously affect another. By combining critical realism and systems theory, a new conceptual model is proposed, opening the black box of an interdisciplinary system and focusing on qualitative matters of causality. The chapter also introduces the empirical research methods subsequent chapters will analyse, including what, why, and how case studies were selected in Uganda, Ethiopia, and Zimbabwe.

5.1 Critical realism and systems thinking: a methodological framework

Travel philanthropy is a socio-economic phenomenon and as such it is treated as the subject of a social investigation. Based on what is known about a systems approach in fields dealing with complexity and non-linear change – in our case evaluating sustainable development in Africa, it was identified as an invaluable theoretical framework to better understand the behaviour of actors and the place of travel philanthropy within the sustainable development context. At the outset of this research, the meaning, significance, and impact of travel philanthropy through a systems lens was identified as a critical gap in knowledge and the most appropriate means of grasping the bigger picture of this phenomenon, as well as unpacking the details.

In contemporary systems thinking, and especially in the social sciences, we tend to be interested in open systems, so a sound understanding of a system looks *within* the system itself but also at its relationship with the environment (Vandenbroeck 2015). In order to fully master systems thinking, we need to step inside the system to learn about its structure and processes and discover the rules that steer its behaviour (Hummelbrunner 2011). Key elements of a systems approach have evolved to include: (1) *Perspectives* – seeing "the world through the eyes of another" (Churchman 1968:231); (2) *Interrelationships* – the multitude of relationship

DOI: 10.4324/9781003497929-5

types, deep understanding of causality and context (Burns and DeVille 2017); and (3) *Boundaries* – boundary decisions and critiquing those choices (Flood 2010; Jackson 2019). These three elements informed the methods used in this study, such as gathering multiple perspectives, refining boundary decisions, and addressing the matter of causality. Explanation in open systems requires "an account of a multiplicity of causes and explanatory linkages" (Rutzou 2017:9), thus our philosophical position needed to substantiate this ambition.

After careful consideration, *critical realism* (initially called *transcendental realism*) emerged as the most appropriate lens because it is a philosophy for science in which the structures and mechanisms that generate phenomena, rather than the phenomena they produce, are the objects of study (Bhaskar 1975). This scientific process improves competence with respect to open systems. Over the past few decades, critical realism has gained popularity as a philosophical framework for social scientific research, emerging out of the positivist/constructivist 'paradigm wars' of the 1980s (Denzin and Lincoln 2018). Some critiques of critical realism, especially those informed by Guba and Lincoln (2005), argue that critical realism is too close to post-positivism, naïve realism, and critical theory.

However, many other sources that stem from Bhaskar (1975) emphasise a far greater element of interpretivism, being a social science. The essence of critical realism is an inquiry into the nature of things, a combination of explanation and interpretation (Archer et al. 2016). Since Bhaskar, critical realism has been further developed by scholars such as Sayer (1992, 2000), Archer (1982, 1995), Collier (1994), Lawson (1997), Elder-Vass (2010) and identified as:

> *An alternative paradigm both to scientistic forms of positivism concerned with regularities, regression-based variables models, and the quest for law-like forms; and also to the strong interpretivist or postmodern turn which denied explanation in favour of interpretation, with a focus on hermeneutics and description at the cost of causation.*
>
> (Archer et al. 2016:1)

This research aligned philosophically with Bhaskar's (1975) original ideas about critical realism and more recently with Mingers (2014:11), who specifically links "critical realism as a significant philosophy of science able to underpin the systems disciplines". A core focus of critical realism is causality, but regularity determinism or the hypothetico-deductive view of science behind the actuality of causal laws is disputed (Bhaskar 1975:58). It maintains that knowledge is always fallible, and functions as a methodological framework rather than a set of distinct methods. Its position is ontological realism and epistemological relativism (Archer et al. 2016) and holds that we must separate epistemology from ontology.

The ontological focus of systems thinking is on explanatory mechanisms, emergence, generative powers, and causality, and thus, there is a symbiotic relationship between the fields of systems thinking and critical realism (Mingers 2014). Mingers (2014:28) formally brings together critical realism and systems thinking, arguing, "Critical realism clearly embodies systemic and holistic themes at its very heart with concepts such as totality, holistic causality, emergence, open systems, autopoiesis, and levels of stratification". Furthermore, "these concepts have their own history of development within the discipline known as systems theory or systems thinking but there is almost no reference to this literature within Roy Bhaskar's (or other critical realists') writings" (Mingers 2014:28). He also argues that systems thinking can gain from critical realism philosophically as a position between positivism and interpretivism by providing a more rigorous philosophical underpinning that systems thinking lacks, and by its development of concepts, such as absence/negativity.

Critical realism argues that positivist models lack a holistic conceptual engagement with complex causal environments and often draw on a narrow evidence base of observable regularities of 'facts' ignoring, for example, where patterns arise from (Kurki 2007).

The research methodology behind our study is therefore designed around a synthesis of critical realism and systems theory, contributing to the science of open systems research, which had yet to inform any theories of travel philanthropy. Viewing the world as an open or closed system is what Bhaskar (1975) says leads to totally different conceptions of science. Events (change) in open *and* closed systems may be explained; however, the antecedent and consequent of lawlike statements are only viable under entirely closed conditions where predictions are possible. Open systems characteristically involve multiple mechanisms of different kinds combining to produce effects, and in such systems, we cannot deductively predict events. In open systems, especially the social world, causation therefore cannot be understood as a closed equation of isolated events and regularities like "in the controlled confines of a laboratory", but 'in the wild' as a chaotic and disordered series of contingent constellations and conjunctions in which things often appear in 'impure forms' (Steinmetz 2004:338; Bhaskar 1986:110). The real world interpreted as an open system means:

> [Seeing] *the world as it is, where what occurs is a consequence of the interaction between a number of different powerful particulars, and the operation of any individual particular is modified and restricted in various ways because it's operating in relation to all sorts of other powers.*
>
> (Bhaskar and Callinicos 2003:3)

In open systems, therefore, "outcomes might be the result of many different causes and the same cause might lead to different outcomes" (Patomaki and Wight 2000:18). For these reasons, the chosen methodological framework aids the appreciation of travel philanthropy as a complex open system of the social world.

5.2 Exploring travel philanthropy: methodological perspectives

Reflecting upon critical realism and systems thinking as the most appropriate methodological framework to inform this study of travel philanthropy, this section explains the research methods and how they unfolded as the study advanced. The research objective was to discover mechanisms we understand may already exist, but we do not yet know:

> *As a piece of philosophy we can say (given that science occurs) that some real things and generative mechanisms must exist (and act). But philosophical argument cannot establish which ones actually do; or, to put it the other way round, what the real mechanisms are. That is up to science to discover.*
>
> (Bhaskar 1975:42)

5.2.1 A qualitative research approach

Findings from the literature review provided evidence that addressing the qualitative aspects of travel philanthropy would be the most appropriate first step in progressing our theoretical understanding of the travel philanthropy phenomenon. Abstracting causal mechanisms and immersing in research contexts across multiple case studies to aid the discovery of both patterns and contextual influence (Bhaskar 1975; Danermark et al. 2002; Fletcher 2016) was key in our research.

Methods were designed around observation, questioning, and listening, and an inductive/abductive analysis of data as the research unfolded, allowing theorisations to gradually emerge and be shaped into coherent new elaborations. In line with Patomaki and Wight's (2000) thinking, as a first step, it was necessary to discover *what* had happened, before we could ask *why*, so it was critical to discover relevant aspects/events associated with the travel philanthropy phenomenon, before investigating the reasons behind its existence as a whole (Adamides et al. 2012). An element of the 'what' began to emerge through the review of extant literature (see Chapter 3), but in order to avoid pre-conceptions biased by the literature, an empirical assessment was deemed necessary. We decided to do so through relatively unstructured qualitative interviews and, by exploring multiple perspectives, with the 'system' as the object of inquiry rather than the research participants themselves, the *what* began to emerge (Senge 1990; Gates 2016).

'Perspectives' were considered essential to enable a broader and more inclusive sample of local stakeholder voices, both for ethical reasons and for our own theoretical efficacy, enabling a more robust understanding of travel philanthropy with practical relevance informed by a bigger picture view. Meanings of travel philanthropy as a system can differ by perspective, such as the purpose of 'travel philanthropy' for a donor vs. beneficiary, and because the role meanings are known in other social systems to influence how a system functions (Meadows 2008) it is important to discover how travel philanthropy is viewed from different angles and how that influences stakeholder behaviours. Ultimately, our research intended to create a qualitative interpretation of the travel philanthropy system, informed by diverse stakeholders' lived experiences and perspectives, analysing different meanings and realities in-depth (Senge 1990; Checkland 2000; Ulrich 2005).

More recently, scholars such as Maxwell (2012a) and Bennett and McWhorter (2016) supported Bhaskar's (1975) view that we need qualitative approaches and methods to make valid and useful claims about 'what works' without subscribing to regularity theory. Whilst critical realism does not reject quantitative methods such as statistics, it does emphasise the importance of examining deeper processes at work in the world by abstracting causal powers or mechanisms of an object being investigated and thinking conceptually about how it operates (Roberts 2014:2). However, the 'critical' in critical realism is that the social world is 'open', which implies that researchers can follow a realist methodology but nevertheless produce competing causal accounts of the same social phenomenon (Roberts 2014). There are criticisms in social science about working with 'abstract' conceptual causes e.g., reasons, but critical realists argue that due to the inherently unobservable nature of most social causes, we must capture them through conceptual models (Kurki 2007:13). This approach to qualitative research informed our research, which prioritised causality and explanation.

5.2.2 On the matter of causality

To identify and understand travel philanthropy as an open social system, it was critical to move from '*what*', to '*why*' something happened. Applying critical realism to methods inspired this study's fundamental aim of explanation, answering what caused events to happen, continually asking why, and identifying underlying social structures and mechanisms (Easton 2010). As travel philanthropy is part of the social world, methods could not be designed to look for law-like antecedents (Van der Riet 2008), rather '*impacts*' or change could be statements about how a causal mechanism *tended* to operate (Bhaskar 2008).

While causation remains one of the most contested concepts in the philosophy of science, and especially social science, debates continue around the meaning of causality as a concept, the reality of causes, and methods for causal analysis (Kurki 2007). A major reason for the rejection of causation in social science, particularly by qualitative researchers, has been the long-standing positivist account of causation, generally known as the 'regularity' or 'successionist' theory of causation derived from David Hume denying any knowledge of causality that goes beyond the observed regularities in the association of events and rejecting unobservable mechanisms (Maxwell 2012a, 2012b). Qualitative researchers embracing the constructivist paradigm, led by Lincoln and Guba (2005), assume social action is not caused, rather it is constructed through an ongoing process of meaning making that occurs within and across individuals, and that the complexity of social life can only be captured by thick description, rejecting the notion of causality on ontological grounds (Donmoyer 2012). Critical realism, however, engages in explanation and causal analysis, not thick empirical description of a given context, looks beyond the observed, and is thus useful for analysing social issues we find in travel philanthropy and identifying change solutions (Fletcher 2016).

For this study, to add theoretical and practical weight to the understanding of the travel philanthropy phenomenon in the context of sustainable development studies, qualitative methods able to deal with causality in a scientifically robust manner were imperative. Critical realism was therefore embraced to inform methods and provide an appropriate way to address the causality issue found in social sciences, offering an alternative to positivism/empiricism and interpretivism/constructivism debates (Sayer 1992; Archer 1995; Layder 1990; Putnam 1990; Pawson and Tilley 1997; Baert 1998). This standpoint informed both our data collection (e.g., the questions asked) and analysis (e.g., systems thinking and map) in order to achieve a well-reasoned *qualitative* contribution of causality to the travel philanthropy debate. Our study is therefore based on a non-positivist model of social scientific causal analysis to overcome the binary of the other approaches to research (Kurki 2007), addressing the complexity of causal relations in the social world without explanations being constrained by empirical regularities (Rutzou 2017), but arguing the reality of causal processes leading to outcomes (Maxwell 2012a).

Social reality was understood as structure and agency, but 'critical' in rejecting structural determinism (King 2009; Bates 2006). Both structure and agency were examined in our study, because in critical realism structures and agents have a mutually influencing relationship, in which the actions of agents are both constrained and enabled by existing structures. The actions of agents in turn reinforce, reproduce or transform existing social structures (Bhaskar 1989; Lennox and Jurdi-Hage 2016; Fletcher 2016). Although structure is unobservable, it is considered real because it has an impact on observable events (Lewis 2000). In open systems, such as travel philanthropy, structures of society are changeable (Bhaskar (1975)). However, individuals also follow certain rules (Pleasants 1999), so the approach taken in our research was to provide an account of the multiple webs of causality in this respect, including both structural enablers/constraints, in which individuals operate, and also that of agency and structural transformations (Burns and DeVille 2017). While there is not a specific section on structure-agency in this volume, it is an aspect that emerges throughout the discussion of our findings offered in the chapters that follow.

It was important to realise that agents may not be aware of structural causes or effects (Bennett and McWhorter 2016), so the explanation of the system was based upon the interpretation of actor narratives, which included individual experiences, perspectives, and decisions (Dev Singh 2015; Giddens 1993). As structures and causal mechanisms tend to exist beyond the consciousness of actors in their narratives, analysis started with identifying empirical tendencies and then assessed deeper and broader issues informed by a critical realism methodology (Lennox and Jurdi-Hage 2016). Our research therefore considered perceptions and descriptions surrounding empirical events to identify possible mechanisms that may have produced those events, and examined conditions that triggered those (Zachariadis et al. 2013; Fletcher 2016). As part of the causal analysis, academic theories from the social sciences were combined with empirical data to help frame these interpretations, give unobservable cause–effect relationships meaning (Lennox and Jurdi-Hage 2016) and help determine the nature of mechanisms or structures (Danermark et al. 2002) of the travel philanthropy phenomenon.

Our ultimate goal was to open the black box of causality, looking for relationships between agency and structure, from the domain of empirical to the domain of the real, and understand events and mechanisms in different contexts. To do so, a conceptual system map understood through qualitative causal explanation informed by critical realism was to be identified. A case study method was chosen as the most adequate method encompassing appropriate qualitative data collection techniques and analytical processes.

5.2.3 Case study method

Case studies are found to be appropriate for "peering inside the 'black box' of correlation and exploring causal pathways" due to their richness of detail (Weller and Barnes 2016:2). Examining a context for causal processes is a way to open the black box that quantitative studies may leave closed when they rely on statistical analyses and the law of large numbers (Bennett et al. 2016). Critical realism also offers a philosophically informed account of science suitable for empirical investigations using small-n case comparative analysis, combining explanation and interpretation (Archer et al. 2016).

A small-n case study approach therefore framed our empirical enquiry in Africa. Travel philanthropy in this study is not something we are measuring, we are rather trying to critically understand. This meant the case studies demanded a richer understanding that could not be provided by large scale quantitative methods, and an open-eyed approach to what data might tell us. This subject required a process of in-depth research to understand what things are and why things are as they are, and because case studies are suited to explanatory rather than frequency-based questions, they are the most useful method for disentangling the complex sets of factors and relationships identified previously (Easton 2010; Yin 1989; 2003).

Stake (1995:2) proposed that a 'case' is an 'integrated system' with emphasis placed on holistic qualitative research, and although separated from systems and complexity literature, synergies have been identified with his approach to case study research, systems theory, and critical realism. Thus, the research methods in our study also drew upon some of his ideas. The case studies were designed flexibly, which meant using a set of refined research questions to inform and structure the initial case studies, but then allowing for changes to be made as knowledge of the subject progressed, partly due to limited information available at the outset, and because research and analysis took place simultaneously (Stake 1995).

Although flexible, the research was framed around a bounded context (Merriam 1998) in terms of 'travel philanthropy' as a phenomenon, spatially in terms of tourism destinations, and by individual research participants. A critical realist case study approach is suited to "relatively clearly bounded, but complex, phenomena such as organisations, interorganisational relationships or nets of connected organisations" (Easton 2010:123) and whilst travel philanthropy has fuzzy boundaries as a social system, it has many of the other characteristics. In systems thinking it is also important that methods make it possible to see a situation from different perspectives and reflect on boundary decisions, because choices have to be made as to what is relevant, what is included and by whom; thus, identifying a boundary to a phenomenon and assessing the consequences of those boundary choices was paramount (Hummelbrunner 2011).

This research aligned with Mingers' (2011:320) thinking in that a system is not something given in nature but 'defined by intelligence', it is the concept of a 'system' existing within an 'environment' that is the foundation for systems theory, but to define a system we have to define its boundary, and what separates it from its environment. However, demarcating the system with properties from its environment can be contentious when dealing with complex social systems consisting of components, such as ideas or beliefs that may not be physical (Mingers 2011). In the context of our research, we also had to select from an infinite number of relations between things, because without patterns identified, we would not be able to interpret "what otherwise might be a meaningless cavalcade of arbitrary events" (Mingers 2011:320). Building a better understanding of what is considered within the bounds of travel philanthropy in this respect was therefore a key part of our research method.

Research was conducted across multiple contexts (Yin 1984), including five rural tourism destinations in Africa across three countries – Ethiopia, Uganda, and Zimbabwe, to allow richer comparative perspectives between countries, cultures, types of philanthropy and tourism encounters/experiences, beneficiaries, etc. This produced comparative evidence to set stronger foundations for this under-researched subject currently informed theoretically by isolated and anecdotal evidence. Differences, similarities, and patterns across data, within and between cases, aided theoretical development and contributed to the knowledge about causation. Examining a context as a case study for causal processes is a way to open the black box of causality that quantitative studies may leave closed where they rely on statistical analyses and the law of large numbers (Maxwell 2012b). There is ample evidence across the literature that context is intrinsic to causal explanation; therefore, it is the richness of case studies and the ability to examine different contexts for causal processes that led to the multiple case study method for this research (Maxwell 2012a; Weller and Barnes 2016; Mahoney 2001; Pawson and Tilley 1997).

5.2.4 Research site selection rationale

Previous knowledge about the destinations, access, relevance, and budget constraints played a relatively important role in the choice of the case study locations. Uganda and Ethiopia were chosen out of the entirety of Africa partly because – in addition to the specific criteria listed below – we had previous knowledge, experience and connections in these countries, which would aid access to research participants in some locations, as well as general awareness of travel philanthropy practices in some of their tourism destinations. Zimbabwe was selected as the third case study country partly due to the collaboration with a corporate partner considered to be leading the way in terms of travel philanthropy interventions in the region, a possible benchmark in the overall assessment of the phenomenon, and a way to gain insights into corporate practices.

The following criteria guided the selection of specific research sites from a spatial standpoint in each of the three countries:

- Rural/small town tourism destinations in Africa linked to protected areas
- High levels of perceived poverty in local/host community
- Emerging/growing tourism sectors representing different niche products
- Presence of travel-philanthropy-related activities
- Different contexts in terms of geography, culture, national, and local political structures, policies, tourism encounters, philanthropic activities, tourism operators, etc.

Although limiting the case studies to three countries and five destinations (see Table 5.1) excluded other parts of Africa, it is worth noting that the main purpose of our study was to provide a theoretical standpoint and conceptual elaborations based on empirical evidences, which could indeed be furthered through research in other countries. The selected number of cases provided sufficiently rich data to advance new knowledge through this small-n in-depth study.

5.2.5 Establishing the research boundaries for data collection

Following an initial conceptual boundary of travel philanthropy informed by existing literature, the first empirical research boundary decision made was the geographical location (spatial) of the case studies previously discussed. This was followed by a network analysis with a range of key local stakeholders conducted in the field aimed at identifying which actors were considered relevant to the system and help clarify the geographical boundary from a local perspective. For example, at an early stage of the field research in Uganda, a number of participants were separately asked to free-list all the people and organisations (actors) on blank sheets of paper (using post-it notes for each actor), which they identified to be connected to giving and receiving donations in the specific case study area. This evolved into a type of network map (Figure 5.1) and a visual representation creating a space for dialogue (Van der Riet 2008). They were then asked to categorise them into specific groups to understand how local stakeholders viewed actor groups and roles in the system. The method incorporated ideas from social network analysis to map formal and informal interactions between different actors (Schiffer and Hauck 2010). An understanding of network flows was also probed including the interconnections between actors concerning resources, information, power, conflict, etc. The precise definition and boundaries of travel philanthropy at the start of this research was blurred, so, in order to avoid instilling bias within research participants, rather than inquire about 'travel philanthropy' *per se*, the exercise began with discussing a more general perspective of 'giving and receiving donations'. During the network mapping, an informative discussion took place to understand participant thought processes and clarify any queries.

The findings were then cross-checked with the previously conducted analysis of the extant literature for further interpretation, as well against other research

Table 5.1 Case study countries and locations of research

Case study code	*Case study country and locations*	*Overview of the area research site*
	Uganda	Predominantly a nature-based tourism destination. Across country driving safaris in 4WD private or small group with itineratires mainly between national parks to see wildlife. Most foreign leisure visitors tend to stay on average about 2 weeks in total and 2 nights in each destination. Mixture of budget to high end tourists. The main attraction of the country is primates particularly mountain gorillas and chimpanzees, as well as birds. Community/cultural products tend to be by-products
U1	**Bwindi Impenetrable National Park**	**UNESCO WHS and national park. Bwindi is area of high-altitude mountain forests demarcated for biodiversity conservation. The main tourist product is nature-based – hiking in the rainforest to see mountain gorillas.**
U1_1	Village 1	Two high end lodges and one low end, gorilla tracking gate entrance and various craft shops – all close proximity
U1_2	Village 2	Locally run mid/low end lodges (approx 4) within larger village located between the two gorilla tracking briefing points and an entrance point for gorilla tracking – all close proximity
U1_3	Village 3	Mid/high end lodges (4 operating at time), craft chops, gorilla tracking gate entrance – all close proximity
	Zimbabwe	Predominantly a nature-based tourism destination. Fly-in destination (to Victoria Falls) – the market has shifted in recent years from backpacker to high end tourism and this destination is now predominantly combined with a multi-country trip including South African destinations. Visitors are often in larger groups, mid/high end market and transported to/from activities in coaster buses run by one leading company. Itineraries tend to be tight and pre-arranged. Community/culture is a by-product and if included as a tourist activity it is formerly organised as a tour
Z1	**Victoria Falls National Park**	**UNESCO transboundary WHS (with Zambia) including the waterfalls, river system and some surrounding land areas. The main tourist product is nature-based – seeing Victoria Falls, or adventure activities in the nearby river system.**

(*Continued*)

Table 5.1 (Continued)

Case study code	*Case study country and locations*	*Overview of the area research site*
Z2	**Hwange National Park**	**Largest protected area in Zimbabwe and approximately 2 hours' drive from Victoria Falls, particularly well known for elephants. Various high-end hotels are located within and surrounding the national park. The main tourist product is nature-based – typical wildlife safari.**
Z2_1	Village 1	Remote village receiving daily tourists for village tours and linked to a tourism operator travel philanthropy programme – no tourist accommodation
Z2_2	Village 2	Area around the main entrance gate of the national park where the majority of mid/high end lodges are located – relatively spread out
	Ethiopia	Predominantly a cultural/heritage-based tourism destination. Cross-country destination circuit tourists tend to follow, almost always including Lalibela and more recently the South Omo area has increased visitor volumes due to infrastructural changes – roads and tourist accommodation. Air links have also expanded nationally in recent years so there is now a combination of visitors either flying or driving between destinations. Most visitors arrive on a pre-arranged itinerary tour led by a local ground handler in 4WD private or small group tours. The market is mixed but dominanted by mid/high end visitors due to the cost of tours. Wildlife is a by-product.
E1	**Hamar South Omo (Southern Ethiopia)**	**UNESCO WHS protected area as prehistoric site. Hamar people are a tribe located in the Omo Valley area of Ethiopia. Visiting and interacting with cultural traits of Hamar is the dominant tourist product such as attending a coming-of-age ceremony, visiting markets, and individual homes. Main product cultural-based**
E2	**Lalibela (Northern Ethiopia)**	**UNESCO WHS protected area for rock-hewn monolithic churches and intangible heritages of church practices. A village/small town located in the mountains of Northern Ethiopia in the Amhara region. Visiting the churches is the dominant tourist product, as well as by-products such as visiting the market. Main product cultural/history based**

Table 5.1 (Continued)

Case study code	*Case study country and locations*	*Overview of the area research site*
EN	**National perspective**	**Research was also conducted in Addis Ababa amongst tour operators operating in both Hamar and Lalibela as they were not permanently present in the destination but key stakeholders to fill gaps in knowledge about the role of the tourism industry in travel philanthropy – they have been coded separately as a location to reflect the general view of the system in both destinations**

participants. This helped develop system boundaries by including or omitting actors identified by some participants. For example, a government agricultural department was listed, but whilst that actor might have been somewhat interconnected with the system of giving and receiving in travel philanthropy, it was found to be a weak influence when cross-checking with other participants, so certain boundary decisions eventually had to be made by the researcher.

By combining findings from individual and group research tasks of this nature, patterns became clear and stakeholder/actor categories for the research participants were refined and selected. This process informed the interview strategy to ensure representative voices from the identified groups. Purposive sampling was then combined with snowball sampling (Cohen and Arieli 2011) to ensure participants represented a thorough cross-section of individuals/organisations. These steps ensured that rather than having a preconceived criteria and assumptions biased by an outsider worldview, it was open to continuous learning as the research unfolded.

In Zimbabwe and Ethiopia, similar methods were used, but in the interest of time and to avoid unnecessary repetition a simplified version of a network mapping process was used, as very similar patterns for stakeholder groups were identified. Therefore, after initial conversations, rather than providing blank sheets, participants were probed for missing actors or categories. In some of the case study locations, listing every single actor was also unachievable due to the size of the destination, and full knowledge or access of an entire geographic area proved too challenging for the blank sheet method. The information gathered through the in-depth and simpler approaches provided solid grounds for the interview strategy and was successful. Table 5.2 outlines the taxonomy of travel philanthropy stakeholder groups resulting from this research method.

5.2.6 Multiple perspectives: qualitative interviews

Having identified the taxonomy of stakeholders, qualitative interview methods were selected. These do not "bury the voice of research subjects beneath piles

Table 5.2 Stakeholder groups

Case study code	*Actor category*	
	Organisations	*Description*
_TB	**Tourism business – management**	A hotel/lodge located in the destination or a tour operator which visits the destination as part of a national tour or more rare, based in the destination
_NGO	**International NGO**	Non-governmental charitable organisations owned and managed by non-local actors
_CBO-F	**CBO – formal**	Non-governmental charitable organisations owned and managed by local actors – formerly registered and relatively larger in size compared to informal CBOs
_CBO-IN	**CBO – informal**	Small organisations or informal groups established and managed by local actors either with charitable purposes or self-group 'development'
_B-INS	**Local institution**	Schools, businesses, traditional authority, etc facilitating or receiving donations
	People	
_INTER	**Local intermediary**	Also called connectors/implementers/ go-betweens, these are local resident (in destination community) individual actors intermediating information, resources, and people between donor and beneficiary
_TE	**Tourism employee/ local guide**	Working formerly or informally in the tourism sector as a guide or hotel employee – not identified specifically as an 'intermediary' of donations
_B-COM	**Community beneficiary**	Individual donation recipient
_D	**Donor**	Tourists, religious organisations, other non-national donors

of anonymous standardised data" and so a combination of semi-structured and unstructured techniques was used to encourage respondents to talk freely, particularly around emotionally loaded topics, to gain insight into how people feel and think about the phenomenon under investigation (Roberts 2014:3–4). In line with Churchman's (1968:231) thinking that "[a] systems approach begins when first you see the world through the eyes of another", it was absolutely critical to capture different viewpoints, experiences, and meanings of travel philanthropy through these qualitative interviews and to continue until data saturation was achieved, within the time-frame and resources available. This data saturation point differed slightly according to each specific case study location, but this was attained through a similar number of interviews in each location.

Seventy in-depth multi-stakeholder interviews were conducted across the three Africa countries from August 2018 to February 2019. This exact number of interviews was not a pre-planned element of the study design, rather it was found that a type of '360-degree view' of the system was reached at this point, from a stakeholder perspective, following the simultaneous research/analysis case study method by Stake (1995). The saturation of data included stakeholder types and system perspective themes that were emerging. This did not mean that full analysis of all data was conducted in the field, but an iterative process of listening and reviewing field notes enabled the identification of this saturation point – i.e., when interviews were not yielding new information, and a qualitative judgement could be made with confidence (Brod et al. 2009).

Interviews were detailed with an average length of 1.5 hours, adjusted depending on participant knowledge, experience and time constraints. Table 5.1 provides detail of the ID codes used for each interview, also used in the chapters that follow. Stakeholder categories in *Table 5.1* formed the framework for research participant selection, allowing for some flexibility through a snowballing technique suggesting other key informants as research progressed (Cohen and Arieli 2011). A range of people/organisations considered most highly active in 'travel philanthropy', including those who had previous experience regarding gift-giving/receiving, as well as more periphery stakeholders, were included to vary perspectives. Some actors could be located within more than one category (e.g., tourism employee and intermediary). However, the category that they most represented in their interviews was selected for participant code identification.

5.2.7 Interview technique and framework

Interviews conducted within these defined case studies (bounded by space and stakeholders) offered a means to progress new understandings of the system boundary of travel philanthropy, *conceptually*. With permission from participants and the agreement that findings would be anonymous, all interviews were audio recorded to effectively capture rich data for later analysis.

This empirical phase of the research was intended to learn about the nature of the system, rather than assess it. The interviews aimed at gathering data about the *what, who, where, when, how*, and *why* of giving and receiving donations in different contexts, and the broader system this was nested within. The interview questions were left flexible, pivoting around personal stories, events, experiences, and perspectives of the system. The aim was to probe answers in terms of cause–effect, observable, and unobservable mechanisms/structures, and the decisions/behaviour of people/organisations. This would help to identify what produced and shaped the travel philanthropy phenomenon as a system. Stories of travel-philanthropy-related experiences were collected, asking participants to think back to what happened during particular events, and then used as tools in the dialogue to uncover unobservable mechanisms such as beliefs, values, motivations, feelings, attitudes, and choice processes. The people/organisation network was also questioned on interconnections regarding information, resources, knowledge, people, ideas, etc., and insights into the types of relationships between different actors in the system were gathered. Overall, the research questions aimed at enabling the elicitation of information about events and causal mechanisms relating to gift giving and receiving in different contexts (Pawson and Tilly 1997, Roberts 2014).

Whilst a relatively unstructured interview approach was chosen in practical terms, there was a rough guiding framework for the discussion, as there were certain bounds to the subject of interest (Merriam 1998, Mingers 2011). Like in the previous phase of the empirical research, this framework was developed as the research progressed (Stake 1995).

Whilst this interview framework outlined the type of information potentially useful to understand travel philanthropy from this methodological standpoint, utilising an informal conversational technique throughout interviews and a predominantly unstructured approach to questions enabled greater rapport with participants than following more structured interview methods. Most importantly, this method also allowed data to surface by being open to the unknown or unexpected, as well as being a suitable interview style in such typically 'informal' research settings. The authors' previous experience in Africa, in addition to experimentation during fieldwork, informed these decisions. A central aim of the research was to understand frank perceptions of the system and experiences within it; therefore, using such in-depth and anonymous interview techniques helped participants be more open (Sinwell 2008).

As our aim was to "piece together clues and diverse bits of information" (Rutzou 2017:9), the interviews provided insights into some of the 'pieces' of the travel philanthropy system in the specific destinations from the perspective of different stakeholders, at a specific time, and aided the mapping of the bigger picture. However, quickly it was found that actors could not or did not view the whole system, and they often looked inward, as they were interested in the way the system worked around them. Therefore, whilst similar themes across interviews emerged, different perspectives indicated key, interrelated elements that others did not see. Thus, each interview contributed to the identification of parts of a bigger picture, which later defined the conceptual boundaries of the travel philanthropy phenomenon.

The reasons that actors gave for their/others' actions and indications of structural causes conditioning them to act in certain ways contributed to the later causal analysis (Kurki 2007). While it was important to gather and uncover the 'narratives' behind issues by describing the impact they had on individuals, it was necessary to go deeper and assess their impact on broader society (Lennox and Jurdi-Hage 2016). Different from interpretivist approaches to research, the result in this study was the identification of one or more mechanisms that could be regarded as having caused the events (Easton 2010). The goal of explanatory mechanisms constantly informed the questioning and listening in the field.

During the empirical phase of the research, as knowledge advanced by interviewing and observing, the initial ideas about understanding the 'impact' of travel philanthropy (at the outset of the research), such as its aggregate impacts, needed revision. Redefining what 'impact' meant became necessary, as to judge the 'impact' or 'good' of travel-philanthropy-related 'interventions' would, for example, demand in-depth investigation into the impact of 'formal education' in Africa in a broader socio-economic system after discovering that most travel philanthropy donations were allocated to fund children's education. Investigating this aspect would have demanded substantial empirical evidence beyond the scope of our initially planned research.

Furthermore, it was also realised that it would have been necessary to undertake *direct comparisons* with other forms of development aid interventions to 'judge' the value of travel philanthropy in the development field from an interventionist perspective. One question that emerged was for instance why The Bill & Melinda Gates Foundation, Oxfam, United Nations Development Programme (UNDP) or other non-governmental organisations (NGOs) might be considered differently from travel philanthropy once the 'giving' reaches the recipient. Another early discovery was that many beneficiaries or local people did not differentiate between aid models or donors – a donor was a donor. Both these considerations posed problems for investigating the merit of travel philanthropy interventions *per se*, especially since much research already exists on NGO interventions, and this overlap made it difficult to theorize the specific merits of travel philanthropy. Therefore, exactly where to set the boundaries for this research conceptually and how to interpret 'impact' identified as a gap in research was increasingly a challenge growing mid field work. Finally, a process of logical deliberation led to the realisation that a broader understanding of the system clearly had to be researched first and it was the impact of the *travel philanthropy system* as a whole (not specific interventions) that was important at this stage, to be able to set the foundation for these types of questions in the future. This was a turning point in how to progress with defining what impact meant in the context of this research, so to align with the theoretical and practical elaborations of travel philanthropy across the case study settings.

5.2.8 Data Analysis

As the empirical research progressed, the researcher fully transcribed electronically all audio-recorded interviews (approx. 500k words total) leading to a rich data set. The study then followed an inductive/abductive research process, moving backwards and forwards, evolving in a continuous cycle of research and reflection (Easton 2010).

In the first phase of analysis, the researcher coded data within each individual interview. The coding avoided a rigid approach to avoid preconceptions and distorting observations (Saldaña 2016). An extensive list of codes was initially drawn from the interviews (approx. 200), and through a series of coding phases, they were refined into useful categories (initially 20). This was done manually, with Nvivo used to manage the process and store the data. The key code themes from the first phase resulted in several categories. These included 'behaviour-action-doing' identifying empirical events and behaviour of people/organisations in the system, data about specific actor characteristics, cause–effect, network flows, perceptions, properties of the gift, sources of influence on the system, such as beliefs, emotions, motivations, reasons, and anything specific to define travel philanthropy exchanges. This reflected the interview framework outlined in Figure 5.2.

Code themes from Nvivo for each country were exported separately for the first phase of analysis. Quotes under coded themes were printed, read carefully, and marked with highlighters and notes, which provided data for a set of visual cause–effect diagrams and mind-maps. The diagrams represented sub-themes within the broader code categories for a greater level of detail, and created a visual tool used for analysis. This process was carried out for each country over a period of approximately three months. As part of this process, reflecting the critical realist ontology, data analysis began by searching for demi-regularities at the empirical level of reality, and looked for tendencies, or events and rough patterns in empirical data (Danermark et al. 2002; Fletcher 2016).

Key characteristics emerging from data collected in each country were then identified in the systems/mind maps as tools. Following this, various cycles of analytical processes were carried out, for example identifying key characteristics from the systems/mind-maps and noting them on post-it notes placed on blank sheets of paper, which were then later further organised into simplified themes. The goal was to keep refining the patterns to reach the fundamental essence of the findings.

At this stage of the analysis, an unexpected key finding began to emerge. When the essential characteristics were abstracted from the data for each case study, a great deal of similarity was discovered across all the cases. While the initial intention was to create a system map for each country for comparison, by viewing each country case study separately and then comparing and contrasting, some clear themes found across all cases began to emerge. Therefore, simplifying the data to the essence of the findings allowed one system map to represent all cases. This process of the analysis revealed the key elements of the system, specifically those found to be necessary for the system to function across all cases:

- A resource – money and other tangible objects
- Stories
- Beliefs
- Emotions
- Inequality or need perceived
- Technology
- Giving-receiving processes
- Actors and roles (active/passive – donor, storyteller, story audience, intermediary, recipient of resource, affected by recipient of resource)

Post-it notes linked to each of the above listed characteristic were then organised under each of these themes (resource, stories, beliefs, etc.), combining the three countries, but still identifying them separately by colour for further comparison. Different system functions/purposes (or perspectives) also began to clearly surface from the data at this stage, with the travel philanthropy system emerging as an *exchange or transaction economy*, exchanges of tangible (e.g., money) and intangible objects (e.g., stories and emotions). This top-level outline of the system elements were then followed by an iterative research process of returning to the interview data, disaggregating codes, revisiting the system/mind maps created earlier and back again, to gather further information on each of these core themes. This in-depth analysis led to carrying out a further phase of data coding to find more granular details of beliefs, principles and norms, stories and dominant narratives, emotions, tangible and intangible objects exchanged, types of exchange, etc.

Data was extracted based on the revised codes, studied in-depth and bullet-point notes compiled with the findings from analysing coded interview quotes. By way of illustration, all the *rules* were identified in giving-receiving processes (by individual country). The findings of this are presented in Chapter 8. These lists of rules were then studied to identify emergent themes from that data and then organised into sub-themes within 'rules'. Within the Ugandan data, for instance, there were approximately 150 rules identified, which were then consolidated into 22 themes, and further analysed to conclude 7 "principles/rules governing behaviour in exchanges". This was carried out for each case study and then the cross-cutting themes were concluded. These included:

- Justification of exchange (e.g., copying someone else)
- If and how the exchange happens (e.g., rules of refusing a gift)
- About what is exchanged (e.g., authenticity of object)
- Creating and maintaining relationships (e.g., turning a blind eye)
- Expectations (e.g., making promises)
- Power (e.g., othering)
- Ethics (e.g., rights and wrongs)

Theoretically, complexity emerges from the collection of agents following simple rules (Burns and DeVille 2017), and it became evident that such rules were standing out in the data. They were emerging as a foundation of the system, part of the unobservable mechanisms.

A similar manual process was carried out for each of these categories (beliefs, emotions, stories, etc.) for each country.

This summary and visualisation tool contributed to the methods used to generate the findings regarding the belief system and informal rules underlying the travel philanthropy system discussed in Chapter 8, as well as the other findings in Chapters 9–11. These findings then informed the decisions regarding the subsequent analytical approach to understanding the 'travel philanthropy' system, found in Chapters 8–10. However, it became evident that the findings generated through the methods described so far had not entirely represented 'the system', but rather some of the parts. The research had certainly revealed a combination of unobservable mechanisms (activated and inactivated) such as beliefs and empirical events such as gift-giving, some of which contributed to the emergence of the system and some shaped how it functioned, with a mix of structure and agency. They were indeed all key elements of the system; however, this was not the complete picture of travel philanthropy as an interconnected system and could not outline the whole system story. It needed further organisation and refining. We particularly needed to look towards emergence to gain clarity on the system (Meadows 2008). Therefore, further analysis took place that concentrated on the structure of the system and in particular delved deeper into causality to explain the processes within it.

The use of transitive verbs such as producing, generating, driving, pushing, and pulling lie at the root of the notion of 'cause' in critical realism (Bhaskar 1975:111), so "to ask for the cause of something is to ask 'what makes it happen', what 'creates' or 'determines' it, or what 'enables' or 'leads to' it" (Sayer 1992:104). Critical realism is characterised by its *retroductive* mode of inference, shifting from knowledge of some phenomenon existing at any one level of reality, to knowledge of mechanisms at a deeper level or stratum of reality (Lawson 2003). Therefore, this research would generate knowledge of those "mechanisms of nature that produce the phenomena of our world" that may or may not be triggered but endure (Bhaskar 1975:37). Informed by this stratified theory of reality, the domains of the real, actual, and empirical (Bhaskar 1975), the data analysis then evolved to identify the observable phenomena at the empirical level, and beyond, to causal mechanisms at the level of the real – i.e., the 'demi-regularities' or patterns observable as events (Lennox and Jurdi-Hage 2016; Danermark et al. 2002).

The role of theory distinguishes critical realism from purely inductive approaches, which lack the ability to develop abstract and general theories (Lennox and Jurdi-Hage 2016). In this study, theories were formed using rational judgement to help get closer to reality by aiding the identification of causal mechanisms driving the events or phenomena (Fletcher 2016). Abduction guided the analysis, generating explanatory hypotheses from data similar to induction (Mabsout 2015), but taking the "middle ground between induction and deduction" (Patokorpi and Ahvenainen 2009:126) by generating ideas, or 'theories', as hypothetical explanatory concepts (Thomas 2010). This meant 'moving backwards' and asking, "What must be true in order to make this event possible?" (Easton 2010:123). Mechanisms or tendencies of relationships among elements theorised to cause events were then studied (Easton 2010) as well as *events* and causal pathways (Bengtsson and Hertting 2014) and knowledge about the behaviour of mechanisms in different contexts (Danermark et al. 2002; Benton and Craib 2001).

The iceberg analogy is often used in critical realism and systems thinking to shift our focus from horizontal cause–effect relationships to vertical hierarchy. The visible tip of the iceberg – the observable events of repeated cause–effect patterns – takes on meaning against the background of structural factors, which may include unspoken beliefs, values, and standards that guide how they function (Vandenbroeck 2015:10–11). In the hierarchy of explanations of events, patterns, and structures, the lowest levels give context and meaning to the highest levels. At this stage, the research needed to discover the higher and lower-level elements of the system. An iterative process from the tip to the base of the iceberg took place, which meant returning to the data.

5.2.8.1 Unobservable mechanisms and causality

It proved fundamental to understand the travel philanthropy system by allowing ideas, meanings held by individuals, beliefs, feelings, intentions, reasons, norms, social structures, and so on to be equally real to physical objects and processes, because causal impacts on agents and behaviour were evident (Bhaskar and Callinicos 2003; Kurki 2007; Maxwell 2012a; Mingers 2014). As said by Bhaskar and Callinicos (2003:11):

> *That an illusion such as belief in witchcraft, if it's unreal – which I'm sure it is – is not true of an object, but it's real in the sense that it's causally efficacious and so part of the world ... ontology actually includes everything – it includes contradictions and mistakes – there's nothing that's not included within ontology.*

Without conceptualising the unobservable structures beneath patterns, it would arguably have been impossible to grasp a realistic understanding of the travel philanthropy system, as so much seemed to rest on such unobservable ontological objects in this social world, such as the belief that giving is good.

The other useful influence from critical realism was considering causal powers in terms of 'absence'. Even an 'absent' mechanism is still an ontological entity that might affect us in some way (Roberts 2014). Mingers (2011) argued that the idea of absence is an important topic not generally discussed or considered in the contemporary literature on systems thinking. This research helped fill the gap identified by Mingers by considering what alternative possibilities could conceivably have occurred and inquiring about those absences in the travel philanthropy system.

5.2.8.2 Mapping the system

In Chapter 3, we identified the growing interest in systems and complexity amongst international development professionals working in an interconnected world of wicked problems. However, the question "what actually is a system from a methodological point of view?" emerged. The word 'system' derives from the Greek *synhistanai*, which means 'to place together' or 'to stand together' (Ison 2008). The 'system principle' refers to 'the organising principle' characterising intra- and interrelationships existing in and between a system and its environment (Flood 2010). *Systems thinking* describes phenomena as a system, thinking *systemically* is about 'making sense of the world' (Hummelbrunner 2011:397), the approach enables a *holistic* view on research issues (Peric and Djurkin 2014) and an understanding about interrelationships and patterns (Bowman et al. 2015; Flood 2010). In its early days, whilst systems theory considered 'systems' as 'objects' in the world, physical entities that were fixed, our study aligns with later concepts interpreting systems as representations not to be confused with the real world. Thus, as the famous line goes, the map is not the territory (Korzybski 1931). This informed the conceptual model presented in Chapter 11.

A further question emerged, "what actually constitutes the travel philanthropy system and how could it be mapped?" More than the sum of its parts, a system is defined as "an interconnected set of elements that is coherently organised in a way that achieves something", including elements, interconnections, and a function or purpose (Meadows 2008:11). This study applied systems theory to examine how 'the whole' of 'travel philanthropy' was made up of processes and structures (Paliwal 2005), what its patterns were and the sources of system behaviour that had formed over time (Peric and Djurkin 2014) and analysed the systems' functions/purposes and outputs (Meadows 2008). As extant literature called for more work on understanding the inner structure of complex systems and the roots of their dynamic behaviour (Burns and DeVille 2017), particularly in international development, this study also considered complexity theory. This helped make sense of 'emergence' found in systems and to challenge assumptions about predictable behaviour and only linear change (Mingers 2014).

Drawing on a combination of system thinking methods by Sherwood (2002), Senge (1990), and Johnson et al. (2019), for example, a series of systems maps drawing on themes identified in interviews were then sketched, re-sketched, and repeated. Data analysis was now moved to an electronic platform (using Kumu 2020, systems mapping software) to manage the empirical data, which was increasingly complex to manage manually. The goal of this process was to identify emergent elements of the system and to learn its interconnections and dynamics. Chapter 8 of this book elaborates on this aspect of the findings.

The key to systems thinking is to take a holistic view and understand chains of causality, so the research goal was to identify cause–effect relationships that underlie the travel philanthropy system (Sherwood 2002). Informed by critical realism (as established above), chains of causality can also be vertical, from the domain of the empirical to the real (Bhaskar 1975). In systems mapping, causal loop diagrams are one of the tools to show variables identified that are connected by key causal relationships displaying the systems cause and effect behaviour, and the diagrams intend to turn complex elements into a simple format that is easier to understand. Figure 5.1 illustrates the meaning of such diagrams. There are only ever two types of links, which can be referred to either as 'same', 'S' or '+' as it is in the Kumu (2020) programme. This is further discussed in Chapter 3. If a causal loop diagram shows a closed continuous loop it is known as a feedback loop, which can be either reinforcing or balancing. Reinforcing feedback loops show when change is amplified, and balancing feedback loops show change in a counteracting effect. All systems have interlinking networks of causal loops.

Sherwood (2002:71) argued that good systems maps require deep knowledge of the system, "the insight to see the forest from the trees" and the "explicit articulation of relationships that we all know are present but rarely talked about". The rich data within the cross-case interviews made this possible. Sherwood also urged the consideration of fuzzy variables, which are very important but difficult to measure, making explicit our beliefs about how the world works. Everything starts with a cause and leads to an effect, forming a network of interactions, which is the system model from which to consider an understanding of the dynamics, interconnections and relationships (Banson 2015). Circularity in systems language effectively expresses interrelatedness within a system and every interrelationship can be both cause and effect, working both ways. This study examined these types of feedback loops and the consideration of fuzzy variables carefully. To master systems thinking, it is important to step inside the system to learn about its structure

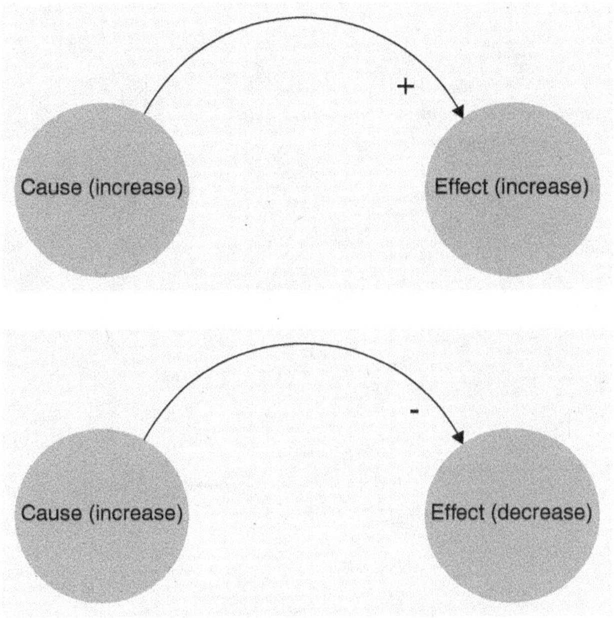

Figure 5.1 How to read interconnections in the system map.

and processes and to discover the rules that steer its behaviour (Hummelbrunner 2011:402–403).

To begin developing this system map for travel philanthropy, the different events and interconnections identified as similar across cases (from the various phases of coding as well as referring to the original mind maps) were uploaded electronically. This allowed the study of all the elements and interconnections by extracting or highlighting (using Kumu) certain relationships and abstracting causal loops. The system story began to emerge. Figure 5.2 visually illustrates this complex phase of the system map. This of course makes no sense when looking at it in detail and does not provide a useful communication tool to discuss travel philanthropy as a system theoretically. Therefore, deep reflection, further thematic analysis of cause–effect abstractions and the use of theory, led to a simplified version of the map (see Chapter 9).

5.2.8.3 Opening the black box in an interdisciplinary system

The aim of mapping the system was to identify causal processes that resulted in particular outcomes including the unobserved factors that lie between – i.e., opening the black box. The analysis involved iterative cycles of reflection between theories in academic literature, data, and propositions in an effort to achieve analytical stability about the mechanisms (activated or unactivated) characterising an event or outcome (Zachariadis et al. 2013:12).

Figure 5.2 Method: developing the travel philanthropy system map – the details.

Note: This diagram shows part of the process of developing the travel philanthropy system map (found in Chapter 9, Figure 9.1) – tendencies of causal interconnections were identified in detail before refining the final system map which includes only key elements – this is not expected to be read in detail rather represents part of the method of dealing with complexity and producing a more useful understanding of the system.

Abstraction was a key aspect of research and analysis when continuing to ask the question *why* (Easton 2010:123). The analysis of the travel philanthropy system followed Bhaskar (1975), who disregarded the use of existing theory as a starting point for empirical research. Thus, theories were applied at this later stage, informing interpretations and enabling explanations by conceptualising structures and causal mechanisms (Lennox and Jurdi-Hage 2016; Fletcher 2016). As a complex social system, it became evident that different details of its behaviour could actually be constructed using theories from a wide set of disciplines, whether economics, psychology, anthropology, biology, history, etc. To explain the complex system that was emerging from the data, an interdisciplinary theoretical approach was fundamental. See Chapter 7 for the resulting theoretical lens that helped contribute to the explanation of data. With time and space limitations in this study, boundaries needed definitions.

This systems approach informed by critical realism followed Aristotle that 'nothing comes from nothing', believing all events, processes, objects, and agents arise out of some conditions and influences, from a pre-existing causal context (Kurki 2007:5), which are objects of knowledge affecting people's actions (Lewis 2000). Drawing on Pawson and Tilley's (1997) realist approach, causal analyses considered contexts, mechanisms, and outcomes to identify relationships. This led to a probabilistic understanding of causality based on the idea that outcomes cannot be determined a priori by knowing the type of mechanism that is at work, because the context they interact with is inconsistent (Trampusch and Palier 2016). According to conditions, the same mechanism may produce different events, and conversely the same type of event may have different causes (Sayer 1992). This study addressed the relations of causal mechanisms, which produce complex entities, and in the most simplified way possible, identified cross-case themes of cause–effect (Rutzou 2017, Flood 2010) discussed in the system map in Chapters 9 and 10.

5.2.9 Ontology, epistemology, and the changing social world

System models have become more common in the global development field in recent years, and currently advocated for in sustainable development debates, so this was an appropriate method aligning with the field of travel philanthropy, as outlined in Chapter 4. Nevertheless, it must be noted that in a world full of complexity and uncertainty, such maps are never completely correct, because they are always changing, and socially produced knowledge is always fallible (Bhaskar 1975). Representations of knowledge do have limitations, so even though the thought process of expressing structures and ways of acting means knowledge can be justified, this is only as conceptual frameworks (Bhaskar 1975; Archer et al. 2016). This methodology should therefore be considered as enabling a model of the travel philanthropy system to open further discussions and debates, with this process of thought being conceptual, rather than a representation of the actual reality.

This study also drew on critical realism and systems theory to delve into the ontological realm of travel philanthropy, holding that "objects apart from our descriptions of them" exist, which appear as generative mechanisms and structures (Bhaskar 1975:239–240). 'Travel philanthropy' is also socially produced and individuals will continue to transform this system. Thus, epistemological relativism is for critical realism "the handmaiden of ontological realism and must be accepted" (1975:241). This understanding is inherent to the findings in this study, in that is produced the best possible conclusions and insights about the system and its outcomes from the sources and methods drawn upon at this specific time and locations. Therefore, our understanding of this phenomenon as a system is not conclusive.

5.3 Research ethics

Good practice for social science research is fundamental and understood in the context of specific research projects (ESRC 2021: online). In line with the UK's Economic and Social Research Council (ESRC – funding this research), we followed various principles and projects that included: maximising benefits and minimising risks and harm, respecting the rights and dignity of all people, that research participation should be voluntary where possible, with responsibilities clearly defined, and that research should be conducted with integrity and transparency. Key ethical considerations for this study are summarised in Table 5.3.

Table 5.3 Key ethical considerations

Participant age	All participants were over the age of 18, so although issues regarding children are discussed in the findings, those below this age were not interviewed for ethical reasons.
Payment	Interview participants were not paid. Interpreters were paid a fee reflecting local wages for the time hired. The majority of interviews did not require interpreters because many individuals linked to travel philanthropy tend to speak English. When interpreters were required, they were fully briefed on the ethical protocols and quality translation, such as confidentiality, accuracy, integrity, working together practically, gender equality, etc.
Voice	The 'local communities' in rural Africa were considered a relatively marginalised population from a social inequality perspective, in the global and regional context. Researcher positionality was therefore reflected upon continuously during the research design, fieldwork, and in analysis to ensure power issues were addressed – economically, socially, culturally. Applying the systems thinking approach and methods was beneficial for reducing issues of power and equality through enabling multiple perspectives, listening carefully to what participants considered important over researcher bias through relatively unstructured interviews (see 5.2.1) and overall generating a greater voice to such marginalised participants within the global travel philanthropy debate. Key stakeholders, which were eventually identified as research participants, were also significantly informed by local people to reduce outsider and pre-research bias, in who should have a voice.
Consent	Participant information was delivered in written and oral format in English or translated where necessary. A consent form was provided as an agreement to the terms and conditions between the project and the participant with all contact details available for follow up concerns. Participants were made aware of their freedom to immediately terminate (or pause) their involvement in interviews.
Anonymity and confidentiality	Names of individual participants were not considered important for the presentation of this study because the interest was in overall patterns of behaviour, systemic structures, etc. All participants remain anonymous throughout. All personal data collected has been treated with strict confidentiality and whilst themes, situations, and insights may be raised between interviews, no individual names were discussed specifically.

Table 5.3 (Continued)

Sensitive issues	The research was not used as a vehicle to expose specific issues such as corruption nor aimed to deconstruct such practice. However, even though they emerged through interviews and systems mapping cause–effect discussions, the focus was rather on the broader issues of travel philanthropy, and any individuals/organisations mentioned by name remained anonymous.
Researcher experience	The researcher had experience living, working, and researching in remote rural communities in Africa and had learned how to adapt professional and personal behaviour in specific cultural contexts, which aided the field research methods.
Quality of knowledge	An honest and respectful account of data has been presented. Methodological decisions and positions, including epistemology, are clarified in the remainder of this chapter.

This chapter has presented the research methodology guiding our qualitative study, explaining its philosophical position and methods. A synthesis of critical realism and systems theory has been introduced as a framework, and the relevant literature reflected upon. This included matters of causality in social science, such as the critical importance of understanding unobservable and absent mechanisms. The case study method, research site selection, as well as interview, data analysis, and systems mapping techniques followed. The following chapters will present the empirical findings of this study. Chapters 7–11 provide an in-depth analysis of the findings, leading to a better understanding of the travel philanthropy system at the time of research being conducted in the three mentioned African countries, inspired by critical realism as a philosophy of science.

6 Actors, roles, and perspectives

shaping the system

This chapter identifies key travel philanthropy system actor groups/stakeholders (the people and organisations) and their dynamically shifting roles, compared across case studies. Although across all case studies assessed in our study, system roles were found to be the same, occupants of each role differed and were not static. Here, systems perspectives are analysed, and a taxonomy of five system functions is identified, leading to the introduction of the business of poverty concept and its significance.

6.1 The system stakeholders – people, organisations and roles

Stakeholders are actors that can affect or are affected by the system; they have a stake in travel philanthropy (Freeman 2015). This study began by unpacking stakeholders' opinions, experiences, and perspectives within the travel philanthropy system and understanding that they may differ. In this chapter, we introduce the network of stakeholders identified, contributing to the contextualisation of the travel philanthropy phenomenon in the cases studied in Uganda, Ethiopia, and Zimbabwe. A key aspect that emerged was the relevance of identifying stakeholders in two ways: (1) *actor groups* defined by people/organisation categories and (2) *roles* defined by an actor's function in the system, as categories. Stakeholders will be referred to in both ways depending on the context of the discussion.

Stakeholder groups in this system are not static social objects but rather have changing interests and unfixed landscapes of power (Lyon et al. 2017). Whilst across the three studied countries and cases, stakeholder categories and actors directly involved in travel philanthropy were almost identical, a key finding of our study was that there are very similar patterns of stakeholders' *roles* and *functions*. However, the notable difference between countries (not between cases *within* countries) lay with the roles and their occupants, which might perform similar functions but varied in behaviours, producing context differentiation and influencing a variety of outcomes.

DOI: 10.4324/9781003497929-6

6.1.1 Key actor groups – people and organisations in the travel philanthropy system

The identification of travel philanthropy stakeholders was an important task, as omitting any of the actor groups such as travellers/tourists or local community actors, and focusing only on tourism companies, for instance, would be erroneous. In Chapter 4, we raised an issue of inconsistency across the literature regarding *who* was considered part of travel philanthropy. In this respect, our research highlighted that, even if a tourism company was 'directing' a gift-giving pathway, due to the complexity of the system, one cannot explain the system by simply identifying and discussing tourism companies alone. Firstly, all stakeholders summarised in Figure 6.1, followed by the development of an empirical taxonomy (see Section 5.2), were found to be highly interconnected in this system. Figure 6.1 illustrates the actors at the local, national, and international levels that have emerged from our empirical assessment.

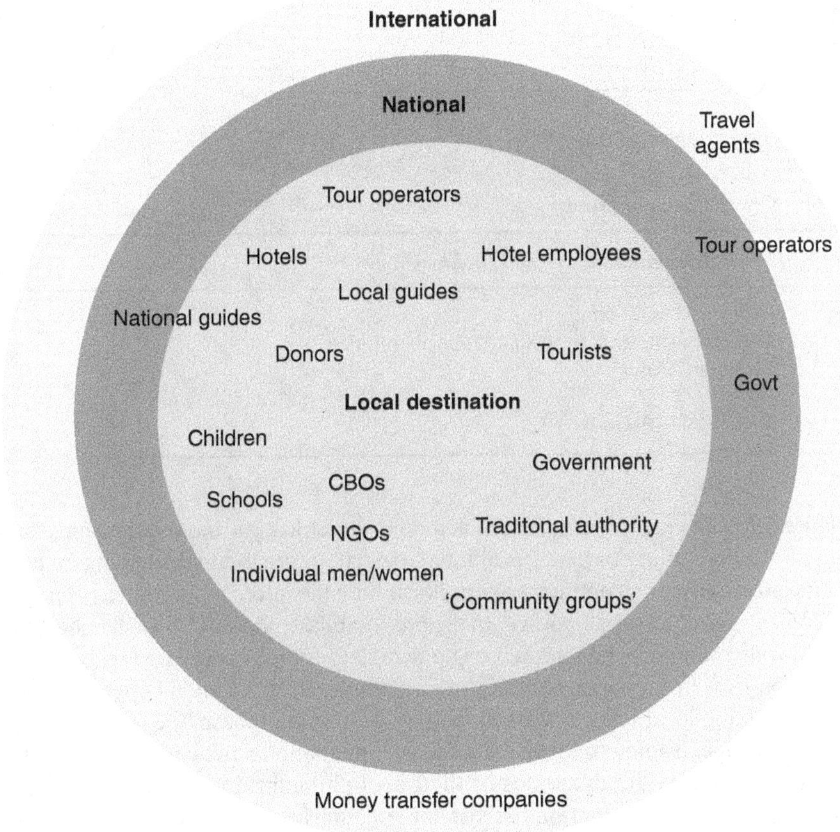

Figure 6.1 Local, national, and international travel philanthropy stakeholders.

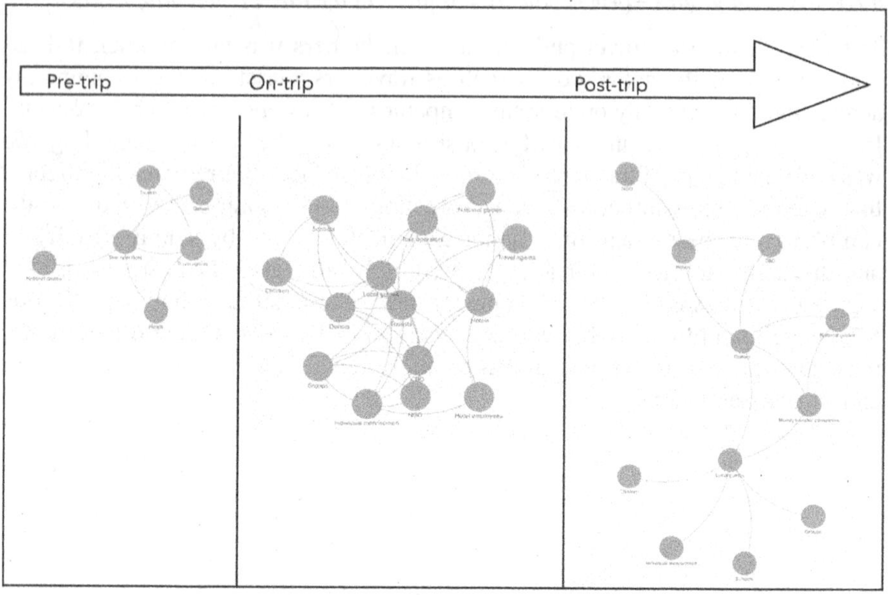

Figure 6.2 Emerging actor network.

Table 6.1 Key stakeholder groups

	Key in-destination active stakeholders
1	Guides and hotel employees
2	Tourism businesses: hotels and tour operators
3	CBOs and NGOs
4	Tourists
5	Schools and children

Figure 6.2 is an example diagram of actors' communication and relationships, and because in reality numerous patterns of interconnections are found depending on context, this only indicates a common pattern illustrating the interconnectedness of actors at different stages of a tourist journey. In the pre-trip phase, there are few stakeholders, and the local community has limited or no access to tourists compared to when the tourist is on-trip. In the latter phase, a more complex web of interconnections predominantly at the local level is formed, and in the post-trip phase, the actors that are connected change, leading to local, national, and international network connections.

As we focused on the dynamics of the travel philanthropy system in five specific African tourism destinations across three countries, the majority of actors of our study were at the destination level. However, it should be noted that more peripheral or indirect stakeholders, such as international media, airlines, banks, etc, were interconnected with the system and have influence in, for example, exposing new places to travel philanthropy, enabling online financial transactions, and shaping beliefs and norms. However, for the purposes of our study, the actors that were identified as influential and therefore most relevant are listed in Table 6.1, supported by a full description in Table 6.2.

What emerged is that different destination contexts influence the relationships each actor has with each other, such as local individuals with tourists, which causes differences in aid pathways and the characteristics of exchange. For example, it was mentioned that:

> *[In Zimbabwe] the person with the relationship with the client is going to be the lodge... guides are just moving them from A to B... it's more transport than a guide.*
>
> (Z1_TB_29)

By analysing the different cases, it became evident that the 'community' contexts are undoubtedly not a static backdrop, confirming South et al.'s (2019) view. In fact, significant diversity was found within each stakeholder group, as individuals and organisations did not hold the same beliefs, behave in a uniform manner, or have the same capabilities. To illustrate this, a tendency emerged amongst tourism business stakeholders in Uganda, contrasting with many of the tourism businesses in Zimbabwe, which had established more formal giving programmes. In Uganda and Ethiopia, the management of tourism businesses was much less inclined to be directly involved in the giving–receiving pathway and occupy a different system role. This reiterates the point that to fully understand travel philanthropy, we need to look beyond the assumption that tourism businesses are the principal actors and that network structures are consistent across destinations.

As philanthropic philosophies across the tourism industry are also not uniform, they do not necessarily influence the travel philanthropy system by choice. Whilst the literature reviewed led to questioning assumptions about the homogeneity of tourists and tourism businesses, such as idealistic visions of travellers seeking sustainable change in the places they visit, embarking on ethical consumption (Lindkvist 2011; Hunke and Koot 2012), this study confirmed that the heterogeneity of tourists and tourism businesses is part of the systems' complexity, as highlighted by some tourism businesses:

> *I don't think I would have the skill-set to go in and run that, I mean we are hired for a specific purpose and that purpose is not running the community, or running donations and that kind of thing you know... It would be a nightmare. You can't be for the community and also deliver a certain level of comfort, because they almost totally go against what you are doing, you are spending money on frivolous things and then trying to manage all these other little projects which are so different to that, I mean I don't see how it would work properly. Certainly not here. Certainly not with me doing it.*
>
> (U1_TB_3)

Table 6.2 Stakeholder groups descriptions

Key in-destination active stakeholders' description		*Indicative quotes*
Guides and hotel employees	In the travel philanthropy system, guides and hotel staff play key roles, spending significant time with tourists and often gaining their trust. This trust can position them as intermediaries in the philanthropy system, helping to manage donations and, in some cases, select beneficiaries. Whilst local guides deliver the core tourist experience, they may receive clients from national guides or hotels. Occasionally, hotel staff, depending on guest duration and setup, also participate in this process. Not all guides work directly in tourism; some, like traditional authorities or teachers, may serve informally in this role.	"…when you are a guide you have much more interaction with the client… so those are connected to donations" (U2_INTER_9) "our guides are quite principled but I think some guides are not so principled … there are very professional guides … and you have got the people that want to get the money" (EN_TB_67) "so they collect some stationary materials, teaching aides and so on and distribute to local schools in remote areas like in Turmi, so we encourage clients if there are tour leaders the tour leader will do this" (EN_TB_65) "this is the weakness of our local guide" (E1_TE_46E1_TE_46) "when they come to the school it is the teacher who takes them around the school but when they come to the village, first and foremost it's me, they hand them to me, I will do the tour and it's me to say go to that village" (Z2_1_B_INS_38)
Tourism businesses: hotels and tour operators	Tourism businesses vary greatly in size, goals, and involvement in travel philanthropy. Key players include hotels, lodges, tour operators, and occasionally international agents. Lodges differ significantly in their approaches, from active fundraising and community projects to merely advising or funding CBOs or NGOs. Some adopt a hybrid model with attached non-profit entities. Tour operators, often based in cities, are less involved in local philanthropy due to their spatial detachment, generally providing contacts for clients, whilst guides lead most giving activities individually.	"they are profit oriented and just benefitting their company, they are not much in for the community" (U1_CBO-F_2) "we all have different drives for our companies... we are all different… marketing different products, having a unique selling points" (Z2_2_TB_41) "just that is not our purpose you know… the responsibility of the private sector is running tourism and business" (EN_TB_70) "we feel this is part of our business, this is part of what we do and we feel responsible" (E2_B-COM_64) "we feel very strongly that we have a role to play" (Z1_TB_20) "a community guide takes them around, then they may feel of donating something, without my concern" (U3_TB_19)

CBOs and NGOs	Community-based organisations (CBOs) are locally owned entities, often unregistered, with activity levels varying across countries. In Uganda, CBOs are highly competitive, striving for roles in the travel philanthropy system, whilst stricter regulations in Ethiopia limit their numbers. Some informal, guide-led entities operate in Ethiopia, and Zimbabwe's regulations are between Uganda's and Ethiopia's in terms of ease of establishment. NGOs, defined as non-local-owned non-profits, include both tourism-linked and independent entities, though their engagement with tourism is generally weaker in Ethiopia due to government restrictions and local norms. Rather than transparency or participatory goals, CBOs often focus on economic opportunities through donations.	"someone doesn't have a project but I mean rents a house and then writes a very big poster and says orphans and vulnerable children" (U2_CBO-F_8) "if these people keep going to one project they definitely benefit one person, they benefit one, than all these other communities" (U1_CBO-F_1) " it's their fat cat thing ... they are kind of like a weird little maffia in their own right" (U1_TB_3) " there are people who run these community organisations entirely, they kind of take them to be like charity organisations or community organisations helping he community but they also rely on them for their livelihoods" (U2_CBO-F_8) "there are 100 or 200 guides but the guides who have charities is not more than 30 or 40" (E2_TE_63) "they don't have the right training on how to promote their project" (EN_TB_65) "we are a bridge between knowing and working with communities and seeing what the needs are and having all of these people on our doorstep visiting with money and sometimes with an interest to put it towards something useful" (Z1_NGO_21)

(Continued)

Table 6.2 (Continued)

Key in-destination active stakeholders' description		*Indicative quotes*
Tourists	International leisure tourists vary widely in demographics, interests, and experience, influencing their behaviours in travel philanthropy and their relationships with other actors. In these sites, tourists generally fit into high-end or mass adventure categories, with most on bucket-list trips, staying 2–3 nights per destination within a 2–3 week guided driving tour. Visits focus on protected nature or cultural sites, with community experiences typically arranged during spare time on the trip.	"people getting more connected to tourists who come to these ends" (U2_INTER_12) "some foreigners have different principles they have different behaviours so some have good behaviours and some have bad behaviours" (E1_B-INS_47) "they are in a way out of their comfort zone" (EN_TB_65) "it is written in the guide books ... they know the situation but when they get to the ground from their hotel to the restaurant from their hotel to the church in the meantime the children within a short period they change their mind, they have the capacity to change their mind" (E2_NGO_61) "they spend the rest of their time at the hotel so it's very in and out and unless the hotel has got information about it in their rooms" (Z1_TB_29)
Schools and children	Schools are central to village tours, with children often a primary attraction in travel philanthropy. Hosts like teachers or village heads typically receive no payment, instead hoping for tourist donations. In Uganda, some village tours are CBO-run and paid, aiming to inspire tourist giving. School visits are a key part of these tours, with children sometimes performing for tourists at hotels. In Ethiopia, school visits are rare, and interactions with children occur casually in public spaces. In Zimbabwe, children play a significant but indirect role in tourist donations, serving mainly as fundraising symbols rather than direct participants.	"see these older white people all smiling at them they know that they have got to smile and do a song and then they'll get stuff" (Z1_NGO_21) "we don't allow them to give out to the kids because will create a culture of begging, we don't allow them, but if it's a if a guest want to specifically give to a child they will have to consult the teacher" (Z2_1_B-INS_31) "they want to see how the kids are taught, they want to see what is lacking, make arrangements they are bringing gifts" (U3_TB_16) "one thing is a negative part is the tourists come here and give money to kids, this money is really not good" (E1_B-COM_69) "I saw the chidden becoming happy so it was good for them and I cannot be angry of that when the children are happy." (U2_INTER_12)

> *It's just ethical stuff we don't believe that you just drain and drain resources from a place and you do not plough back, so this is our nature.*
>
> (Z1_TB_28)

Conversely, whilst there was diversity found amongst stakeholder groups, the institutional monocropping of non-governmental organisations (NGOs) argued by Evans (2004) as a flaw in the international aid system was also identified amongst community-based organisations (CBOs) linked to travel philanthropy, as well as some of the tourism businesses and hybrid models (businesses with a formalised charitable arm, see Section 10.2.1). In travel philanthropy, rather than big donors pushing blueprint approaches and driving the top-down giving trend (Stirrat and Henkel 1997), interventions have been emerging from the bottom up – i.e., CBOs or local actors. However, mechanisms of stakeholder imitation of established Western paradigms in local settings did not challenge the status quo.

Our research questioned idealistic post-development theories concerning stakeholder categories (Escobar 1995) and the assumed efficacy of bottom-up approaches. Whilst Sustainable Travel International (2008) recommends projects receiving donations to be designed and managed by 'locals' rather than tourists to ensure they are worthwhile, it ignored the significant complexities in this assumption. Donors may not be directly dictating 'worthwhile' projects or needs, but CBOs and businesses are often responding to expected needs of tourists, confirming Novelli's (2015) view that travel philanthropy mirrors some of the mistakes of traditional aid models. Whilst Honey (2011) argued that 'community organisations' need to be the drivers of 'philanthropy-based development initiatives', 'local' stakeholders are not neatly aligned with idealistic notions of '*local* means best'.

Despite this monocropping, it became evident that diversity within stakeholder groups (see Table 6.1) also contributes immensely to the complexity of the system, in which interactions between actors can be very unpredictable. There are nevertheless characteristics that define the diversity of actors in a 'group' as a stakeholder category, with relatively distinct actors' *roles* emerging, which, although similar across cases, felt the influence of contextual factors. It is for this reason that an analysis based on stakeholder groups only would not suffice. Discussions in our study therefore draw either on actors as stakeholders, such as a guide or tourism business, or by reference to roles, such as donor or intermediary.

6.1.2 Key roles in the travel philanthropy system

A significant finding in this research was that across Uganda, Ethiopia, and Zimbabwe, very similar patterns emerged with regards to actors' roles and their *functions*. To understand the travel philanthropy system, the research highlighted the importance of understanding which actor occupies which role, with inconsistencies found across different contexts. Actors' behavioural characteristics

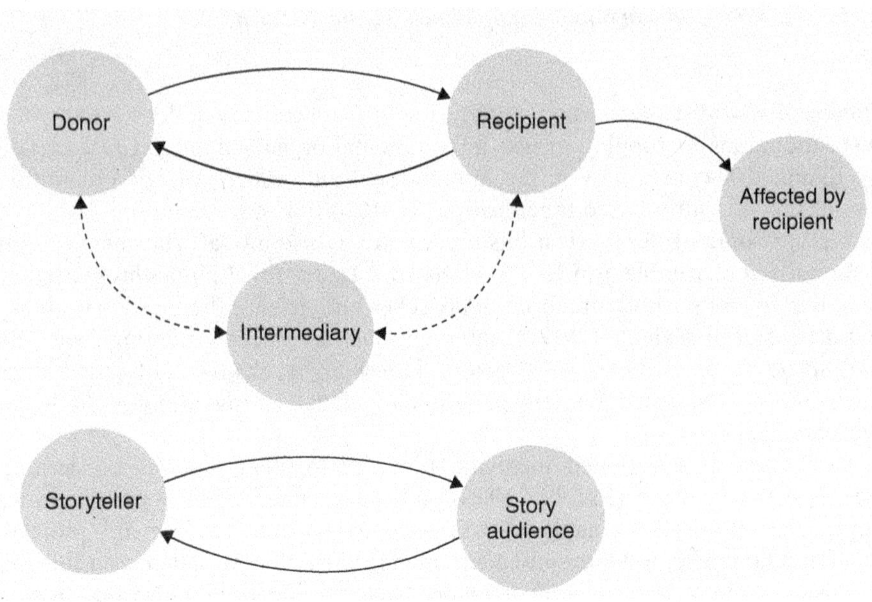

Figure 6.3 Travel philanthropy system roles.

and purposes can differ, but what those in specific roles do was largely comparable across cases. These roles are key elements in the system, and without actors occupying these roles, the system would simply not exist. Gaining clarity about the roles in the system helped disentangle some of the complexities in decision-making, related aid chains, and governance. Whilst Figure 6.3 introduces the *role* categories in travel philanthropy that emerged from the system, *Table 6.3* defines each role's function in the system.

Each role emerged as pivotal for the system to function as well as being interdependent. For instance, whilst a donor cannot exist without a recipient, a recipient cannot exist without a donor; an intermediary can only operate if a donor and a recipient exist, with donor and recipient often relying on an intermediary to facilitate the exchange. Equally, a storyteller cannot tell a story without an audience, and an audience is not an audience if there is no story to listen to. The research found that travel philanthropy's exchanges involved between two and six key roles (Table 6.2), of which two are fundamental – the donor and the recipient, which are further explored in Chapters 8–10.

The holistic approach of this study did not reveal a straightforward linear aid chain or a simple transaction pathway motivated by altruism that begins with the tourist or tourism business as a donor ending with a local beneficiary recipient. Rather, it was found that travel philanthropy is a complex system of

Table 6.3 Definitions of key roles identified in the system

	Role definition: functions in the system
Story teller	Tells stories and communicates the travel philanthropy narrative to the story audience
Story audience	Listener and spectator of stories
Donor	Transfers a tangible object (monetary or non-monetary) to another actor (in the form of a gift/donation)
Intermediary	Responsible for channelling/brokering tangible objects from donor to recipient – managing resource distribution As the go-between intermediaries can link, connect, manage, monitor, communicate, report, advise, bridge, channel, distribute, and often select recipients. Their roles varies.
Recipient	Receives a tangible object from a donor – monetary or non-monetary (gift/donation)
Affected by action of resource recipient	Affected by recipient of resource behaviour – public or private consequences

diverse stakeholders occupying different dynamically shifting roles within a relatively informal economy of exchange. For instance, whilst a guide at first may occupy the role of storyteller, they may shift into the intermediary, donor, or recipient role in the succession of exchanges. A tourist or tourism business may also perceive them as an intermediary, for instance, yet from another perspective, such as by someone in the local community, the guide has become a local donor, or recipient. We therefore cannot fully understand the system unless we study interactions from a context and role perspective. As one local participant in Uganda said:

> *When they get the chances of meeting these people, the tourists, let's say the guides, when they get the tourist and they tell them about the community, the challenges of the school fees etc etc etc, they end up being the go-between because now the tourist gets connected to them, they exchange contacts and they keep giving updates the other side it comes to donating, they donate sometimes through these people.*
>
> (U2_INTER_12)

Part of the rationale for focusing on roles rather than actors was also because many local people did not differentiate amongst specific stakeholders. For example, a foreign donor is a 'donor', whether a tourist or an individual representing an established NGO.

> *To them a donor is a donor.*
>
> (Z2_2_NGO_45)

> *For him he is not a real donor, he's not a donor, but he fundraises from donors, then he manages donations, he's a businessman.*
>
> (U2_B-COM_13)

This evidenced not only the social construction of an identity that is useful to recognise, but that such beliefs, perspectives, and interpretations of stakeholders can have real implications on behaviours in the system as part of the system's causality.

Table 6.4 presents overall differences and parallels of the most active roles and occupants across cases, with a general overview followed by patterns identified in each country. The 'role occupant rank trend' indicates stakeholders that tend to occupy each role in each country, with the most common at the top. A particular value in understanding these details of the system is in the context of exchanges. A full analysis for each role and country can be found in Tables 6.5–6.8 – indicative contexts and mechanisms influencing role patterns are discussed.

In the travel philanthropy system, emotions also significantly influence roles. The most evident emotions found were *fear*, *happiness*, and *compassion*, with evident implications on the gift pathway. If *fear* was activated, this was either 'fear of being involved' or 'fear of not being involved'. Considering tourism businesses as a main driver for their level of involvement as an intermediary or donor, the fear of *not being involved* was generally caused by a fear of not meeting industry or societal expectations, informed by global or local social/business trends, leading to fear of not gaining (Fastenrath et al. 2018). The fear of *being involved* could stem from a fear of losing business or personal status because of politics of mistrust, jealousy, envy, and loss of community reputation, or by fear of greater pressure beyond business duties associated with occupying that role. Businesses avoiding involvement provided greater freedom for other actors, guides as intermediaries, direct giving, etc.

Similarly, ample evidence existed amongst non-business local actors who feared 'touching the money', thus involvement in the system too, which can cause wilful blindness (Heffernan 2012). On the other hand, fear of losing the chance of being the recipient of a donation appeared to at times drive competition to occupy a role in the system. Overall, emotions played a strong influence on the relationship amongst actors, their behaviour, and the system roles they occupy, confirming Hanoch's (2002) views.

The role of the recipient is a significant issue. McNamara and Morse's (2004) 'romanticisation of the local', shaped by the naivety of foreign donors to hidden local power dynamics, is found to be very much alive in travel philanthropy. The identification of 'deserved beneficiaries' mirrored similar processes found in critical narratives about NGOs (Gardner 2015; McEwan et al. 2017). The slight difference with travel philanthropy is that in many circumstances (except Zimbabwe), traditional authority or other formal leaders tend not to be involved as much in this aid chain, at least not on the level of traditional NGO aid chains as argued in the literature.

Travel philanthropy appeared to be a good example of development programmes that aggravate elite capture (Arcand and Wagner 2015), as there are

Table 6.4 Roles and role occupants by country (summary table)

	Storyteller	*Donor*	*Intermediary*	*Recipient*
Role definition: functions in the system	Tells stories and communicates the travel philanthropy narrative to other actors in the system	Gives a tangible object (monetary or non-monetary) to another actor in the form of a donation or gift	Responsible for channelling/brokering a tangible object from donor to recipient and managing resource distribution. As the go-between, intermediaries can link, connect, manage, monitor, communicate, report, advise, bridge, channel, distribute, and often select recipients.	Receives a tangible object from a donor – monetary or non-monetary
General across case findings	**It is how the story is delivered that differs, and who the storyteller is, not the story content**	**Tourists overall are the main role occupant for donor**	**The most influential role in the system – whichever actor occupies this role, they follow similar patterns and functions**	**Children most common end recipient**
Ethiopia overview role occupants	**Storytelling in more informal spontaneous settings, less tourism business involvement**	**Tourism businesses are extremely rare donors – tourists are the main donors**	**Either very informal or no intermediary**	**Guides and children – direct and indirect**
Role occupant rank trend (Ethiopia)	1. Local guide 2. Children (hasslers) 3. National guide 4. Hotel employee	Tourist	1. Guide 2. No intermediary	1. Children 2. Guides 3. Women (Hamar only) 3. Household donation (Hamar only)

(*Continued*)

Table 6.4 (Continued)

	Storyteller	*Donor*	*Intermediary*	*Recipient*
Uganda overview role occupants	**Would-be storytellers compete for the audience – guides control the story**	**Tourists dominate donor role**	**Businesses fear the intermediary role, and locals strive for it**	**Children but indirect**
Role occupant rank trend (Uganda)	1. Local guide	1. Tourist	1. Guide	1. Children
	2. National guide	2. Tourism business	2. Any local individual	2. CBO/School
	3. Hotel employee		3. CBO	3. Intermediary
	4. CBO		4. Tourism business	
	5. School teacher/ proprietor			
	6. Tourism business			
Zimbabwe overview role occupants	**Tourism businesses are attentive in controlling the story and the audience**	**Tourists fund hybrid models – tourist and tourism business – as donors**	**Those businesses with capacity dominate the intermediary role**	**Children and schools indirect**
Role occupant rank trend (Zimbabwe)	1. Tourism business	1. Tourist	1. Tourism business	1. Children
	2. Hybrid NGO	2. Tourism business	2. Hybrid NGO	2. School
	3. Guide		3. Traditional authority	3. Traditional authority/village
	4. Traditional authority/ village tour host			
	5. School teachers			

Table 6.5 Storytellers' roles and function in the system by country

Storyteller	
Role definition: functions in the system	**Tells stories to audience and communicates the travel philanthropy narrative to other actors in the system**
General across case findings	**Indicative quotes**
It is how the story is delivered that differs, and who the storyteller is, not the story The storytelling in tourism is shaped more by who tells it than by the story itself, with local guides – both formal and informal – being primary storytellers through community-based experiences. Regardless of the storyteller, narratives often revolve around themes like 'what is lacking', 'my vision', or 'you can change the world', aiming to inspire tourists as potential donors. Factors like personality, language, and access enable certain individuals to become storytellers, influencing audience beliefs and potentially positioning them as intermediaries or recipients. Tourists, too, act as both active and passive storytellers, bringing their own stories or beliefs to the experience and seeking stories that align with their perspectives. Different contexts enable or constrain certain actors to become storytellers.	"it's very difficult because they get information from the guides or the tour companies so it's not easy to access tourist before they come" (U2_B-COM_13) "those people can anyway influence the tourist, they can easily guide them on how to donate their money" (U2_CBO-F_8) "there are tourists they visit that village and then they ask problems, why kids they do not go to school what is the problem and they ask problems from the village" (E1_TE_52)

(*Continued*)

Table 6.5 (Continued)

Storyteller

Ethiopia overview role occupants – storyteller	**Indicative quotes**	**Role occupant rank trend (Ethiopia)**
Storytelling in more informal spontaneous settings, less business involvement With limited tourism business interest in philanthropy and few CBOs or NGOs, Ethiopian guides and children often take on the storyteller role. Unlike Uganda and Zimbabwe, where project visits or formal village tours are common, Ethiopian storytelling is typically informal, focusing on individual life stories encountered on the street rather than structured projects. Stories in Ethiopia often unfold through personal interactions on-trip, with follow-up exchanges occurring post-trip via email or social media. Children, prominent storytellers, now represent themselves directly, moving away from being depicted solely as passive recipients in NGO narratives.	"…when they go to the village they saw kids and they are like very lovely they love the people and then the tourist operator start working on this, for me like the influence is from the operators, to support the kids" (E1_TE_52) "he's emailing like talking that one of his wife had cancer or cut legs" (E2_TB_56) "the children within a short period they change their mind, they have the capacity to change their mind" (E2_NGO_61)	1. Local guide 2. Children (hasslers) 3. National guide 4. Hotel employee
Uganda **overview role occupants – storyteller**	**Uganda indicative quotes**	**Role occupant rank trend (Uganda)**
Would-be storytellers compete for the audience – guides control the story Guides, hotel employees (not management), and CBOs often compete for access to tourists to shape storytelling, with tourists typically directed to these actors during free time. In Uganda, hotels generally prefer not to control narratives beyond their premises. 'Community walks' in Uganda resemble those in Zimbabwe, where tourism businesses are minimally involved in design, instead facilitating structured interactions aimed at inspiring donations through 'seeing is believig'. In some cases, tourism businesses in Uganda do act as storytellers, not to fundraise, but to promote these walks, aligning with responsible travel expectations and strengthening local relationships for social licence.	"once they walk out the door .. we have no control or knowledge… if he forms a bond with the guest then that's his prerogatives" (U1_TB_3) "when we give them the story, I have seen some in tears crying" (U2_INTER_12)	1. Local guide 2. National guide 3. Hotel employee 4. CBO 5. School teacher/ proprietor 6. Tourism business

Zimbabwe overview role occupants – storyteller	Zimbabwe indicative quotes	Role occupant rank trend (Zimbabwe)
Tourism businesses are attentive in controlling the story and the audience In Zimbabwe, tourism businesses exert greater control over tourist movements and experiences than in Ethiopia, especially during 'free time' outside pre-arranged itineraries. They often manage the storytelling directly, with a strong focus on formal village tours and organised interactions aimed at fundraising or promoting responsible tourism. Unlike in Uganda and Ethiopia, Zimbabwean businesses closely shape the narrative, often embedding stories at various touch points like online platforms, room materials, and guided experiences. Guides in Zimbabwe avoid areas they cannot oversee to maintain control over the story. During village tours, storytellers – such as guides, traditional leaders, or school staff for greater authenticity – typically remain in their roles without shifting to intermediaries, a notable difference from other destinations.	"the guide is very very influential, the guide will, the guide can even change rules on the way and say ahhh but I think it is better for... take instructions from the guide" (Z2_1_B-INS_31) "they are making an effort either whether it's through marketing, some flyers, some envelopes in the room or hosting a community tour or something like that" (Z1_NGO_21) "I can't just drive a guest into an area where I am not trusting... where they will eventually have funny emotions, we are not putting on a show, but we take them in an area were they will understand how locals are living how locals are living day to say life" (Z2_2_TB_41)	1. Tourism business 2. Hybrid NGO 3. Guide 4. Traditional authority/ village tour host 5. School teachers

Table 6.6 Donors' roles and function in the system by country

Donor	
Role definition: functions in the system	**Gives a tangible object (monetary or non-monetary) to another actor in the form of a donation or gift**
General across case findings	**Indicative quotes**
Tourist main role occupant of donor Viewing exchanges in this economy through different lenses suggests that locals can occasionally be seen as donors by beneficiaries or intermediaries. However, this study simplifies 'donors' to mean the initial providers of tangible goods, typically foreign tourists. Some tourism businesses allocate a portion of profits as donations or to cover hybrid entity admin costs, yet most funds reaching communities come from client contributions, either directly or indirectly. Across cases, donor perceptions vary; whilst tourists often donate to communities, they are not always seen strictly as altruists, nor are they always distinguished from NGO-based donors – sometimes, a donor is simply viewed as a donor.	"he is not a real donor, he's not a donor, but he fundraises from donors, then he manages donations, he's a business man" (U2_B-COM_13)

Ethiopia overview role occupants – donor	Indicative quotes	Role occupant rank trend (Ethiopia)
Tourism businesses extremely rare donors In Ethiopia, whilst a few tourism businesses do contribute to selected causes, although usually off the record compared to Zimbabwe and Uganda, the vast majority of donors are the tourists. It is evident that the absence of tourism businesses in travel philanthropy, as donors or intermediaries, has increased the proportion of tourists as donors in this context.	"the hassle the guide is facing and drivers are facing in the village so we take all that information and together we decide, ok we have this problem with the village with the kids and so on and how to solve it so this is something that we set in consultation with the drivers and the guides, so there is no reason for any guide not to abide by that regulation because it is a kind of common decision" (EN_TB_65)	1. Tourist
Uganda overview role occupants – donor	**Uganda indicative quotes**	**Role occupant rank trend (Uganda)**
Tourists dominate donor role Tourists provide most of the funds in the travel philanthropy system of Uganda, but businesses are also occupying this role to some extent. In this study, one of the high-end lodges has a type of hybrid, in which the business donates a proportion of profit to a CBO they are contracted with through a bednight levy model, but it is less of a donation and more of a business contract. However, it is sold as part of responsible tourism in a travel philanthropy context, so will be highlighted here. The majority of additional donations are directly from guests to the community. Tourism business on the whole connects their guests as potential donors to a local actor. Tourists are the main donor occupants.	"we are always encouraging guests that come to stay with us, you know, if you really like the surrounding area and if you find a bond with one of the doctors of the clinic or pastor at the church.. A: or whatever it is .. J: or someone at the orphanage of whatever, they obviously would be thrilled to accept donations" (U1_TB_3) "we're not an NGO, we're a business" (U1_TB_3)	1. Tourist 2. Tourism business

(*Continued*)

Table 6.6 (Continued)

Donor

Zimbabwe overview role occupants – donor	**Zimbabwe indicative quotes**	**Role occupant rank trend (Zimbabwe)**
Tourists fund hybrid models Hybrid models, whilst separate entities legally, are considered tourism business actors because they are dependent on each other, so in this context therefore, tourism businesses are more dominant donors than in the other two country cases. There are many more hybrid models. However, it is rare that donations are generated from profit, as in more standard CSR practice, the funds tend to be generated from tourists, and the businesses and hybrid NGOs are intermediaries.	"if they are donating for school fees and things like that then it still comes directly from the tourist because that is our source of income and without the tourist there is no source of income for us" (Z2_2_TB_43) "it's depends on whether their [tourism business] pot is full and spilling, if it's not they will not" (Z2_2_NGO_45) " unless they bring their clients here it is not easy for us to receive any help so we receive help because people who come here they are the ones who have money … they go out and seek for help to help the schools not that the money they generate with their profits, no they don't channel their profits to here they are only like a vehicle to help us get help" (Z2_1_B-INS_31) "they are ploughing back yes, it's all about the tourist it's not … about [company] but it's all about the tourists … they are facilitating" (Z2_1_B_INS_38)	1. Tourist 1. Tourism business

Table 6.7 Intermediaries' roles and function in the system by country

Intermediary	
Role definition: functions in the system	**Responsible for channelling/brokering a tangible object from donor to recipient and managing resource distribution. As the go-between, intermediaries can link, connect, manage, monitor, communicate, report, advise, bridge, channel, distribute and often select recipients.**
General across case findings	**Indicative quotes**
The most influential role in the system – whoever occupies it has similar functions The intermediary role is highly influential, responsible for channeling resources to recipients. It may be voluntary or involuntary – some seek this role, whilst others feel obligated by donor requests. Often, storytellers transition into intermediaries, though not always. Intermediaries connect donors with recipients, manage resources, and shape donation impact. Control over the resource path is significant but typically informal, relying on trust rather than formal criteria. When tourism businesses step back as intermediaries, direct community donations increase, often through local guides. Whilst the role offers incentives, it also entails pressure and ethical responsibilities, affecting intermediaries personally and professionally. This role is both praised and critiqued. In Uganda, guides usually act as intermediaries; in Ethiopia, it's often guides or none, and in Zimbabwe, tourism businesses commonly take on this role.	"the in-betweens selects, in fact there are 2 different ways, there is when they are like given something, it is when they are bribed, it happens" (U2_INTER_12) "they kind of make decisions on which projects to sponsor, individuals to sponsor so you find the wealthy class has kind of dominated the sponsorship, the donations" (U2_CBO-F_8) "for her she is just donating as if she is the one donating" (U2_CBO-F_8) "for me I want to stay clean, I don't want any sponsorship or NGO or anything like that… I don't want to be in that I am not a charity" (EN_TB_68) "we want to be a bridge between the tourist market and the communities in terms of generating funds to put towards projects" (Z1_NGO_21)

(*Continued*)

Table 6.7 (Continued)

Intermediary		
Ethiopia **overview role occupants – intermediary**	**Indicative quotes**	**Role occupant rank trend (Ethiopia)**
Either very informal or no intermediary In Ethiopia, the intermediary role in the aid chain is often absent, with donors frequently giving directly to recipients, typically children. When an intermediary is present, it's usually a guide, prompted by storytelling dynamics – children may directly share personal needs, unlike Uganda, where guides or CBOs more commonly represent others' needs. Local guides in Ethiopia sometimes manage donations through personal accounts, taking on the intermediary role informally. Tourism businesses rarely engage as intermediaries, and legal CBO and NGO channels are mostly inactive in travel philanthropy. Some guides, especially in areas like Lalibela, may have informal 'projects' to attract donations, though the overall absence of formal intermediaries leads to direct giving. Many actors are thus pushed involuntarily into the intermediary role, facing ethical concerns about the responsibility and control over resources.	"the driver or the guide... he receives the money and then he give the money to the kids" (E1_TE_52) "the main problem is really they trust the national guide …" (E1_B-COM_69) "in Lalibela there are 100 or 200 guides but the guides who have charities is not more than 30 or 40" (E2_TE_63) "the one thing is no-one who trust, the second thing is … I don't have time" (E2_TB_58) "that is out of our mandate, one, and we don't have any resource to follow up on that issue" (EN_TB_65) "our part is just to give the right contact so that they can pick up and go ahead" (EN_TB_66) "do you shut off the one or two chances that would make a big difference to… I usually say ok and then I end up getting into war because the hustlers are then hassling me and I'm the go-between" (E2_TB_62)	1. Guide 2. No intermediary

Uganda overview role occupants – intermediary	Uganda indicative quotes	Role occupant rank trend (Uganda)
Businesses fear the intermediary role and locals strive for it In Uganda, intermediaries are more common, typically representing the recipient to the donor, unlike in Ethiopia, where recipients often speak directly to donors. Guides, CBOs, or locals with face-to-face tourist encounters usually fill this role, whilst tourism businesses generally avoid directly managing donations, citing capacity, risk, and lack of interest, though they may promote village tours. However, tourism businesses sometimes serve as intermediaries upon client request but tend to connect donors with a CBO. As a result, CBOs are the most frequent intermediaries when formally linked to tourism, whilst guides take on the role independently.	"every person needs a sponsor, and people who are involved in giving these sponsors, like the go-between are blamed, are take as corrupt people by those ones who don't get the sponsors, like they hate us" (U2_INTER_12) "when tourists pass money through us … the community will not see the favour and the sacrifice they will say oh they are eating our money so to avoid that we always connect them" (U3_TB_18) "they want money, most people who start up these projects and who bring the ideas to tourists and other people that they do projects they are targeting some benefits, some money" (U2_INTER_12) "a driver … he was kind of fighting to hold onto the money to become like the treasurer" (U2_CBO-F_8) "for us it's not worth it … it would be just a lot of work unless we are hiring someone to do that, it's a full time job" (U1_TB_3) "as the lodge of course we have to play a role there because ok we link them" (U3_TB_17) "I was the one to decide the beneficiaries" (U2_INTER_12)	1. Guide 2. Any local individual 3. CBO 4. Tourism business

(*Continued*)

Table 6.7 (Continued)

Intermediary		
Zimbabwe overview role occupants – intermediary	**Zimbabwe indicative quotes**	**Role occupant rank trend (Zimbabwe)**
Those businesses with capacity dominate the intermediary role In Zimbabwe, tourism businesses primarily serve as intermediaries, often through hybrid models with charitable accounts or directly if such models don't exist. NGOs, separate from tourism hybrids, are increasingly involved in travel philanthropy, bridging donations to local projects or funding their own. Guides rarely act as intermediaries, though they may be storytellers. Formality is greater in Zimbabwe due to legal requirements and the prevalence of tourism hybrid models, leading to a high proportion of funds channelled through the tourism sector. Some businesses cite capacity limitations rather than avoidance, as in Uganda.	"I am sorry if the monies would come to the committee they could be abused" (Z2_1_B_INS_38) "everything, everyone who wants to do something in the community must go through the chief, that is a requirement you cannot just drive into the community and you just establish something without the traditional leadership knowing" (Z2_2_NGO_45) "there are a few projects that they will show you around and if you are willing to make a donation we encourage that all donations go through [hybrid NGO]" (Z2_1_NGO_37) "we are the ears on the ground, we are the facilitators for anything you want to do ... as opposed to going directly to the communities" (Z2_2_CBO-F_40) "most of these projects still die when they are not supervised but we also try and say we are not in the business of supervising these people, so if you are donating this you are donating this also at you own risk it might be a good outcome it might be a bad outcome" (Z2_2_TB_44) "we do not have the capacity as a company to be receiving all sorts of money and then saying this is how it's supposed to be used and all that because then you need to employ a full set of people to just do that" (Z1_TB_28)	1. Tourism business 2. Hybrid NGO 3. Traditional authority

Table 6.8 Recipients' roles and function in the system by country

Recipient	
Role definition: functions in the system	**Receives a tangible object from a donor – monetary or non-monetary**
General across case findings	**Indicative quotes**
Children most common end recipient Local recipients of travel philanthropy donations are typically near tourism sites, with children often serving as primary recipients, directly or indirectly. Indirect recipients are frequently chosen through personal connections, which raises ethical concerns. Simply viewing recipients as a blanket 'community' can dilute impact and empower those with influence, complicating the aid landscape. Identifying true beneficiaries mirrors challenges in traditional development, as donors, particularly inexperienced ones, may reach only surface-level recipients – often wealthier, well-connected locals. Determining legitimate beneficiaries is complex, requiring knowledge of community dynamics, power structures, and varying levels of access and influence.	"No they don't specify they say community" (U3_TB_16) "there is where it is not clear, sometimes there are primary beneficiaries and secondary beneficiaries, the way we are talking about people in the middle, so I don't know how we can differentiate it... [it matters] whether we mean the last beneficiary or both, or the manager" (U2_B-COM_13) "…we cannot, we are not in a position to get in touch with somebody in America and ask for donations but they can do that on our behalf so we really appreciate that, we really are happy and there people can do that for us" (Z2_1_B-INS_31) "… you choose which are the poor people it's obvious there will be commotion there, friends will choose each other ... after they have gone after the see confusion" (Z2_1_B_INS_39)

(*Continued*)

Table 6.8 (Continued)

Recipient		
Ethiopia overview role occupants – recipient	**Indicative quotes**	**Role occupant rank trend (Ethiopia)**
Guides and children – direct and indirect In Ethiopia, recipients are typically guides or children, linked to donors through direct or indirect relationships. Donations often support school-related causes or provide capital for ventures like hotels or motorbike taxis. Women, particularly in one case study site (Hamar), are also common recipients; however, aside from income from photos and souvenirs (excluded from this study's scope), donations for women generally consist of clothing rather than monetary aid.	"guides mostly get money easily ... they kill their money even in one day they kill a lot of money" (E1_TE_52) "most tourists come one time but those are which are giving money for direct for kids" (E1_B-COM_69) "women are always expecting that thing because Hamar they see one thing once and they are always looking for that thing" (E1_TE_52) "this has been happening for quite some time, so the kids when they were 15 or 16 and getting this money what have you seen when they become like 25 or 30 what do they do, do they stop or they carry on" (E2_NGO_57)	1. Children 2. Guides 3. Women (Hamar only) 3. Household donation (Hamar only)
Uganda overview role occupants – recipient	**Uganda indicative quotes**	**Role occupant rank trend (Uganda)**
Children but indirect In Uganda, recipients tend to be children in education-related donations but indirect via intermediaries. CBOs are also key recipients, predominantly promoting the children and education-related causes but also other themes such as women, HIV, Batwa people, etc. It is also found that non-CBO intermediaries are in reality recipients as well, whether it is the intention of the donor or not.	"when a mzungu says he wants to help, he has his limit or sometimes is helping for one term or two terms and when here, when a parent hears that a child has got a sponsor he expects for a children from the primary 1 to the university and he finishes his education that's where the problem is" (U2_INTER_9) "it's combination of so many things, first of all some donations don't reach them then secondly they decide for them what kind of donation to give them so it never makes any impact, thirdly... that donations never reaches to the ground, never reaches them as beneficiaries" (U2_CBO-F_8)	1. Children 2. CBO/School 3. Intermediary

	" even the locals themselves don't even say anything about the donations but I think one of the major reasons is that they are not aware of the donations and they are not aware of how much they are donated to and they are even confused of which organisation is helping them" (U2_CBO-F_8) "there are no structures, that would follow up such kind of fake beneficiary. First of all the way the money is sent there are no agreements" (U2_CBO-F_8)	

Zimbabwe overview role occupants – recipient	**Zimbabwe indicative quotes**	**Role occupant rank trend (Zimbabwe)**
Children and schools indirect In Zimbabwe, recipients are largely children and predominantly sponsored at school indirect, or schools. There are other projects funded in communities not targeting children specifically and tend to be organised through the traditional authority, such as boreholes/water pumps. Hosts of village tours can also be recipients of small amounts of cash. The largest end beneficiary is, however, children and schools. The villages closest to the tourist accommodation are the common recipients.	"which is the closest school to town that we can work with because we want our clients to be visiting the school" (Z1_TB_28) "I have got an American guy who has sponsored 17 children... this guy came forward ... he was doing it but wasn't sure the fees were getting to where they should go, which they weren't, funny, but no, so we, I said no that's fine, one because we work with the school..." (Z2_1_NGO_37) "they [tourism businesses] tend to come as individuals to do the same project like say we want a teachers cottage, at the end we may end up having 2 teachers cottages because they don't want to put their resources together and work on this particular project. The other we want to bring their own complete resources to do that teachers cottage and the other will want to bring their compete resources to do that same project" (Z2_1_B-INS_31)	1. Children 2. School 3. Traditional authority/village

certain individual characteristics of who can become gatekeepers of the philanthropic exchanges (e.g., access, language, personality, and confidence) that play in their advantage rather than specific evaluated deservedness. This is caused by the informal nature of the exchanges in this system and types of relationships amongst actors. This study provided evidence to support Brest's (2015) warnings about travel philanthropists lacking time to really understand problems and identify solutions and Abernethy's (2011) argument that by being pushed to think fast, tourists, like many other actors, behave instinctively and emotionally as opposed to strategically.

Understanding actors and roles in the travel philanthropy system is critical in clarifying the 'context', hence why our study drew on Yin's (2003) use of comparative contexts, enabling an examination of how different conditions cause different outcomes. The combination of beliefs, norms, principles, and others shapes the system structure and behaviour that happen in specific stakeholders' contexts, thus the setting in which activities occur (Gates 2016). So, the actor–role contexts are part of this, with the boundaries of these 'settings' becoming very blurred (Mutale et al. 2015) in our study too.

6.2 System perspectives: purposes/functions

As an exchange economy, travel philanthropy does not simply function as a purely altruistic system of donations in Africa and/or African recipients. In fact, our study substantiates the theory that actors in the aid chain of travel philanthropy are pursuing different interests (Song et al. 2013). Whilst system perspectives provide the framework for these interests and represent purposes/functions (Senge 1990), our study drew on Senge (1990) and Arensman et al.'s (2017) arguments that when evaluating non-linear processes involving multiple stakeholders, it is critical to understand different interests. Perspectives in development tend to focus on the mental models from the donor perspective (Green 2016); therefore, our research considered multiple perspectives in a system approach, aiming at generating new understandings of how different actors manipulate the same system for different purposes. There is currently not a common discourse despite some efforts of social change (Bebbington 2005); therefore, understanding different perspectives (particularly those from specific destination contexts and cases) is critical to understanding why and how the system operates in a certain way.

One could view travel philanthropy from Korf (2007) and Silks (2004) angles as power dynamics between the donor and recipient of the gift, but whilst the superior self of the donor is sometimes obvious, the *passive* recipient perspective is actually contested in our research. In fact, whilst there were incidences where donors chose to give and decided what to give through a direct philanthropy perspective, with specific implications emerging, there were also common patterns where recipients decided if donors were to give anything, what they would give, and to whom. Power dynamics appeared to be far more complex than the traditional and polarised North vs. South exchange, and in line with Burger and Owen's (2008) stance, it became clear that in travel philanthropy, one cannot assume that altruistic

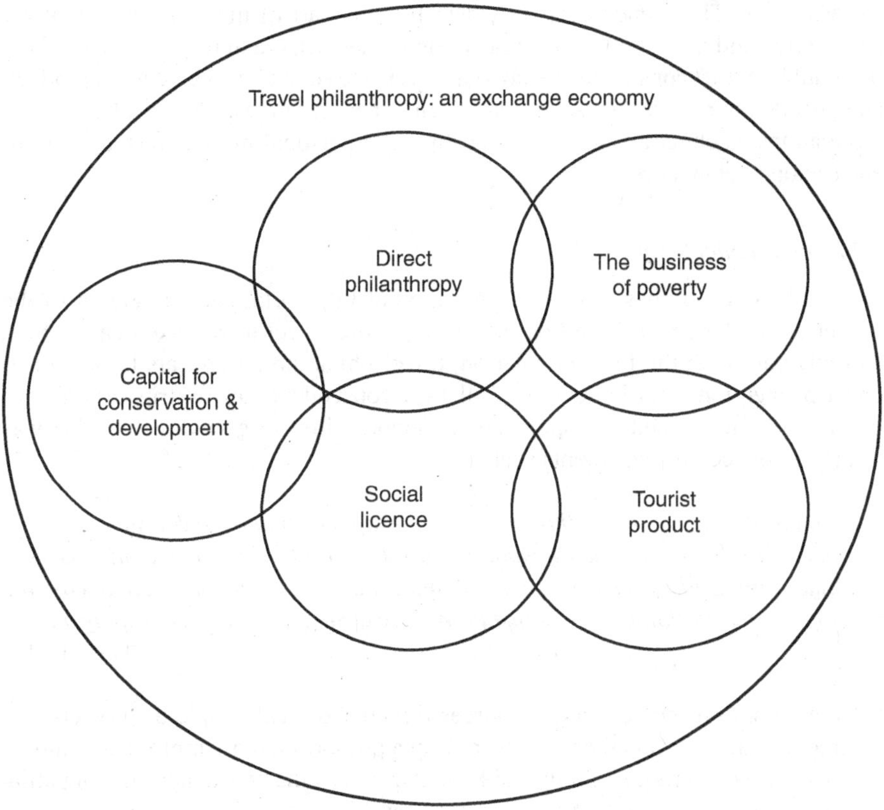

Figure 6.4 System perspectives.

motivations and non-monetary preferences drive the donor or intermediary. On the contrary, apart from what seemed apparent amongst some tourist donors, although a philanthropic self-identity may arguably be a more dominant motivation, the rest of the stakeholders tend to use the system for purposes other than altruism.

Our study highlighted five key purposes (Figure 6.4) in the system, all of which shaped its functions. These are as follows:

1. Direct philanthropy
2. Capital for conservation and development
3. Tourist product
4. Social licence/reward
5. The business of poverty

Ashley et al.'s (2006) and Ham's (2011) theories of consumers (tourists) taking social impact into account when designing their trips and being driven by personal satisfaction of 'improving' the places they visit appeared influential across our study's cases. This trend became evident both as part of the direct philanthropy perspective and as an influence on how companies behave in response. Therefore, one could say that consumers' behaviour affects the social licence, tourism product, and business of poverty, as actors adapt their behaviour in response to others. Other motivations, such as business relationships and personal motivations, also influence actors' behaviours.

6.2.1 Direct philanthropy

Direct philanthropy has always existed around the world, such as giving to the homeless on the street. However, as more people travel to remote areas in SSA and integrate with the local population, travel philanthropy appears to be part of pre-trip intentions and beliefs, as well as a consequence of on-trip experiences involving both planned and spontaneous opportunities to give directly. This was clearly evidenced by statements such as:

> *At times they come aware through their tour agency that once you stay at [lodge name] you have an added advantage to interact with the communities so they come organised with some scholastic materials to the school neighbouring the lodge, some clothing and at some point they even get some kids to sponsor.*
>
> (U3_TB_17)

Some tourism businesses gave resources directly to local people or projects too, whether fundraised from their guests or from a proportion of their profit, as a model of corporate social responsibility (CSR) that involved the corporate entity donating to the beneficiary. For example, a tourism business said:

> *If you don't have donors, we can stick to our budget and say per year we're going support two kids.*
>
> (Z2_2_TB_41)

At the most surface level, some actors viewed travel philanthropy as a form of direct giving and a linear path that attempts to cut out the intermediary, such as international charities, by giving directly to a beneficiary (Silk 2004). The act of giving itself, through a donor-to-beneficiary pathway, was in fact the purpose of the exchange. However, 'direct philanthropy' is often not as direct as it seems.

Figure 6.5 outlines a situation where a local actor and donor might consider a linear, direct-giving pathway to the beneficiary, even though it is not. The *primary donor* may be a donor located overseas, whereby the *secondary donor* – e.g., the tourist, would be responsible for allocating that resource in the destination when they travel (they may or may not be in the aid chain). Similarly, whilst the *intermediary* may or may not be in the aid chain, the *secondary beneficiary* is included because actors sometimes expect the *primary beneficiary* to distribute the resource further down the aid chain – i.e., to other beneficiaries. There are not

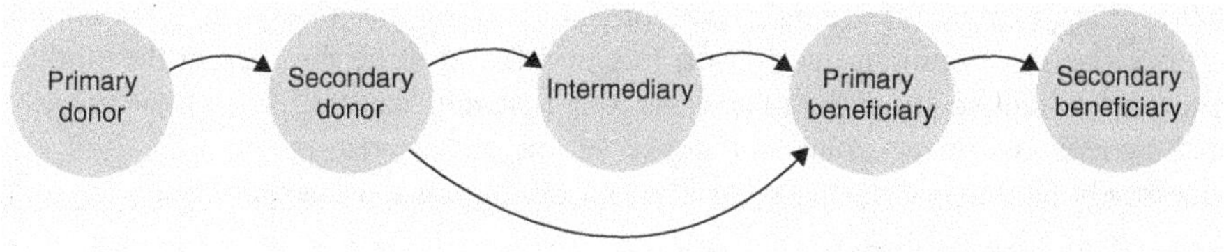

Figure 6.5 Perspective: Direct philanthropy – From donor to beneficiary.

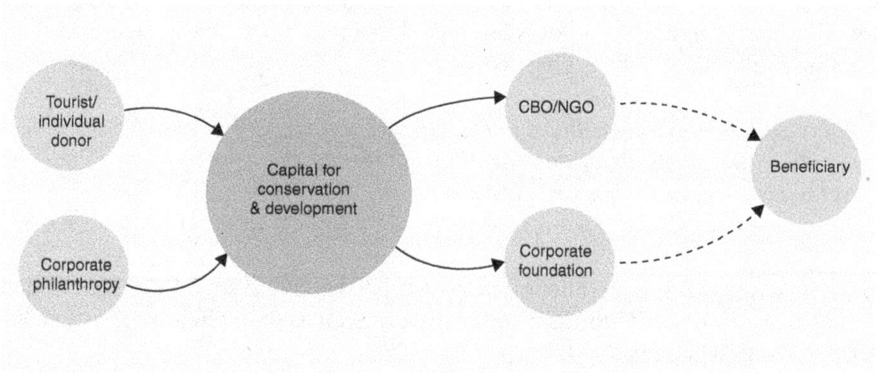

Figure 6.6 Perspective: Capital for conservation and development.

always primary and secondary donors or beneficiaries, but when they exist, actors often overlook them in a presumed direct exchange.

6.2.2 Capital for conservation and development

Capital for conservation and development generally aligned with Burns and Barrie's (2005) conceptualisation of travel philanthropy as turning development needs into a holiday experience and Burgold and Rolfes' (2013) development projects centred around ethical consumption. This perspective (see figure 6.6) was identified amongst a range of actors, including some tourism businesses and local stakeholders. For CBOs, NGOs, and/or other stakeholders raising capital from tourism businesses, perhaps as part of their CSR programme, this 'donation' tends to come with some kind of recognition as an important aspect of the exchange. As a Zimbabwean NGO put it:

> *They have to get something out of it, if they aren't going to get anything out of it why are they going to keep you as a partner.*
>
> (Z1_NGO_21)

Ashley et al.'s (2007), Song et al.'s (2013), and Turner et al.'s (2001) believe that tourism companies are not development agencies and rarely have the choice to prioritise development over profit; this was found to be the case in the studied countries whenever tourism businesses were raising capital. So, whilst tourism businesses may raise capital for conservation and development, this does not necessarily lead to non-intrusive or more sustainable ways to experience a country (Burgold and Rolfes 2013).

From a recipient or intermediary perspective, the purpose of participating in the system could be to raise capital for local conservation and development objectives. Recipients regarded travel philanthropy as a financial resource, fundraising platform, tool, or opportunity within a tourism destination, in which the donors are the tourists or tourism businesses and those raising the capital are locally based organisations or individuals. An NGO linked to a tourism business described their function as facilitating the giving process:

> *What I do among other duties is I get to spend time with guests and eat lovely meals telling them about the work that we do but the whole idea is to try and stimulate them to support conservation work.*
>
> (Z2_1_NGO_33)

The key actors that tend to adopt this system perspective were CBOs, NGOs, and formal charities linked to tourism companies, as well as individuals such as guides. A Ugandan CBO emphasised:

> *The donations are the ones that have developed this orphanage, like the building you have seen the what, the sponsorships, there is no government that is giving us even a single coin.*
>
> (U1_CBO-F_1)

Differently from direct philanthropy, involving active donors, this perspective required active intermediaries, as mentioned by a village headman:

> *We, with the chief, we were talking about it that all companies that are surrounding us they need to assist us when we identify a project.*
>
> (Z2_2_B-INS_42)

6.2.3 Tourist product

A key aspect of travel philanthropy is the *tourism product*, an increasingly competitive tourism experience designed around philanthropy, led by both the demand and supply sides of travel (Phocuswright 2015; Crouch 2011; Sustainable Travel International 2008). Tourism products can be both a cause and an effect of giving – i.e., the purpose of the experience in a community may not be giving, but the experience can cause giving, or an experience centred on giving can become the tourist experience in the community (Figure 6.7), with the overarching goal

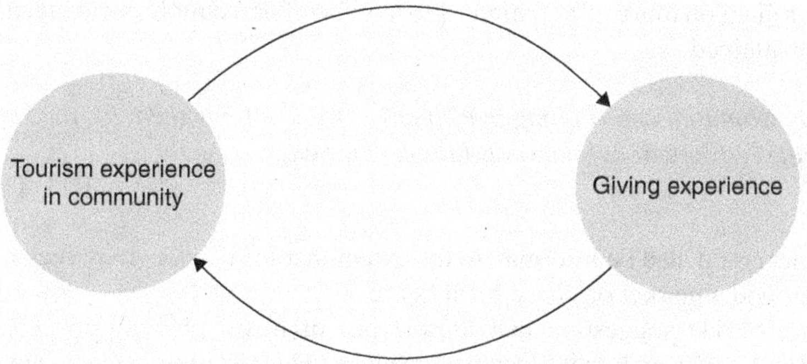

Figure 6.7 Perspective: Tourist product.

being satisfied customers rather than just raising funds. As one Ugandan tourism business highlighted:

> *They [tourists] go for the experience, if they [tourists] don't have gifts, they get overwhelmed [during their community experience], most of them, and they give [gifts] from that.*
>
> (U3_TB_16)

Normally, an experience in the community tended to be secondary to another core activity – e.g., a safari. Seeing and interacting with local people can become part of the trip if free time is available. Visiting local projects or vulnerable groups can thus become a reassuring mechanism that the tourism company is involved in responsible tourism. As suggested by an Ethiopian tourism business:

> *One of the basic things is that people get the experience and they feel it themselves directly, they share the happiness of giving themselves and seeing where their donation is going.*
>
> (E2_B-COM_64)

Given the demand for such experiences, the majority of tourism businesses offered these types of interactive experiences or services designed around the guests' desires and safety, involving villages within close proximity to their hotel or safari location, as part of their packages. In this case, the product tended to be formal, designed by the tourism business, or referred by them to local community-operated options. Visiting children in schools appeared to be firmly on the itinerary, sitting alongside gorilla tracking in Uganda or a big five safari across all the destinations. In Zimbabwe, for example:

> *It's custom, it's sort of standardised, so you go to the school and you see what.*
>
> (Z2_1_NGO_37)

Some hotels invited groups to entertain clients on the property, rarely a paid activity but rather reliant on tourists' donations into a basket. For example, a Ugandan hotel manager explained:

> *We only want that entertaining our clients... we don't pay them for the singing, so whatever [donations] they get, I think they benefit directly.*
>
> (U3_TB_18)

The product could also be informal, to the extent that local actors may seek clients to show around a market or visit a local home.

Honey's (2011) suggestion that these types of travel philanthropy products promoted by tourism companies deepen travellers' understanding about inequalities is largely questionable, as they very often merely provide communities with *direct access* to tourists or they just offer this as another local tourist experience. Some community groups and/or individuals charge for the experience; others offer it for free with the hope of receiving donations as a result. In Zimbabwe and Uganda, the product was similar; more often than not, it involved a tour of a school, interacting with kids, seeing inside a medical clinic/hospital, visiting marginalised indigenous groups – e.g., women or those disadvantaged by HIV, and a strategic selection of projects they would like the tourist to see and hear about, so to stimulate direct donations. In contrast, Ethiopia's products tended not to be standardised tourism experiences geared around giving, especially not by the tourism operators, but they were rather subtle and based around 'spontaneous' encounters and exchanges.

6.2.4 Social licence/reward

In Uganda and Zimbabwe, tourism businesses primarily engaged in a system of exchange in order to achieve or maintain a social licence/reward sought from a number of stakeholders, including the local population, travellers, international agents, and the government, and as indicated in Figure 6.8, there were also other inter-stakeholder social rewards generally sought.

Tourism businesses' approaches appeared as not homogenous: some might act as intermediaries to connect clients as potential donors to the community, whilst other businesses might donate directly themselves. What businesses were looking for in return from the community, for example, was to be licensed/awarded certain business rewards, greater responsiveness, less hassle, or a smoother business environment in which to operate. As one hotel manager said:

> *Ever since then I have never had any complaints from guests about noise at night, that helps us, because they know if we continue making noise, we won't get the services, we won't get these gifts we are given.*
>
> (U3_TB_16)

The influence of the social licence perspective on the system structure became evident in merit to travel philanthropy affecting how actors behave, the type of

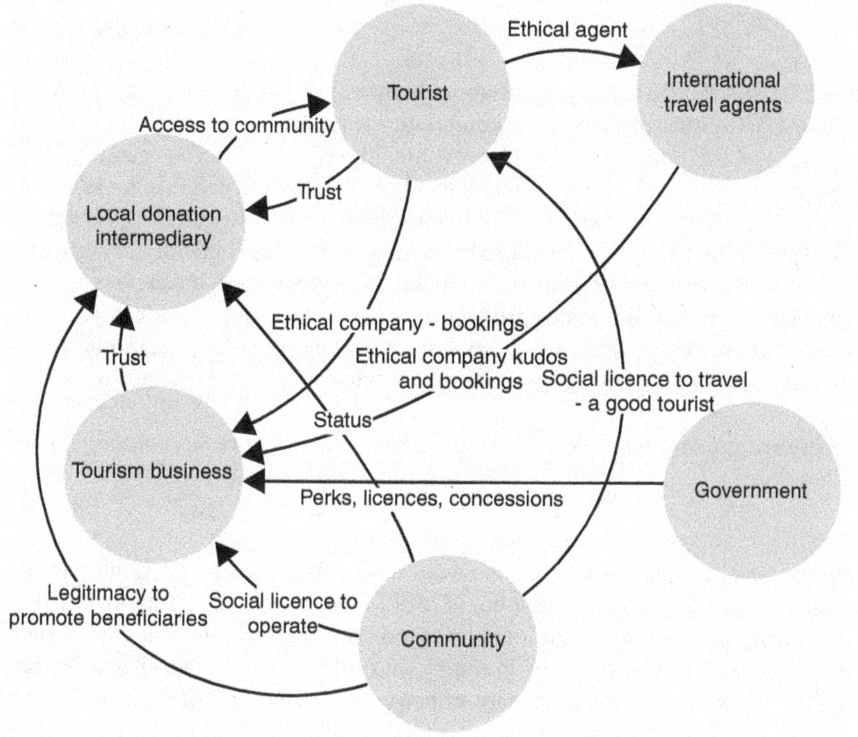

Figure 6.8 Perspective: Social licence/reward.

relationships between actors, and types of exchanges that occur. Both an NGO and school headmaster in Zimbabwe stated:

> *It comes up all the time, so how do tourism businesses work with us to get more business and be seen to be doing the right thing.*
>
> (Z1_NGO_22)

> *They are competing ... to be embraced by the community.*
>
> (Z2_1_B-INS_31)

Travellers and the international agents found it increasingly important to establish the image of an ethical or responsible company in line with industry trends, with travel philanthropy playing a key role. Our findings confirmed McIntyre's (2015) and McEwan's (2017) arguments that most tourism operators cannot afford or do not want to contribute a slice of profit, so they regard travel philanthropy as a means to maintain a local social licence. 'Communities as recipients' were well aware of tourism operators' function in the system:

> *There is this thing that says plough back to the community but the core business of why you are here is because you want to make profit.*
>
> (Z2_1_B-INS_31)

Echoing Gardner's (2015) argument about contrasting moral economies between corporate CSR efforts and the local community perspective, whilst many tourism companies saw their efforts through travel philanthropy as a generous act of giving, thus earning them a social licence, many local recipients regarded this act as something *due* as payback for what the companies take away (Nickel and Eikenberry 2009). The company may be awarded their social licence, but not for altruistic or philanthropic reasons. Rather, they aimed to protect their businesses and to keep the environment – i.e., their product pleasant and pristine. As McIntyre et al. (2015:307) found, companies appeared very dependent on a local community for competitiveness by preventing 'groundswells of discontent', with cases in which:

> *People were being mobilised [to say] we don't want [tourism company name] in the area.*
>
> (Z2_1_B_INS_38)

A certain level of payback was achieved under the banner of philanthropic donations. Engaging in this system could lead to social licences being awarded to intermediaries of donations on a local level too. For example, intermediaries such as local guides or CBOs engaged in the travel philanthropy system because they gained social status in their community, effectively another form of a social licence.

> *If you can really kind of help some beneficiaries to get some donations from tourists, then you are kind of, your status kind of raises up, people talk about you they kind of sing about you that you have helped them to do what.*
>
> (U2_CBO-F_8)

Travellers could also seek social licences, which might include the reassurance to themselves, as well as the local and global communities, that they are responsible travellers, perhaps through the act of offsetting their luxury vacations in often-deprived destinations.

6.2.5 Business of poverty

Evidence showed that many local actors were simply not the *passive recipients of aid*, as many donors often assume or dominant international development narratives portray (e.g., Korf 2007). Rather, local people were part of the businesses and actively engaged in driving them. As two research participants said:

> *They're [hotel employees] like the kings of this village... . if they can get in close with the butlers, they can have a direct route to the clients and potentially their donations, they do hold a lot of sway.*
>
> (U1_TB_3)

> *They [CBOs] are like pastoralists... so for example if I was running a school now, I shift to helping women, or I shift from helping women to helping pygmies, the Batwa, or I shift from that and go and help refugees.*
>
> (U2_CBO-F_8)

The *business of poverty* was arguably one of the most prominent functions of the system that emerged from our research, particularly from the perspective of local actors, which fills an identified key gap in the extant literature. What emerged was that travel philanthropy is not only a form of direct or corporate philanthropy, or a source of funding for development and conservation work, or a tourist product as the end goal (Honey 2011); rather, it is a commercial enterprise based on a system of economic transactions aimed at personal or business benefits, utilising the system as a platform to achieve a specific purpose.

In fact, across all case study locations, local actors openly regarded the donations from tourists and tourism operators as a business. As these participants from across all three countries said:

> *They are seeing them as businesses, they can get much more money from community organisations than setting up a shop.*
>
> (U2_CBO-F_8)

> *I basically sell our problems that's what I do.*
>
> (Z2_1_B_INS_38)

> *So much dependent on incomes generated in such a way so that means they don't have any other alternative that's all, they have become dependent on this.*
>
> (E2_B-COM_64)

The business of poverty was by some regarded as a superior economic opportunity even when other employment or business opportunities existed. Salaries tended to be low in rural SSA, so some choose to leave paid employment to pursue business opportunities through travel philanthropy (U1_CBO-IN_4, U2_CBO-F_8). Others generated high incomes, greater than through government positions (especially in Ethiopia), at young ages and focused their aspirations and livelihoods on the business of poverty as a preferred source of income and/or social status, as two research participants, for example, highlighted:

> *Now my life is a bit better than being in a class... when I was still a teacher ... you find that you are very tired, you grow very fast, you have no money and you find that nothing is at home.*
>
> (U1_CBO-IN_4)

> *The guides get an unusually high amount of money compared to government [officials] or whatever.*
>
> (E2_TB_58)

Establishing entities to legitimise philanthropic business transactions (i.e., CBO or private school) had become commonplace, especially in Uganda. People regarded CBOs as self-employment opportunities as well as ways to employ others, thus a means to power. People sometimes operated them as a pure business, or echoing a social enterprise model for profit and social good (Maltby and Rutterford 2016; Talpalaru 2014). As this community research participant in Uganda explained:

> *The real motivation is a number of things. Number one a person like [anonymous name] who will be wanting to have the project in place will also be wanting to benefit, he knows when a project exists, he just benefits in one way or the other. Number two he wants to help the people in this community, to improve livelihoods, that is the overall goal.*
>
> (U2_INTER_12)

One might arguably sum up the product of these businesses as marketable 'poverty' by making perceived development needs tangible and making poverty profitable. The product could also be a CBO itself, sold as an idea, as it was later added by the same community research participant from Uganda:

> *They want money, most people who start up these projects and who bring the ideas to tourists and other people that they do projects [with] they are targeting some benefits, some money.*
>
> (U2_INTER_12)

Mirroring typical charity organisations operating across Africa, local people had learnt which products are appealing to consumers and sell, using specific words and imagery to generate interest. 'Groups' were formed as a type of branding of marginalised people (e.g., name of an indigenous population, women, orphans, and HIV patients) and competed for funds. Within communities, power was not flat, and certain community groups had become the most saleable products in the business of poverty. In Uganda, the *Batwa* population has effectively become a commercialised brand in development, not only by local actors but national and international too, and the Batwa are very aware of being "the product of their business" (U2_CBO-IN_7), as clearly stated by one of them:

> *They [Bakiga] knew that when we [Batwa] remain in that situation, then they take that opportunity, chance to have friendship from outside, then they told them about the way we are suffering, then they bring them and come and see us, then they earn money, so it's their business.*
>
> (U2_CBO-IN_7)

Equally, there was a general level of disbelief in the pure altruistic motives of international donors and an inclination towards a belief in their rather business-oriented intentions, with connotations of causal powers. This perception of donors had implications for how local actors participate in the system as they respond to expectations they have of other actors. As this Ugandan CBO said:

She was in business...she was helping but she was in business.

(U2_CBO-F_8)

Locals perceived that donors drove the business of poverty, and whether true or not, this role appeared to incentivise local individuals to take the role of intermediary or recipient. Mimicking personal business benefits (from 'donations') that they perceive donors to be enjoying appeared to have become a local norm. This practice has implications on transaction pathways and impacts the meaning and value of objects exchanged, as highlighted in the following response:

When she told me her story she was just an aid nurse in one of the hospitals in the USA and she told me she was working for few hours and getting very little money... when she came to track gorillas she saw a kind of an open opportunity where she could fundraise and raise money ... she was using the fundraising part of it as a business, she was raising money and then she would give us a negligible amount of money and then she would use the money either for herself, or for her family or for any other needs that she wanted.

(U2_CBO-F_8)

Overt cynicism in many interviews indicated that local people understood this subject and willingly played their roles in such a system of exchange as win-win situations. As the vast majority of villages in the researched locations have a history of development interventions, whether tourism-related or not, it is arguably unsurprising that local actors were not 'passive recipients of aid' but rather *active observers and trainees* of the foreign aid intervention scene. The business of poverty emerging from travel philanthropy may just be a replica or an adaptation of previous aid for development scenarios, as showcased in the response by a community member in Uganda:

The way I see it, it is a bit psychological... if he sees you are needy he goes to your home, takes pictures then takes them to Romania or anywhere and starts fundraising to build you a house.

(U2_CBO-F_8)

Tourism businesses are well aware of the business of poverty perspective in the local community, with some also embracing the business of poverty as much as local individuals and promoting 'poverty' or solutions to it for commercial purposes. Uganda and Zimbabwe showed evidence of these tendencies more strongly than in Ethiopia. On their view of local behaviour in this respect, two tourism businesses said:

They just prey on the weak, they know what they're doing ... they are wolves at the end of the day and they know how to pick out the weak links and get what they need from them ... and at the same time make them feel like they are doing something good.

(U1_TB_3)

they know how it works, there is a whole philosophy and game plan involved.
(EN_TB_67)

Figure 6.9 illustrates the basic concept of this perspective. Perceptions of 'poverty' were packaged and promoted for donors as products using storytelling by a recipient or intermediary for business purposes. This is the result of the capitalisation of a resource that is in abundance in SSA.

Whether through theatrical entertainment or a village/school tour, targeted consumers engaged with poverty products through what looked like informal encounters. Photography was a central activity, making giving and poverty tangible for donors. Although entertaining tourists takes time for hosts, they tended not to receive standard payments – i.e., ticket fees. Instead, hosts generally hoped for donations, raising questions about the boundary amongst donations, tips, and payments and about the adequacy of the remuneration for services provided in the business of poverty. Some may say that paying for the experience may lack authenticity from a tourism point of view, or perhaps it may be unethical to put a ticket price on the voyeuristic experience of observing children in a school, as if in a 'human zoo' (Staszak 2009) or a story tale. Nevertheless, there was evidence that local people willingly enabled donors to create their story, whilst unspoken rules appeared to shape these events, with ethical questions amongst actors remaining almost entirely absent.

We rely on local marketing, when a mzungu [term for white person/tourist] comes, that is local marketing.
(U2_CBO-F_10)

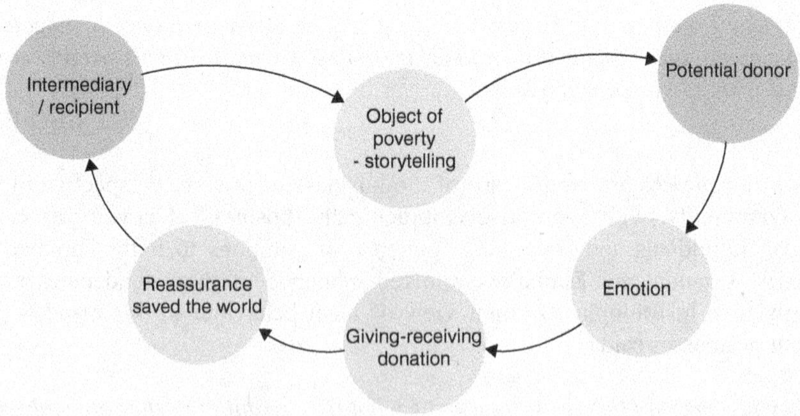

Figure 6.9 Perspective: The business of poverty.

> *There is need for potential people who can fund these projects to see what we are talking about.*
>
> (Z2_2_CBO-F_40)

> *Most of the communities activities we offer they help us to market.*
>
> (U1_CBO-F_2)

Storytelling by local people to generate sales of poverty products was a key characteristic of the system. One community participant explained that:

> *Whenever they see they have a chance of talking to foreigners, people who come in, they have the thinking they have money and they can support different projects so people keep trying, different people, they keep trying different people, mentioning to them different projects.*
>
> (U2_INTER_12)

Children appeared to be the most sought after, approved, or favoured product in the business of poverty, being used as marketing objects in travel philanthropy, with the majority of transactions sponsoring schools' projects or individual children in formal education. This appeared to have led to a range of tourist-donor-ready neat packages. The research did not assume that the business of poverty translates into 'unethical dealings' or 'scams', but some questionable behaviour became evident during the research, such as the covert manipulation of donors vs. recipients taking advantage of an opportunity:

> *They are not doing anything, but they want money from donors, because maybe they take donors to be stupid, at times I would think so because at times donors prove to be that way.*
>
> (U2_CBO-F_8)

> *They will gage what they ask for depending on what they think they can get, and I mean there may be some genuine cases in there, but I think most of them are not genuine.*
>
> (EN_TB_67)

However, right or wrong, good or bad, all perspectives, illustrated together in *Figure 6.10*, arguably raised a number of ethical dilemmas that are important to interrogate. What was clear is the importance of becoming more forthright in appreciating that travel philanthropy is not an entirely altruistic system of exchanges. If we continue to frame it around linear, benevolent transfers from businesses and tourists to communities, we are failing to address critical issues and to understand the real-world system of travel philanthropy in SSA. The importance of surfacing and understanding this perspective is to lay the foundations for understanding the real meaning of exchanges in this system. Thus, we can

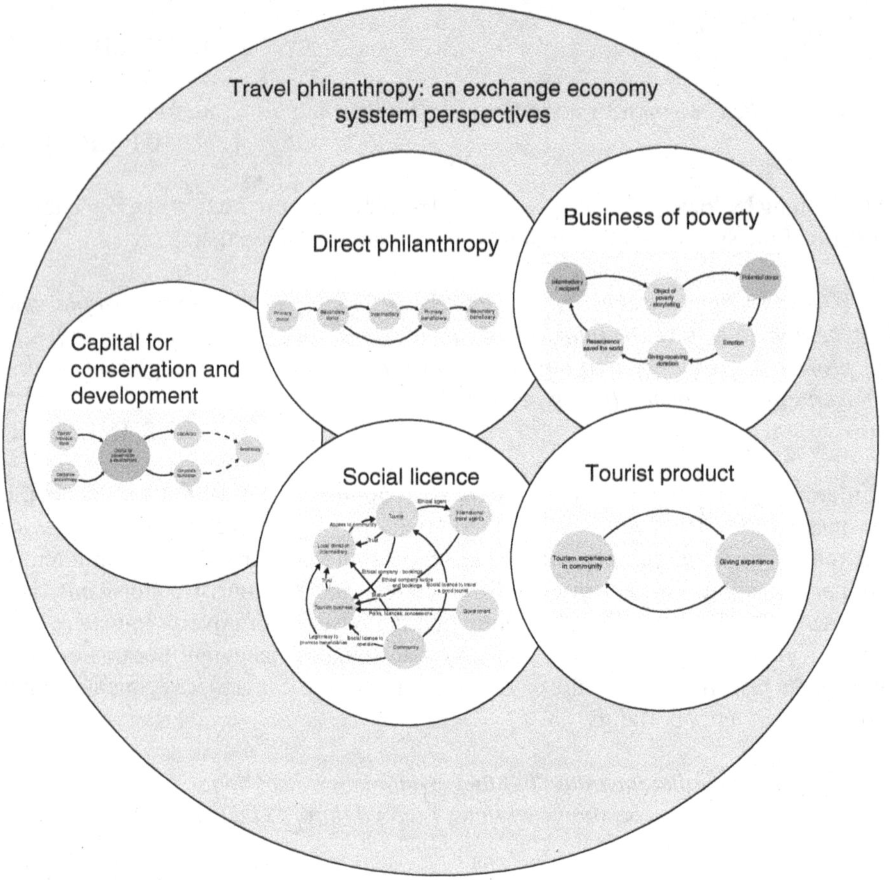

Figure 6.10 Travel philanthropy – An exchange economy and it's multiple perspectives.

understand the effects of those exchanges on who is involved, how actors participate in the system and shape exchanges, and the outcomes and impacts of such exchanges.

This chapter introduces the network of travel philanthropy stakeholders, actors defined as groups of people/organisations, as well as key roles or functions in the system. A combination of the heterogeneity and homogeneity of actors is part of the system's complexity. Key findings highlighted similar patterns or relatively distinct stakeholders' roles across all cases in all countries, but notable differences between countries in terms of roles and their occupants. These were found to be producing context differentiation, influencing different outcomes. Travel philanthropy has now been identified as a complex system of diverse stakeholders occupying different dynamically shifting roles within a relatively informal economy of exchange. Travel philanthropy was found not to be a purely altruistic system of donations in SSA, but rather actors operating in the same system but pursuing different interests, framed as system perspectives. This chapter has contributed a part of understanding what the system is and why and how the system may operate in a certain way. Chapter 8 will build upon this understanding and introduce some other elements which underlie the system.

7 Broadening the theoretical perspective

Travel philanthropy as a social, economic, and moral phenomenon

This chapter advances new understandings of travel philanthropy in relation to human behaviour. Theories within development studies, which predominantly informed earlier chapters, were found to lack the depth necessary to explain broader concepts of human behaviour and the social world, constraining a more comprehensive understanding of travel philanthropy as a system. Therefore, to understand the structure of travel philanthropy as a system, its underlying mechanisms and impacts, this chapter offers a helicopter view of broader issues emerging from the empirical research and frames how to zoom into social systems and human behaviour thinking. Key topics covered are reasoning and bounded rationality, wilful blindness and the ontology of ignorance, bounded roles and responsibilities in a complex world, beliefs, principles, norms, and expectations within socio-economic systems, otherness, sameness, and social influence.

7.1 Inside socio-economic systems

The theoretical framework for this research supports the inquiry into travel philanthropy as a social, economic, and moral phenomenon. The research drew on business and management, economics, behavioural economics, social psychology, sociology, anthropology, and moral philosophy thinking to build a stronger foundation to the critical reflections offered in this book. Theories within development studies, which predominantly informed earlier chapters, were found to lack the depth necessary to explain broader concepts of human behaviour and the social world, constraining a more comprehensive understanding of travel philanthropy as a system.

By building on a theoretical analysis of the domains within which travel philanthropy is located and concepts identified in previous chapters, this chapter is meant to offer a helicopter view of broader issues to then zoom into social systems and human behaviour thinking. The rationale for including these specific disciplines and theories was informed by the inductive research process used in the empirical part of the study, deemed necessary to enable reasoning in the analysis of the research findings. To understand the structure of travel philanthropy as a system, its underlying mechanisms and impacts, these topics were identified as most relevant:

DOI: 10.4324/9781003497929-7

- Reasoning and bounded rationality
- Wilful blindness and the ontology of ignorance
- Bounded roles and responsibilities in a complex world
- Beliefs, principles, norms, and expectations within socio-economic systems
- Otherness, sameness, and social influence

7.1.1 Reasoning and bounded rationality

To understand travel philanthropy as an exchange economy, the literature suggests it is necessary to draw upon theoretical contributions beyond traditional economics. Standard economic theory assumes individuals are fully rational and act in a self-interested manner. The decision-maker has 'well-behaved' preferences over a set of alternatives, represented by a utility function, and choices are interpreted as the result of utility maximisation (Collet 2009). Rationality has long been a topic of controversy (see Friedman 1953) and is historically associated with calculation, as well as justification and competence to have reasons for beliefs, justify beliefs, and justify the validity of goals (Krijnen 2016). An increasing number of economists argue that rationality misrepresents reality (Manski 2004) and presents reductionist and distorted views of human nature (Arjoon et al. 2018). Furthermore, all possible alternatives cannot be known (Levi 1997), and amongst known alternatives, *outcomes* attached are often not known/understood (Hanoch 2002), good outcomes are overestimated (Wilson 2016), moral responsibility may overpower self-interest (Arjoon et al. 2018; Aumann 1997), pure material selfishness is an inconsistent driver (Dunning 2012), and decisions in complex uncertain social structures challenge rationality (Beckert 2013). In Rabin's (2002:658) words, "the more real our assumptions about economic actors, the better our economics".

Economic psychology has been providing increasing evidence against the neoclassical model of individual behaviour, choice, and decisions offering a range of alternative theories that integrate the psychological foundations of human behaviour into economic analysis. One of the key theories relevant to draw upon in this research was bounded rationality (Simon 1956, 1979, 1991, 1994, 2000), which has contributed to the significant expansion of behavioural economics as a field. Herbert Simon led the way into this alternative perspective saying it is only when choices are very simple and transparent, and situations are static, that an individual behaves as a utility maximiser; so, situations that are increasingly complex, dynamic, and uncertain cause "severe demands on [the rational human's] attention" (Simon 1979:14). Choices deviate from perfect rationality, and bounded agents are unable to make optimal decisions.

Bounded rationality shifts analysis from outcomes to understanding underlying processes, such as how decisions are made, the mechanisms used, and other constraints or limitations to rational decision-making in the real world. The idea is that the external environment does not only determine decisions, but also abilities and conflicting priorities of decision-makers. This includes knowledge decision-makers have, their ability to apply it, work out consequences, deal with uncertainty, like the reactions of others to their decisions, and navigate their competing wants (Simon 2000:25). In this age of complexity, humans are required to calculate non-linear utilities (Heiner 1983) and are fallible since we can only see a small part of the universe (Wilson 2016). So, the intellectual climate being created by complexity theory is conducive to bounded rationality (Manzini and Mariotti 2014).

Bounded rationality proposes that humans have limited cognitive capabilities, limited or imperfect access to information, or limited will to search for information in making decisions, and search processes are bounded rationally, with 'good enough' (satisficing) choices being made as we are unable or unwilling to maximise utility (Simon 1978). Choices become mental short cuts (Earl 1983), search rules rather than decision rules (March 1978), or they may be the only choice if the optimal outcome is not well defined (Katsikopoulos 2014). Problem-solving consists of searching for alternatives or improvements on the alternatives we already have, but capacity to generate facts and the ability to draw inferences from these observations can differ (Hanoch 2002). The theory also allows for learning and adaptation as mechanisms (Hanoch 2002; Simon 1956), the impact of education because we are not born rational (Wilson 2016), tastes changing throughout life (March 1978), and bounded willpower as people often make choices that are not in their long-term interests (Aumann 1997).

Substantial research exists on decision-making in the context of bounded rationality. Some of it covers *convention* – linking ideology, ethics, values, beliefs, and behavioural norms (Koumakhov 2009); *emotions* – diverting to a more urgent objective affecting our hierarchy of goals, which can conflict, a mechanism arguably omitted by economists when concentrating on ends rather than processes (Hanoch 2002; Elster 1999); and *intrinsic and extrinsic desires* – biasing emotional expectations (Stutzer 2013). Other research addresses how people have different intuitive beliefs about *what makes them happy* (Loewenstein and Schkade 1999), *experience and values* (Matthiesen and Klitmøller 2019), and *stories*, as later discussed (Fastenrath et al. 2018). Some research addresses ways *personality* biases stakeholder relations (Hu et al. 2018; Hambrick and Mason 1984), and the cost of information acquisition in searches (Sims 2003). Others focus on information asymmetry or rational inattention (Schlaile et al. 2018; Dubbink et al. 2008), 'mindlessness' in that people see what they want to see leading to 'rational fixation' and mental rigidity (Langer 1989), and culture and different imaginaries (Ehrnstrom-Fuentes 2016; Smaldino 2014). Finally, some research focusses on risk and willingness (Roehrich 2014), exploited biases by charities to follow moral intuitions over considered judgements (Woods 2015), or *transparency* and opaque processes (Roehrich 2014).

Incorporating bounded rationality into the travel philanthropy system helps us zoom in to understand some of the system dynamics and understand what is enabling and constraining the decision-making processes of actors and their willingness to trigger searches for information, as well as implications for responsibility in an economic, social, and ethical sense. With much of the current travel philanthropy discourse emphasising the role of emotions in this phenomenon (e.g., Ham 2011;

Bertini et al. 2014; Burgold and Rolfes 2013; Honey 2011), a deeper look at the literature in this respect follows.

7.1.1.1 Emotion in decision-making

> *Sometimes a boisterous passion hurries our thoughts as a hurricane does our bodies, without leaving us the liberty of thinking on other things, which we would rather choose.*
>
> (Locke 1846:154)

Simon (1983) argues that for a robust theory of human rationality, we must understand the role of emotions. They are part of the information processing mechanism found to work in conjunction with rational calculation, serving as an additional tool to supplement the inefficiencies of reason, restrict the range of options, focus upon certain variables, start/stop evaluation processes, generate values and preferences, and therefore work as a 'satisficing' mechanism central to bounded rationality (Hanoch 2002; Simon 1982). Emotions are subjective rather than objective in nature, therefore considered irrational in traditional decision-making theories. However, authors increasingly indicate emotions aid mental deliberation *differently* to rational calculation of pros and cons (Hanoch 2002).

Emotions can affect our goals and cause us to pursue more urgent objectives – positive or negative depending on context – because rational does not mean right (Hanoch 2002). The complex interaction of cognitive and emotional responses to external stimuli can result in dysfunctional outcomes (Brunnermeier and Parker 2005). Emotions are found as mechanisms for how we think or act *fast*, with the risk of making mistakes, and *slow*, with more precision (Hanoch 2002). It is argued fast and frugal decision-making heuristics, or gut-level responses aiding rapid assessments, may save us from predators but can also lead to suboptimal judgements and decisions (Woods 2015; Smaldino 2014). Mind scientists call these 'System 1 processes', as opposed to 'System 2 processes', which represent a more deliberate mental activity, which Woods (2015) draws on in a philanthropy context, suggesting System 1 thinking tends to choose non-utilitarian reasons and System 2 tends towards consequentialism.

Stutzer (2013) highlights that if people can better understand events, affective reactions tend to be weaker, so if an event's novelty, unexpectedness, variability, uncertainty, and explanatory incoherence are higher, affective reactions are stronger. Feelings and emotions are therefore 'ideas' produced by situations (Wilson 2016a), but, adding to the complexity, emotions are culturally affected and are cultural statements about the morality of a situation (Siltaoja and Lähdesmäki 2015). Affect is the feeling that something is good or bad, when we assign meaning to information, which drives our behaviour by informing our intuitive judgements about how to act, so affect in decision-making and how we behave is prominent (Hart et al. 2018; Matthiesen and Klitmøller 2019). There is much evidence that emotions are linked to spontaneous action (Picard 1997), and as reactive attitudes to others (Smaldino 2014). Expectations are also theoretically linked because emotional values attached to imagined future situations affect present decisions and therefore behaviour (Ben-Ze'ev 2000).

It is beyond the scope of this study to consider all emotions in decision-making; however, some authors (Matthiesen and Klitmøller 2019; Arendt 1994) reflect on the relationship between empathy and moral judgements justifying the shortcomings of empathy as a moral compass. This includes the issue that those we empathise with are culturally embedded, so demands of knowing the other and what is right are not a straightforward judgement because while some human preferences are universal, others, such as ambition, vary from culture to culture (Wilson 2016a). Dunning (2012) argues that emotional factors are more significant than economic in decisions to trust, such as giving money to strangers, so we must consider how one will feel by trusting, meaning the decision to trust is both instrumental and expressive. Envy influences decisions either as *malicious*, intention to diminish others so they lose the object or advantage, or *benign*, admiration and motivation to improve oneself – thus envy turns into emulation (Siltaoja and Lähdesmäki 2015; van de Ven et al. 2009; Elster 1991; Goel and Thakor 2005). Jealousy is mainly a response to protecting a valued relationship from a perceived threat (Siltaoja and Lähdesmäki 2015). Fear of not following the crowds has been found to have significant impacts to decision-making in organisations (Fastenrath et al. 2018). In fact, despite growing evidence, business ethics and international development thinking may be arguably weak in brushing aside decisions that are emotionally embedded, significantly influencing situations (Siltaoja and Lähdesmäki 2015).

Emotions in philanthropic decision-making include sympathy in giving and prosocial decision-making (Hart et al. 2018), the effect of 'sympathetic education' (Rorty 1993), and the cognitive and affective processes in donation decisions (Dickert et al. 2016). Woods (2015:91) argues that playing with moral intuitions may be an asset for organisations, prompting significant transfers of aid by harnessing "our strong intuitive reaction to the suffering of others, including feelings of guilt, empathy, moral outage, and indignation". However, this is arguably limited in tackling the roots of problems, instead addressing symptoms and providing an imperfect guide for policies. Such complex matters cannot be resolved by intuition, as judgement biases can mean suboptimal resource allocation, caused by deliberate distribution of resources based on the reaction of audiences. He goes on to argue that moral intuitions may be powerful in soliciting donations but can mislead, often being problematic and costly. Judgement biases proposed by Woods (2015) include moral outrage, psychic numbing, and naive realist thinking. *Moral outrage*, for example, crowds out deliberative consequentialist thinking, causing people to overcare about a particular scenario, overriding deliberate thought. Our moral intuitions are found to be subject to biases, such as cute faces over statistics, and exploitation by organisations soliciting donations may introduce new biases. Woods argues we should rather aim for 'moral intuition education' and avoid asking people to blindly follow their intuitions but ask people to evaluate and check in a more carefully considered way, understanding the limits of our intuitions about actual social and economic conditions.

7.1.2 Wilful blindness and the ontology of ignorance

Much like staring into the sun, reality is often difficult to face directly.
(Heyer 2015:643)

The theory of bounded rationality partly recognises informational and cognitive constraints in decision-making that are not necessarily from choice, but rather from *wilfully blind* decisions, whether conscious or unconscious (Heffernan 2012). To evaluate moral obligations or responsibilities of actors in travel philanthropy and/or international development, including blameworthiness, people can decide to be strategically ignorant (Wieland 2017b). However, wilful blindness is also not necessarily an overtly deliberate attempt to unconsciously avoid knowing, with ideas such as 'rational fixation', 'mental rigidity', or 'mindlessness' explaining how actors sometimes see what they want to see, preventing them from exploring alternatives and reaching favourable decisions (Langer 1989). Knowledge about the *Other* in this respect, for instance, is found in post-colonial, feminism, and critical race theories, identifying epistemological ignorance on a broader scale (Croissant 2014; El Kassar 2018).

On the other hand, wilful blindness or ignorance is a *conscious* inability to know, when we know we don't know. It is an attitude towards truth and untruth (Mallard and McGoey 2018), and Heyer (2015) studied deliberate behaviour of turning a blind eye to keep facts and reality conveniently out of sight for fear or popularity of that 'truth'. Related to this is Croissant's (2014:11) identified 'agnotology' emerging framework, based on nictwissen, negative knowledge, non-knowledge, ignorance, and agnosis used across research, and defining these as "intentionally produced agnoses". She highlighted theories of absence and differentiated ignorance, which can be *epistemological* – cognitively inaccessible because we do not yet know – or *ontological* – things that are unknowable as a matter of property because of causal indeterminacy or choice. Earlier work investigating the phenomenon by sociologist Michael Smithson (1989) included describing neglect of information as an economic commodity, with the 'power of knowledge' sometimes superseded by the 'power of non-knowledge'.

Knowing and not knowing within theories of ignorance needs to be distinguished, because if an actor thinks ignorance is bliss, that can include the bliss of refusing accountability (Jungkunz and White 2013). A simple example is when a charitable donor chooses to maintain their feeling of warm glow and remain wilfully ignorant of the effectiveness of that contribution (Niehaus 2014). Traditional economic models are again critiqued for their inability to explain economic decision-makers *ignoring* information (Grossman and van der Weele 2017). These theories also enable the moral dimension of decisions into one's analysis as well, considering ignorance in terms of responsibility acceptance or avoidance (Garcia 1997). Silences have also been found to sustain ideologies, because there are things that go without saying in the social world, or things that are unspeakable, forbidden, and/or taboo (Bourdieu 1977 in Croissant 2014). For instance, we are fallible and there is a known disposition for people not to ask questions (Wieland 2017).

Davies and McGoey (2012) argued that ignorance is actually knowledge, and we should treat it as a sociological phenomenon distinct from knowledge, rather than the binary opposite or absence of knowledge. 'Antiepistimology' – the study of non-knowledge, or how knowledge is deflected, covered and/or observed, and embedded in political and social processes – has therefore grown. McGoey (2012) urged a shift from politics of knowledge towards politics of ignorance, that knowing the least amount possible can actually be the most indispensable tool for managing risk and absolving blame. Wilful ignorance is discussed on an individual as well as social level, including "social silence surrounding unsettling knowledge" (Davies and McGoey 2012:12). A common example is the global financial crisis, where individuals in the system maintained deliberate social silence, and warning signs were actively ignored or concealed. Wilful ignorance can mean deliberately 'avoiding inconvenient information' about the social impact of one's decisions, rather than external constraints (Wieland 2017:115), so suspicions are deliberately omitted (Williams 1961). Lynch (2016) referred to the wilful embrace of ignorance, clarifying that it should be possible to find out the truth with reasonable risk and costs. Complexity and ignorance are here combined, because uncertainty is sometimes used to mean 'unknown' (Croissant 2014). Behind many crises that seem surprising, it is argued that individual actors in a system, consciously or unconsciously, collaborate to maintain an impression of helplessness and lack of foresight, yet mediated strategic ignorance, social ignorance, and denial of knowledge are often used as assets rather than liabilities (Davies and McGoey 2012). In business, some organisations produce ignorance through compartmentalisation (Croissant 2014).

Across the academic field, a spectrum of discussions includes ignorance, wilful blindness, and whistle-blowing, particularly with regard to corporate organisations (Wettstein 2012; Bouville 2008), wilful blindness in the global financial crisis (Davies and McGoey 2012), climate change ethics (Croissant 2014; Robichaud 2017; Grossman and van der Weele 2017), corruption (Sundstrom 2019; Shivji 2006), international politics (Mallard and McGoey 2018), and public administration (Taylor 2018; Lavena 2016). The role of truth as an object of contention in organisations appears the most dominant (Munro 2016). Research covers wilful ignorance and self-image (Grossman and van der Weele 2017), media commentary and post-truth (Mallard and McGoey 2018), strategic ignorance (McGoey 2007), and self-deception (Mallard 2014), and 'bullshit' (Spicer 2012). Other studies have focussed on social unknowing (Thrift 1985), agnotology (Proctor 2008), ignorance as structural (Alcoff 2007), moral, and factual ignorance (Wieland 2017), wilful blindness and self-deception (Lynch 2016), recklessness and negligence (Sarch 2018), silence in the non-governmental organisation (NGO) discourse (Shivji 2006), higher education (Marginson 2011), and negative knowledge (Croissant 2014).

Within the debate, there are also whistle-blowing (Heffernan 2012), messengers of the truth (Bouville 2008), or muted criticism (Marginson 2011), and the norms of silence (Morrison 2014). Whistle-blowing is a behaviour in response to the norms of silence about wrongdoing (Taylor 2018). It involves 'challenging the status quo' (Arjoon et al. 2018), can include disclosing illegal, immoral, or illegitimate practices (Near and Miceli 1995), and exerting power to change behaviour (Wettstein 2012), and its purpose is to expose wrongdoing, stop it, and prevent it (Francis et al. 2015). Staying silent instead contributes to the norms of silence, and morality can sometimes hide that a choice was not external (Bouville 2008), with actors being defined as 'morally mute' (Menzel 1999). From an ethical viewpoint, Bouville (2008) explained that whistle-blowing is mandatory, unless other moral obligations can override this and it becomes a tricky moral situation because one can be culpable for action or in-action. It is indeed a complex subject because there are often negative consequences for whistle-blowers in society, as the actor can be seen as loyal or a traitor, and the action may be ineffective in fostering change or in producing benefits or could even harm society. Thus, understanding conditions under which whistle-blowing is effective is critical (Wettstein 2012). Whistle-blowers can be in a position of vulnerability because their actions may threaten a belief system that enables the exposed behaviour (Francis et al. 2015). Firing or blacklisting employees (Bouville 2008) can trigger accusations of disloyalty, secrecy breaching, or leaking sensitive information (Bok 1981). It can also cause harassment, bullying, or redundancy (Munro 2016). The singularity of the act is often under played because of the personal risk in speaking truth to power (Contu and Girei 2014; Kaulingfreks and Kaulingfreks 2013).

Adding to the complexity is understanding what actually constitutes wrongdoing (Taylor 2018). In the context of climate ethics, for example, Robichaud (2017) highlighted that it is not clear if treating individual actions as the relevant unit of analysis of moral evaluation is sufficient, because if individuals are blameworthy but did not know, they are not blameable for acting wrongly; however, at what point are they morally responsible for their ignorance? Robichaud (2017) suggested that blameworthiness rests on the quality of will, failing to inform oneself, or insufficient lack of concern, and therefore argued for a lack of goodwill as moral responsibility, rather than belief mismanagement. These issues are relevant to understand issues within philanthropy and international development.

Discourses in philanthropy and development aid are permeated with the ideas of wilful blindness. One example is corporate social responsibility's (CSR's) "romantic vision of self-reliant peasants made happy with fishing rods", replicating standard NGO unspoken norms of neoliberal capitalism in merit to empowerment, with beneficiaries providing upkeep for development gifts despite increasing critiques of its outcomes, just ignoring 'morally muddy terrain' (Gardner 2015:16). Meanwhile, 'development speak' allows fuzzy and feel-good words that admit no negatives and cause no disagreement (Cornwall 2010), with development interventions resonating with Western consumers and employees perceptions of 'good business practice' (Rajak 2011) and offering rhetorical power, but absolving the company from long-term commitment (Gardner 2015). Discourse of the 'Third World' represented as incapable of pursuing 'developments' on its own (Chakraborty et al. 2017) may prop up an industry wilfully blind to the truth and continue development programmes by consultants embracing the NGO discourse and ideological myth that they are the third sector, non-political, and non-profit (Contu and Girei 2014). Certain psychological processes affect our ethical judgements, leading to bounded ethicality or a 'blind spot' (Arjoon et al. 2018) and other decision processes knowingly uphold false outlooks (El Kassar 2018). Wilful blindness seems invaluable to better understand the development sector operating systems in poverty-stricken areas of Africa and the grey area of responsibility under imperfect knowledge that may allow people to shuffle out of accountability (Schlaile et al. 2018; Weber 1919), which is further discussed next.

7.1.3 Bounded roles and responsibility in a complex world

Building upon the above theories regarding bounded rationality, wilful blindness, and the use of ignorance in decision processes draws light on questions regarding acceptance or avoidance of responsibility. Roles in social systems are an important part of the moral landscape and seen as a source of obligations for role occupants (Dare 2016). It is a complex subject as while some say a role cannot be ascribed a moral responsibility because it is a social construction, the role occupant may have direct responsibility in actions whilst performing roles. Yet, identity is constantly changing, influenced by roles enacted and social expectations, so without a stable element to evaluate moral responsibility, it is relative (Rozuel 2011).

We tend to be accountable for what we are doing rather than what we are not doing (Wettstein 2012), so when causality is ambiguous and unclear, we are left with bounded morality, where different actors find it acceptable not to take responsibility (Schlaile et al. 2018; Scheffler 2001). In an earlier assessment, Scheffer (2001) had suggested that while *backward-looking responsibility* is blame, *forward-looking responsibility* is a duty or obligation to correct mistakes, arguing that failure to act is an omission. He pointed out that failure to participate in solving structural problems may not be seen as a failure to act, unless instrumental in causing the specific harm at stake, thus may be beyond their call of duty. Young (2006) is another key author of forward and backward-looking models. Differently from bounded rationality but aligning, Schlaile et al. (2018) argued that bounded morality does not require explicit reference to rational choice theory. They focus on decisions that involve a moral issue and attempt to take into account systemic forces external to individual agents such as evolved properties of social networks and culture, as well as internal limitations, such as dissonance and confirmation bias.

Responsibility can have different meanings (Schmidt and Seele 2012). As the world becomes more globalised, interconnected, and complex, there may be no clear allocations of responsibility, but we need to go beyond the principle of doing no harm (Wettstein 2012). Similarly Rozuel (2011) argued that compartmentalisation limits our ability to be moral agents, and Wettstein (2012) that fragmentation of authority fragments responsibility. This is particularly poignant today, as private actors move more into the space of public function, blurring authority, thus blurring responsibility (Wettstein 2012), but places more emphasis on personal responsibility (Allik and Realo 2004). This shift in the development landscape demands a new perspective on morality between individualism in modern ethics and the need to assign moral responsibility for collectively produced harm, including questions about those who have the *capacity* to contribute to viable solutions (Wettstein 2012).

The individual-collective responsibility nexus is debated in the NGO sector, as when unintended consequences on a wide range of stakeholders can paradoxically make good intentions cause harm to the group they intended to protect (Fassin 2012). In philanthropy, effective altruism contributes to the moral responsibility debate, the moral requirement to donate efficiently, but conditional obligations may also generate incentives not to do any good to avoid being 'required' to do more (Sinclair 2018). In business, the moral responsibility of managers has been questioned regarding their capacity to control outcomes (Michaelson 2008) and the growth of CSR is a reflection of a shift in the responsibility debate, the impact of actions, and decision-making on other stakeholders (Fassin 2012). Fassin (2012) provided an interesting discussion on reciprocity of responsibility, with Schlaile et al. (2018) arguing that responsibility is with all stakeholders, not only with the business, including consumers (consumer social responsibility), and calling for greater attention to social and economic responsibility, as the ecological one still gets most attention. The big focus on ethics in businesses needs to shift towards other actors in the transactions primarily viewed as acted upon to the shared responsibility of all societal actors (Tammelleo and Lombardi 2014). However, it is suggested for sustainable development that opportunities need to be created for consumers to be able to act in a responsible way (Antoni-Komar et al. 2012).

From a systems perspective, the collective reproduction of structures that underlie and perpetuate social damages is recognised not only through individual action, but because of a contribution to emergent phenomena by all action (May 1992; Kutz 2000; Young 2011). From a consequentialist perspective, in the context of social interconnectedness of responsibility, Schlaile et al. (2018) called also for a shift beyond idealists, who bear nothing in mind but the immediacy of goodwill, yet pass on responsibilities for their actions to others. Consequentialism is the philosophy that challenges notions of responsibility most profoundly, including both outcomes one fails to prevent as well as those directly generated. Having the capability or *power* to act, not just blame, is also becoming a significant topic (Schlaile et al. 2018) and while corporations often point to a lack of appropriate platforms to provide an opportunity for participation, Held (1970) argued that a rethinking of excuses is needed by stepping up and organising collective efforts. Furthermore, Wettstein (2012) asked if *can* means *ought* in understanding moral responsibility in business, suggesting corporate responsibility needs to be reframed as collaborative responsibility. Decision responsibility can be more complicated than *what you do* or *ought to do*, because choice depends on the decision in its context, so some actions are only good if followed up in the right way (Louise 2009).

7.1.4 *Beliefs, principles, norms, and expectations within socio-economic systems*

7.1.4.1 Beliefs

Fodor and Pylyshyn (1981) proposed a theory of propositional attitudes and suggested that beliefs are mental items with semantic properties, with a causal role in the production of behaviour. Forming beliefs is a basic human function, an attitude of true or false, or good or evil (Schwitzgebel 2006). The concept of a belief presumes a subject (the believer) and an object of belief (the proposition) and so truth as a propositional form has been divided into *belief-that* and *belief-in* (MacIntosh 1994). Epistemologists place their interest in belief-that, so the belief in knowledge that is true, and from a realist perspective, the ontology of belief-in and belief-that is of interest, as well as understanding beliefs and their causal properties (Bhaskar 2008).

Belief systems consist of stories we tell ourselves to define our sense of reality, a mechanism to make sense of the world around us. They form a structure of norms, but part of an evolving open system, and they have an impact on our world (Usó-Doménecheh and Nescolarde-Selva 2016). Societies are founded and changed due to belief systems (Dunning 2012), and identifying central components is an important part of scientific inquiry (Brandt et al. 2019). Beliefs arise through and are interdependent with reason and experience, which together create dynamic contexts, causing beliefs, ways of reasoning, and experience to perpetually change (Usó-Doménech and Nescolarde-Selva 2016; Dunning 2012). Emotion and cognition in conjunction also produce beliefs and affect decisions (Cold-Ravnkilde et al. 2018; Dornschneider 2019). Beliefs can be 'antecedents' or 'consequents' of decisions, interconnected with other beliefs, but are often overlooked by macro-oriented approaches, despite offering useful analytical frameworks to better understand complex social situations such as warfare, terrorism, and environmental protection (Dornschneider 2019). The influential systems thinkers, Maturana and Varela (1980), highlighted that within a living system, if the organisation of beliefs changes, the belief system as a whole changes. Beliefs are a critical component of systems and complexity thinking, not only because they affect human behaviour within systems, but they represent complex adaptive systems too.

Individual, social, or political belief systems can be interpreted as ideologies or worldviews (Brandt et al. 2019). A socially shared belief system emerges from creating shared realities by interacting with others, so an ideology, which is "networks of concepts embedded into networks of people" (Homer-Dixon et al. 2013:344). Therefore, according to Homer-Dixon et al. (2013), ideological change requires restructuring the network, the beliefs in the minds of individuals, and also the new emergent reality and cognitive change at the group level. Complexity theory and cognitive science can tackle multi-level causality, to integrate our understanding of mechanisms regarding the emergence of ideologies, their acceptance and spread through societies, ideology change and stability across individual and social groups, which remain patterns difficult to explain rationally.

7.1.4.2 Norms

In development aid, norms are often assumed to spread in a linear fashion as static entities whose content remains stable as they move from context to context. However, global and local norms can in fact be very different, and so the common efforts to establish and articulate global norms (i.e., gender equality) and frame development activities utilising codified international norms are highly problematic (Cold-Ravnkilde et al. 2018). Such approaches also tend to disadvantage already marginalised voices in international development. It is also evident that in any social system in which these interventions take place, norms are a critical component for understanding behaviour and behaviour change.

Norms are *standards* and *rules* in a society, defined as "a standard of appropriate behaviour for actors with a given identity", and as "sets of interlinked regulating principles" (Cold-Ravnkilde et al. 2018:78). Norms are the principle or rule of behaviour understood by members of a group, guiding or constraining social behaviour without the forces of law, especially relevant when laws are not easily applied (Cialdini and Trost 1998; Baron 2018). As a basic requirement of conformity, social norms call out individuals when they deviate and inform obligations or permissions an individual has to a larger social system (Tomasello 2020; Boella and Torre 2007). Norms are found to play a critical role in human societies and they are generally referred to as 'social' norms because they are rooted in human interaction and expectations about other people's beliefs and behaviours (Legros and Cislaghi 2020). Norms are therefore interpreted as informal unwritten and socially enforced rules, expected behaviour towards a specific situation, and shared by members of society (Mahmoud et al. 2014; Legros and Cislaghi 2020). Social-norms theory investigates why people might comply with certain practices, so it aids an understanding about what can be done to change behaviour (Legros and Cislaghi 2020). Guesses about why people do what they do in policy making are common, so by understanding the normative framework and realising behaviours are regulated by normative principles, such as trust, it is argued one can transcend simple rationality (Taylor-Gooby 1998).

Norms can be studied in a number of ways. Tuomela (1995) distinguishes different kinds of norms as *rules* (e.g., taxes paid by all), *social norms* based on mutual belief (e.g., do not spit), *moral norms* appealing to one's conscience (e.g., one should not cheat), and *prudential norms* (e.g., having to maximise one's expected utility). Descriptive norms are behaviours or opinions that are popular in the group (Cialdini 2007), rules that are expected to be pursued (Mahmoud et al. 2014), and an important component of an intersubjective culture (Gao et al. 2015). Norms have also been divided into injunctive norms (people's beliefs about what has to be done), fundamental norms (driven by injunctive norms and activating attitudes), and descriptive norms (influence behaviour directly) (Mahmoud et al. 2014). Coleman (1990) identifies two main norm categories: *conventions* and *essential norms*. Conventions are natural norms that emerge without any enforcement, when everyone conforms to a desired behaviour, such as driving on the same side of the road. Essential norms solve collective action problems when there is a conflict between individual and collective interests, such as not littering, an act that benefits everyone in the society. Global norms include concepts such as human rights (Cold-Ravnkilde et al. 2018).

Moral principles are different from social norms because they do not come with enforcement, but from values, although they overlap (Baron 2018). Virtuousness, for example, is said to be the internalisation of moral rules that produce social harmony, a constellation of virtues in the aggregate manifested by collectives of people, a virtuous economy therefore guided by practical ethical principles (Arjoon et al. 2018). Values represent beliefs and attitudes that shape a culture and indicate the way things are done through appropriate goals, behaviour, and roles (Manz et al. 2008). People hold principles explicitly and implicitly, but as a group coordinated individual principles become norms, and normative principles refer to what ought to be done, categorically or hypothetically (Henderson 2002).

Norms therefore have behavioural consequences, often referred to as guiding prosocial behaviour, their evolutionary advantages, and an encouragement of positive externalities and discouragement of negative ones (Baron 2018). There may be less emphasis in the literature on social norms that are harmful, such as encouraging unhealthy behaviour or discrimination. However, a constraint on behaviour tends to be part of its definition, in agreement with Legros and Cislaghi (2020). The concept of norms has been applied to a range of behaviours in social science including norms of reciprocity (Koumakhov 2009; Tomasello 2020; Contu and Girei 2014), honesty and trustworthiness (Allik and Realo 2004), corruption (Kyriacou 2015), trust (Dunning 2012; Reginster 2013), and lying (Croissant 2014; Jahn and Bruhl 2018). *Corruption* and *lying* as norms, for instance, can be driven by anticipated personal gains or expectations about how others will act (Sundstrom 2019; Croissant 2014; Jahn and Bruhl 2018). *Trust* is a constellation of beliefs about the extent to which others will be concerned about one's best interests; so trust may be reciprocated or violated, depending if people believe others are untrustworthy or trustworthy (Dunning 2012). Kyriacou (2015) argued that there can be cultural variations in how a norm manifests in a society whether it is individualist or collectivist, influencing whether the norm to trust in-groups and out-groups is generalised or particularised. Communication and complexity theories have been used to understand norm creation and lifecycles, including emergence, feedback, social learning, how norms spread socially, and tipping points (Coleman 1990; Gao et al. 2015; Sen and Airiau 2007; Hollander and Wu 2011; Legros and Cislaghi 2020; Mahmoud et al. 2014).

7.1.4.3 The communication of beliefs and norms through stories

Narrative stories preserve culture and customs, giving members of a culture common referents (Smaldino 2014). Narrative frames in African development can be morally or politically problematic, such as frames of suffering, or of racism, because they can become accepted beliefs or visions constructed by and through white privilege, for instance (Jungkunz and White 2013). The development gift is often critiqued for being based around a narrated success for international audiences and their beliefs, having an effect on actual interventions (Gardner 2015). Woods (2015) argued that human rights campaigns today still rely on the same methods of belief persuasion through telling stories of individual suffering as the earliest ever campaigns did, stories that provoke feelings of sympathy, guilt, and outrage, successful in sustaining norms. Imaginative storytelling is at the heart of global tourism as well (Salazar 2013): narratives with colonial nostalgia underpinning tourism experiences (Scott and Selwyn 2010), culture simplified for tourism consumption, often to confirm Western concepts of the Other. Salazar (2013) argues local tourist guides have now actually become cultural brokers, and stories once 'imagineered' for foreign Others during the colonial era are now increasingly accepted by local people as narratives about themselves. In organisation and management studies, stories are found as mechanisms for expectations, which become norms in the form of socially shared myths about what is rational to do, when the content of the story matters less than the number of story followers (Fastenrath et al. 2018).

7.1.4.4 Expectations

Expectations can be informed and defined by experiences that happened in the past and also by what we have been told or by subconscious beliefs that we hold (Stenger 2019). Expectations are considered the foundation for economic decision-making (Manski 2004) as a predicted outcome (Fastenrath et al. 2018), or in comparing what we imagine rather than calculate what can happen (Shackle 1961). People are not necessarily forward-looking rational economic agents basing expectations on objective beliefs but show biases in their beliefs, sometimes overestimating good outcome probability. Such ideas are found in behavioural economics, opposing expected utility theory (Brunnermeier and Parker 2005; Kahneman and Tversky 1979).

Expectations and behavioural impacts are discussed across the literature from a variety of angles. See Table 7.1 for some examples.

Causal concepts regarding how actors form expectations and why expectations change range from rational and adaptive expectations (Fastenrath et al. 2018) to the inability to calculate – i.e., socially constructed 'mental representations' (Beckert 2016; Rona-Tas and Guseva 2001). In the international development context, despite the importance of expectations for economic decisions and predicting behaviour, elicitation of expectations is rare as a data set, usually because poor, illiterate individuals are assumed incapable of understanding probability, which is incorrect (Manski 2004). For example, social labelling can cause people's behaviour, living up to a behavioural or characteristic label bestowed on them, positive or negative, which creates expectations (Kraut 1973). People can make decisions based on how others expect them to perform, such as by showing confidence or doubt (Maxwell and Dornan 1997). Self-fulfilling prophecies are found when beliefs cause people to act in certain ways, with expectations bringing the thing that was at first a figment of their imagination – e.g., the Placebo or Golem effect – thus psychological phenomena become relevant too (Merton 1948; Maxwell and Dornan 1997; Stenger 2019).

Stories and narratives are central to expectation formation – time and previous story-follower numbers impacting their effect, shifting from economic to social over time, content mattering less with time, so when the majority believe in the story or unquestioned 'myth', it reflects the expectations of the majority, thus remaining actors follow regardless of any rational cost-benefit calculation (Fastenrath et al. 2018). Two distinct causal mechanisms operating under different

Table 7.1 Example theoretical themes of expectations as behavioural impact

Expectations in predicting future utility	Brunnermeier and Parker (2005)
Social labelling and self-fulfilling prophecies	Merton (1948)
The role of stories in the formation of expectations	Fastenrath et al. (2018); Beckert (2013)
Relationships between social and economic expectations	Fastenrath et al. (2018)
Expectations locked-in	Dunning (2012); Fastenrath et al. (2018)
Trust	Dunning (2012)
Role models	Morgenroth et al. (2015); Liu et al. (2019); Wilson (2016); Rajak and Dolan (2016)
Emotions such as expected happiness	Stutzer (2013); Wilson (2016)
Ethical expectations in stakeholder theory	Jahn and Bruhl (2018); Gardner (2015)
Moral economies	Gardner (2015)
Obligations	Buchtel et al. (2018); Motilal (2015); Tomasello (2020); Sinclair. (2018)
Legitimacy	Ehrnstrom-Fuentes (2016); Siltaoja and Lähdesmäki (2015); Contu and Girei (2014); Suchman (1995)

conditions have been proposed – 'calculating gains' and 'fearing social losses' – leading to different beliefs, actors with economic expectations and actors with societal expectations. The *fearing social losses* mechanism causes imitation as a social process (Fastenrath et al. 2018). Myths are reinforced through increased adoption of a practice in a story and people believe the majority view (Meyer and Rowan 1977). The more people who adopt an idea, the less it is questioned and assumed legitimate, and expectations of what is considered rational become locked-in (Liebowitz and Margolis 1995).

Role models and expectation theory align with story imitation discussed above (Morgenroth et al. 2015; Liu et al. 2019). The allure of entrepreneurial possibility by youth in Kenya was discussed by Rajak and Dolan (2016:3), identifying expectations "operating at the borders of hope and despair". Expectations in the context of children have been highlighted by Wilson (2016:143) in discussing Locke, whom he quotes:

> *Being indulged when they are little, it grows up with age, and with that enlarges itself to things of greater consequence. And has ruin'd more families than one on the world. This should be surpassed in its very first rise.*

Locke believes children are "highly plastic, whereas adults are comparatively rigid. Once the critical age is past, the machine, with all its skills and associative habits is fully formed, and change is difficult".

Gardner (2015) applied the concept of 'moral economy' as a heuristic device to frame different motivations, expectations, and responses of givers and receivers within a CSR and community development context. Different moral economies can interact and conflict and need to be better understood. In their Bangladesh study of a mining corporate, gift-giving became a paradox with clashing expectations amongst 'the poor' community that the company 'should do more', as compensation for use of their natural resources, with an entirely different ethos of detachment underlying the corporate gift. Ehrnstrom-Fuentes (2016) also discussed CSR and the issue of local norms being incommensurable with expectations of the corporation, arguing the need to identify other realities beyond the modern world, moral legitimacy from the perspective of multiple realities in different contextual settings, also acknowledging that within locations there may be multiple and hybrid imaginaries too.

7.1.3 Otherness, sameness, and social influence

7.1.3.1 Otherness

Human societies tend to define themselves along dimensions of difference and sameness, identifying with whom they are grouped, producing categorical boundaries and social meanings, both consciously and unconsciously, which inform decisions and affect behaviour (Powell and Menendian 2020). We are always 'others' to somebody (Deifelt 2007). Mead (1934) said that social identities are created and adjusted through our ongoing interaction with other people and there is a subsequent self-reflection about who we think we are according to these social exchanges. It is known as the 'looking glass self' (Cooley 1902). Over time, these ideas form cumulative, shared social meanings, and 'othering' becomes structured in the world through institutionalised or culturally embedded processes (Powell and Menendian 2020). These common social imaginaries are complex systems of meaning and shape the expectations we have of one another (Ehrnstrom-Fuentes 2016). Staszak (2009:1) provides definitions of *Othering* as "transforming a difference into otherness so as to create an in-group and an out-group"; *Otherness* as 'characteristic of the Other'; *Other* as "member of a dominated out-group, whose identity is considered lacking and who may be subject to discrimination by the in-group"; *in-group* "a group to which the speaker, the person spoken of belongs"; and *out-group* "a group to which the speaker, person spoken of, does not belong".

Alterity is a philosophical and anthropological term meaning 'otherness', other than sameness, often referred to in the construction of 'cultural others' (Levinas 1999). Alterity is an ontological separateness of 'myself from others', said to emerge as a relational being in its projection into the Other, and with the prevalence of cultural diversification and pluralism in the world, we are confronted more with alterity and strangeness than in early Western philosophy (Kovacs 2017). Paradoxically, otherness represents the mirror in which one understands his/her own being through awareness of differentiation (Kovacs 2017). Spivak (2012) identified four modes of being that create alterity: nationalism, internationalism, secularism, and culturalism. In the 21st century, group-based difference and 'othering' as a set of processes and structures causes hostility, inequality, and marginality through religion, gender, socio-economic status or class, skin tone, etc. (Powell and Menendian 2020). Human beings have a tendency to make categorical distinctions, such as race, using socially constructed meanings that are circular and reinforcing, and the product of complex collective and social processes (Powell and Menendian 2020). Otherness is central to sociological analyses of the way social identities are constructed and how social categories shape our ideas about who we think we are (Zevallos 2020).

Another important concept for this study was *The Gaze*, a concept regarding alterity that has been built upon across disciplines (Sartre 1943; Foucault 1977; Levinas 1999; Said 1978), encapsulating perceptions of the other and self. In Levinas' philosophy of identity, the Gaze represents openness, and the encounter of the Gaze refers to the authenticity of the Others personhood, so a metaphysical difference is made between the Gaze and observation identified, in which viewing as staring is not the Gaze. Rather, one could say, "I directly glance at my own alterity within the Gaze of the Other" in a face-to-face encounter and "the Other manifests before me as a Gaze" (Kovacs 2017:85–86).

Otherness, according to Staszak (2009), is actually less about the difference of the Other, and more to do with the point of view and discourse of the person perceiving the other as such. Staszak (2009) explains *difference* belonging to the realm of fact, such as biological sex, and *otherness* (or *othering*) belonging to the realm of discourse, such as gender, in which individuals are generally classified into two hierarchical groups – 'us' and 'them'. Otherness and identity are inseparable because the Other only exists relative to the Self and vice versa. Thus, to identify that we are different from others requires an interaction with others and understanding how others see us shapes our view of who we are (Allik and Realo 2004). If otherness is about interacting with other people, geography is therefore considered an effective producer of otherness, which is seen in post-modern and post-colonial literature (Staszak 2009).

Hall (1992) in his 'West and the Rest' discusses the Western as rational and civilised, and Other as located on the dark side. Racist representations that circulate in society can be infused with good or bad intentions, like the 'good' intentions of helping subordinated groups. Ideland and Malmberg (2014) argued that repeated representations and processes of everyday racism shape how we interpret the world, using technologies of differentiation separating Us from Them, building on imperialistic historical narratives. As Ideland et al. (2014) argue, the Other constructed as an object for inclusion and being helped, and Us constructed in a privileged position of helping and inclusive, resonates with debates in international development. However, Giroux (1992) argues the effort to include actually becomes a promotion of the colonial gaze rather than a displacement of it.

Staszak (2009) proposed that otherness and identity for Western society focusses relatively more on binary logic than other societies, using dichotomies such as Black/white, believer/non-believer, and colonist/native. He argues the "anthropologist fiction of races and geographical fiction of continents" allows these categories to be reified and naturalised into a false sense of evidence using these binary forms of otherness. He drew parallels with Said's *Orientalism* (1978) in which the West gains the right if not the duty to dominate the Orient, and to save it from despotisms, superstition, misery, vice, slavery, and decadence. Not all forms of otherness are geographical because stigmatisation to something other than location, such as women and LGBTQIA+, is also prevalent.

Societies around the world may be structured differently in how otherness manifests – concepts of individualism (Binder 2019) and collectivism (Singelis et al. 1995; Binder 2019; Green et al. 2005; Alt et al. 2018; Triandis 1995) have been integrated into this debate, such as conformity with group norms and desire to be part of a collective. Collectivist cultures are found to make strong distinctions between in-groups and out-groups (Triandis 1995), because in-groups refer to people sharing similar beliefs and interests that typically exclude outsiders (Chen and Li 2005) and so influence compliance with social norms and limit self-interested behaviour (Alt et al. 2018). Bounded morality can mean that ethical standards are restricted to in-groups and opportunistic behaviour morally acceptable outside these groups (Kyriacou 2015). Tabellini (2008) provided evidence of the impact of trust in societies and the level of economic development in this context. Individualist societies place less emphasis on in-group and out-group distinctions, have higher levels of generalised trust, take more responsibility for their actions, and are less sensitive to social pressure (Alt et al. 2018); thus individualised countries with greater generalised trust and tolerance of others tend to have greater income redistribution and lower net inequality (Binder 2019). However, this individualism–collectivism dimension is not entirely clear-cut, because an individual can actually access both, be independent and unique and have a duty to an in-group and maintaining harmony (Tomasello 2020).

Ideland and Malmberg (2014) argued that Otherness is equally problematic in more 'positive' contexts because it obscures the diversity within those fabricated groups. Homogenisation is indeed a form of social stereotyping focusssing on salient cultural traits giving iconic statuses (Sharma 2019). Good intentions to 'save the world' therefore demand critique (Ideland and Malmberg 2014), and development professionals need to refrain from creating simple, stagnant narratives of who the other is (Ardendt 1958/1998; Matthiesen and Klitmøller 2019). The Other, reduced to one's own terms and parameters, reflects a dominating and destructive orientation where meaning is ascribed on the basis of self-perception alone rather than in light of the Other and their multiple identities (Deifelt 2007). Ardendt (in 1958/1998 in Matthiesen and Klitmøller 2019) pointed out that when we want to say who somebody is, we tend to say what he is in graspable categories. The relevance of such concepts for travel philanthropy, closely associated with 'encountering the stranger' and the ethics of international aid interventions in Africa, certainly demands an inquiry.

Tourism involves the mobility of people, cross-cultural encounters, and produces social spaces where disparate cultures meet and, some suggest, breaks down boundaries (Jaworski and Thurlow 2010). Within global tourism, the traveller may have the privilege to move to a faraway destination and consider local people and their cultures static. However, it is very important to note that both the tourists and local people experience mobility as they encounter differences, not just the tourist (Sharma 2019). For instance, cultural interaction involves an exposure to different views (Deifelt 2007).

Global tourism is said to be the "quintessential business of difference projection and the interpretive vehicle for Othering par excellence", reproducing stereotypes and categories of ethnic and cultural difference across the world (Salazar 2013:690). Such debates result in travellers presented as an authority to give judgemental accounts of otherness, under the pretence of equality and tolerance (Lisle 2006). In mass-mediated travel products, such as travelogues, othering discourses are created in which an understanding of the world is constructed by those who produce and consume, representing the tourist vis-à-vis the local 'other' (Sharma 2019). In these, the desires, expectations, and worldviews of those being viewed are still rarely reflected (Sharma 2019).

Whilst anthropology is said to have shifted from presenting cultures and societies as passive, bounded, and homogenous entities of Western constructions of Otherness, tourism stakeholders have stepped into their place by objectifying, reifying, homogenising, and naturalising people in destinations, claiming imagined identities and belonging based on place and locality (Salazar 2013). Globalisation has made national and cultural boundaries 'more porous than ever before' but common stereotypes about otherness continue to shape the travel narrative and perpetuate othering discourses (Sharma 2019). Earlier anthropological accounts provided the template for current cultural performances in tourism, as tourists continue to demand 'difference' (Salazar 2013) and Western homes of souvenirs are now becoming display cabinets for modern-day collections (Sharma 2019). Therefore, tourism continues to echo previous tones of anthropology and is still a key instrument for the West to exert control, reproducing colonial ideologies and power hierarchies that maintain privileged ideological positions (Sharma 2019). Westerners increasingly feel entitled to 'visual consumption', such as demarcating "'real, authentic locals' from everyone else" in the "democratisation... [of] the tourist gaze" (Urry 1992:177). In tourism the 'voyeuristic gaze' that represents the 'other' assumes superiority of the travellers' cultural and moral values (Sharp 1999) and ideological domination of the ways the West sees the Other continues (Sharma 2019). In other texts the gaze of the privileged and powerful is said to objectify and trivialise racial identity and to ignore the ways these identities are produced and reproduced through practices of white privilege (Jungkunz and White 2013).

Discourse around representations of otherness has a long history (Said 1978; Hall 1997; Foucault 1980; Castoriadis 1987; Taylor 2004). Said's (1978) seminal work shifted thinking towards critiquing the production of popular representations of 'Others'. Said introduced the idea that difference is understood as inferiority, and a product of social and historical beings unable to see the world through disinterested eyes (Argrou 1999). Representations of otherness in tourism and through the media remain powerful at shaping our imaginations in subtle ways (Jedlowski and Thomas 2017). Staszak (2009) argued that the 'exotic' unknown is really something that has already been identified and marked as picturesque and reproduced for bland consumption. It is less about confronting otherness than having the satisfaction of experiencing the sight of a reassuring version of this confrontation, true to our fantasies, comforting us in our identity and superiority (Staszak 2009).

The gaze of the 'Other' towards the West is rarer (Jedlowski and Thomas 2017). However, by performing identities of 'local' and 'traditional', locals may retain agency to appear different and exotic to meet tourist imaginations motivated by economic gains or resist identities imposed by tourists (Sharma 2019). Local tour guides, as part of the attraction, have learnt that they need to 'exoticise' themselves and present a simplified version of the self that confirms the Western, popularised idea of Other (Salazar 2013). Exoticism has become a significant resource for many countries (Staszak 2009). Salazar (2013) suggested that local tour guides provide a symbolic representation as anthropologists did, but rather than be outsiders disclosing inner cultures, local tour guides use their relative societal position as insiders and gatekeepers to establish contact with interested outsiders. He proposed that there is a growing trend of local people Othering themselves, ironically recycling colonial images and the homogenous, bounded, and unchanging idea of cultures, and promoting images of themselves as anthropological Others with the goal of obtaining "a piece of the lucrative tourism pie" (Adams 2004:124; Salazar 2013). Representation is therefore not a 'unidirectional act', because at times the locals may repackage their identity (Sharma 2019) and individuals have the choice or agency to create their identities based on their own beliefs about the world, but in an encounter of 'others', the negotiation of identity depends on the negotiation of power relationships (Zevallos 2020; Deifelt 2007). Otherness is an event, and it is unpredictable how one human being will affect another because they are surrounded by "a plurality of other things-in-the-world". Thus, sameness and otherness go beyond classifications (Rapport 2015:3).

7.1.4.2 Sameness

There are two angles of 'sameness' highlighted in the literature that are relevant to this study: one is the meaning of sameness as a personal or collective identity and the other is the emergence of sameness in terms of imitation or copying others. As well as people being considered the 'same', *sameness* has been understood as a term that relates to personal identity, allowing for less rigid criteria (Olson 2004). Olson (2004) suggested that identity is a contested anthropological term, but an object of inquiry about who we are in relation to others – on the one hand, sameness with oneself, and the other, collective distinctions of otherness. Identity is found to mean different things such as social self, cultural self, cognitive self, or narrative self (Klein 2014). Ricoeur (1981) defined *self* as permanent characteristics, thus sameness over time. Collective identity is about individuals having shared perceptions about their group's distinct and enduring characteristics and, in organisational literature, for instance, identity is considered in terms of clusters, shared mindsets, social milieu, and sense of belonging (Staber 2010).

Human beings are formed by culture and also shape cultures, and cultures undergo constant change from information gathered through encounters with others

(Deifelt 2007). According to Deifelt (2007:114), in this globalised world cultures are fluid and blend, resulting in new 'cultural hybridity' with blurred boundaries between sameness and otherness constantly crossed, influenced by the past and present. They argue that cultural studies emphasise sameness and increased homogeneity, but such an obsession with identities has meant differences and multiple, constantly changing beliefs get ignored. Individual identities – although we may not notice such changes in the present moment – are in constant flux (Stern 1991) and continual evolution (Shabad 2010). In evolutionary cultural theory, cultural traits are passed on across generations as socially learned norms and behaviours (Richerson and Boyd 2005; Smaldino 2014), and so cultural change accumulates in human populations over time (Caldwell and Millen 2009), seen in individual behaviour, as well as group traits not found at the individual level (Smaldino 2014). The role of narrative in this cultural evolution is an underestimated mechanism linked to sameness (Smaldino 2014; Fastenrath et al. 2018).

Imitation, as part of sameness, is a social process in which future actions are selected by imitating others (Banerjee 1992); it is a commonly observed behaviour in human decision-making (Schlag 1996) and more recently part of socio-economic reasoning debates (Staber 2010; Apesteguia et al. 2004). Most studies focus on imitation amongst children (e.g., Whiten et al. 2009; Wood 2020). However, there is some wider recognition – for example, investigating cultural advantages and disadvantages in complex environments (Apesteguia et al. 2004), the role in business behaviour (Lieberman and Shigeru 2006), armed conflicts (Schlag 1996), consumers and prosocial behaviour (Babutsidze and Chai 2018), and organisational ethics and management (Durand and Calori 2006).

One theoretical framework for understanding imitation amongst human beings is 'mimetic desire', which was proposed by Rene Girard (1977) as a philosophical concept to suggest people tend towards the same objects of desire. Plato first used the term 'mimicry' in the context of poetry, and mimesis has been used in the context of art (Halliwell 2002). Humans survive and thrive not by thinking for themselves but by copying others, especially those they respect; thus, rather than self-reliant innovators, humans are savvy conformists striving for social rapport (Wood 2020). Mimetic theory, outlined by Pearson (2020), poses that human desire is not an autonomous but collective process, and as more people want something and that object remains scarce, it eventually leads to conflict. Girard (1977) argued that all conflict, competition, and rivalry have their origins in mimetic desire.

Various theories relating to *conformity* contribute to the sameness discussion. Conformity is defined as yielding to group pressures (Crutchfield 1955) and three main types of conformity have been distinguished as compliance, identification, and internalisation (Kelman 1958). Conformity has been understood in prosocial behaviours, such as charitable giving, and negative behaviours (Sleek 2016), as well as conforming to expectations of a social role, such as a police officer (McLeod 2016). Mann (1969) added 'ingratiated' to the concepts, a form of conformity to gain favour and acceptance from other people. Similarly, Maness and Cirillob (2016) researched social influence vs. informational conformity, social influence referring to imitation providing direct benefits such as popularity. Informational conformity is when one person lacks knowledge and looks to a group for guidance (Deutsch and Gerard 1955), and there are examples of how policy affects this behaviour (Maness and Cirillob 2016). Culture is also found to influence desires to conform to social pressure or remain independent of the group (McLeod 2016). Conventions are rules that guide behaviour and self-enforcing norms, producing regularity in behaviour (Koumakhov 2009). To be 'social', according to Weber (1922) an actor has to take into account the actions and reactions of others, and there is a difference between a social action where objective opportunities are realised through observing others, and only imitation (Schlaile et al. 2018). In business, while Staber (2010) found that clusters or dense personal social networks have self-reinforcing effects in the development of common strategic dispositions, Siltaoja and Lähdesmäki (2015) established that network density causes actors to become similar, which then constrains the actions of that business.

Richerson and Boyd's (2005) influential concept of imitation identified *frequency-based imitation*, when individuals or organisations copy ideas and behaviours most commonly adopted in a relevant population. Herd behaviour, information cascades, and bandwagon effects are common terms associated with this (Banerjee 1992). Adopters of an idea are driven by the number of previous adopters, creating a low-cost device that does not require extensive cognitive and social investments to collect new information (Staber 2010). Fastenrath et al. (2018) built upon this theory in a corporate context with 'unquestioned myths'. Furthermore, Staber (2010) argued that when people face high levels of uncertainty, they tend to do what others around them are doing and highlighted that this kind of imitation can lead to instability, fashions, and fads (Bikhchandani et al. 1998). In contrast, *trait-based imitation* involves copying others that have a specific trait and requires more deeply social interactionist practices and stronger cognitive abilities (Staber 2010). Social learning is a key area of the literature incorporating both *imitation*, in which an individual learns by copying the action of others, the form of an action, and *emulation*, where they learn and copy *the end result* of a behaviour, or tools involved (Whiten et al. 2009; Caldwell and Millen 2009). Social learning is learning within a social group rather than an individual learning from a single person (Schlag 1996).

As part of the imitation process, role models align with the theory of social learning (Bandura 1977), which involves motivational consequences of observing another individual. Role models have been defined as individuals who influence role aspirants' achievements, motivation, and goals by acting as behavioural models, representations of the possible, and/or inspirations (Arjoon et al. 2018). They can be positive or negative: *positive* motivates similar excellence and *negative* motivates the avoidance of similar distress (Koumakhov 2009). Research has indicated greater power of positive role models over negative, but cultural differences have also been highlighted, such as the Western vs. Eastern, individualist vs. collectivist dichotomies (Lockwood et al. 2005). Evidence suggests that role models can be a positive source of social influence (Arjoon et al. 2018; Babutsidze and Chai 2018) and can be problematic (Thieme 2013). Research in Kenya found that, in the absence of teachers or parents, many of the youths' role models were criminals or local gangsters, representing entrepreneurial characteristics, the self-employed, powerful, and 'cool'. In business, research on strategic leadership in driving ethical behaviour also contributes to this debate (Thomas et al. 2004; Chen 2012), including the influence of storytelling and expectations (Liu et al. 2019; Fastenrath et al. 2018; Bandura 1977). Baden (2014:8) researched the effect of role models from expectancy, proposing self-fulfilling prophecies that "if we believe others will behave opportunistically and further their self-interest at our expense, this can give rise to pre-emptive self-serving or unethical behaviour". They call this 'defensive ethics', the justification for violating rules because we expect others to do the same and everyone does it.

'Sameness' in all its forms evidently influences the behaviour of humans and therefore influences systems. Theories of system dynamics see agents adapting and co-evolving (Mingers 2014), and in international development this growing awareness of constant change in systems caused by people adapting to what others do, rather than by market forces (Mueller 2020), emphasises the relevance of building upon such theories to learn systems better.

This chapter has broadened the theoretical perspective framing the work presented in this book beyond development and tourism studies, to understand socio-economic systems more extensively. Key theoretical areas that were highlighted as particularly important, in the context of this research, included reasoning and bounded rationality, such as emotions in decision-making, wilful blindness, and the ontology of ignorance, as well as bounded roles and responsibility. Theories regarding beliefs, principles, norms, expectations, and stories were also interlinked, as well as otherness and sameness, all emerging as influential domains from the study's empirical findings. Chapter 8 will now discuss core findings of the empirical research which revealed travel philanthropy as an emergent exchange economy.

8 Travel philanthropy

An emergent exchange economy

This chapter introduces the underlying elements of the travel philanthropy system across case studies contributing new understandings about its emergent structure, including a common set of beliefs from which the system was found to be based. A framework of principles and norms that enable exchanges in the travel philanthropy system is discussed, including for example, informal rules guiding exchange relationships. The importance of understanding that both tangible objects (i.e., money and technology) and intangible objects (i.e., emotions, beliefs, and stories) core elements required for the system to function is explained. Furthermore, without certain core beliefs, it is argued that the travel philanthropy system would not exist and understanding its underlying elements as an emergent exchange economy is essential to understanding the system presented in Chapters 9 and 10.

8.1 Introducing the travel philanthropy system: underlying elements

This chapter advances extant understanding of travel philanthropy by focusing on other underlying elements of the system that contribute to its emergent structure, beyond actors and roles. On the surface, it may seem that tangible objects, such as money given and received in exchanges, are the most important elements of the system. However, as shown in Figure 8.1 intangible objects are equally influential in the existence and functionality of the system.

Intangible objects include beliefs, stories, emotions, and services. Technology also emerged as a key element in the system, enabling many of the relationships between actors and exchange pathways, such as money transfers and communication. Finally, inequality, intended as perceived need, is another core element required for the system to function.

This chapter (and following Chapters 9 and 10 which present the emergent travel philanthropy system structure) draws on Meadows' (2008) definition of a system as an interconnected set of elements coherently organised in a way that achieves something more than the sum of its parts, and Sherwood's (2002) direction to look beyond details to the context of relationships in which they are embedded. Travel philanthropy is not a system presented as ontological phenomena. It is rather a way of thinking about the system as an epistemological construct (Gates 2016). Through a critical realism lens,

DOI: 10.4324/9781003497929-8

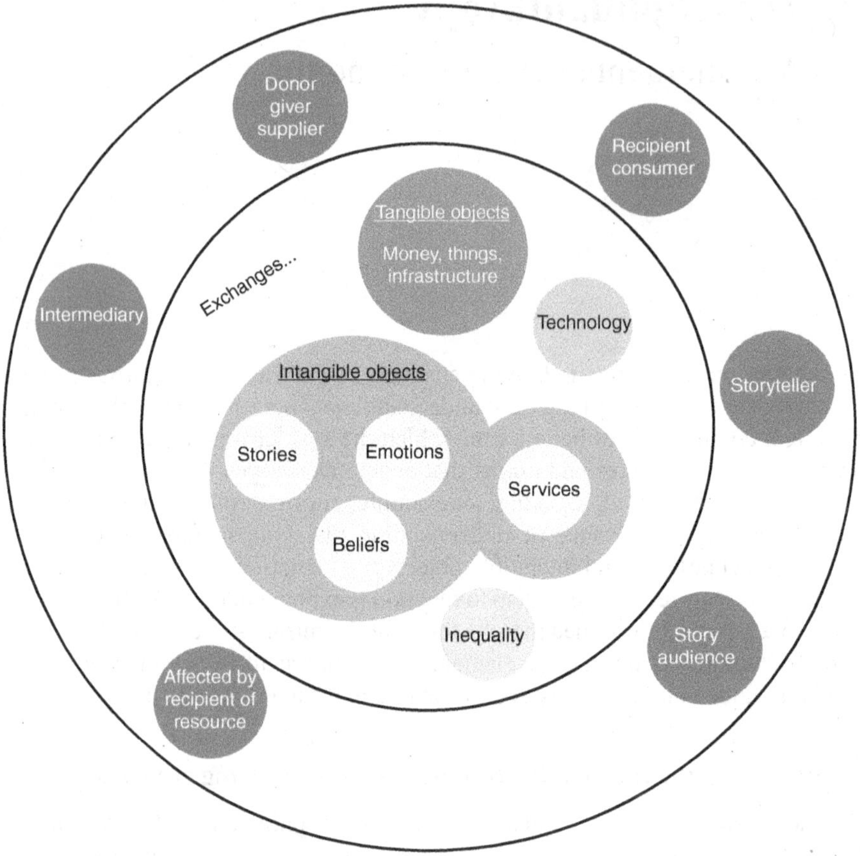

Figure 8.1 Core underlying elements for the travel philanthropy system to function.

this chapter investigates the ontological nature of the observable and unobservable mechanisms that cause the travel philanthropy system (Mingers 2014), with our study taking the position that systems thinking and complexity science can challenge and transform the way travel philanthropy is conceptualised (Gates 2016).

8.2 From a set of beliefs to the emergence of the travel philanthropy system

Beliefs and informal rules emerged as a kind of glue in the system. Informed by Bhaskar's (1975) critical realism, our study shifted from focussing on the epistemological belief that knowledge is true, towards the ontology of beliefs and their causal properties. Whilst the content of a *belief-in* or *belief-that* (Schwitzgebel 2006; MacIntosh 1994) may be true or not, such as *that giving is good* or *in my emotions*, it was found that a belief nevertheless generates effects in the system, and it is real. Beliefs may be hard to measure, and not always explicit, but our study highlighted that the conduct of actors from across interviews emerged to be rooted in their underlying belief system (Usó-Doménech and Nescolarde-Selva 2016). Our study further drew attention to an important set of beliefs, principles, and norms providing the grounds for a travel philanthropy exchange economy to exist.

Travel philanthropy is an emergent socially shared belief system that has been produced by actors interacting over time (Homer-Dixona et al. 2013) and space. Similarly, the travel philanthropy belief system has an impact on our world, as a religious belief system would (Usó-Doménech and Nescolarde-Selva 2016), and so together these beliefs have a causal role in the production of behaviour in travel philanthropy, and in the wider world (Fodor and Pylyshyn 1981). Without challenging certain beliefs, the system will continue to behave in certain (questionable) ways and produce the same patterns that may not necessarily produce positive impacts in the short- or long-term. We therefore need to understand this structure as part of understanding the system.

The travel philanthropy belief system provides a framework to investigate some of the mechanisms of decision-making, and to shape further understandings of this complex social phenomenon, enabling a micro-orientated approach (Dornschneider 2019). The research evidenced that beliefs are not formed by reason alone (Dunning 2012), but they may be *antecedents* and *consequents* of people's decision-making and behaviours, with certain beliefs further triggering others (Dornschneider 2019). Our study specifically focused on *core beliefs*, meaning that without one of them, the system would not exist.

Six themes emerged from our study in merit to beliefs – 'that' and/or 'in', which are core to the travel philanthropy system (see Figure 8.2 and Table 8.1):

1. That giving is virtuous
2. In exchange relationships
3. In money and social power
4. In equality and change
5. In the right to gaze at Others
6. In emotions

Individually these core beliefs are not exclusive to the travel philanthropy system; they are very much relevant in the broader socio-economic, gift-giving, and international development landscape. However, when combined, interconnected and viewed as a whole, this specific set of beliefs generate the travel philanthropy system. Travel philanthropy is a living system and as Maturana (1980) indicated, when a belief changes the whole system changes. For example, if a belief in the right to gaze at others changed, the travel philanthropy system would change too – i.e., if it were not considered a right, or not deemed an unethical act, then tours to schools, marginalised groups, etc. would probably not be facilitated; thus, exchange pathways would shift. As a local participant in Zimbabwe noted:

> *In peak season you would find that the tourist is basically on the ground and almost every day coming to view our area, going to school.*
>
> (Z2_1_B_INS_38)

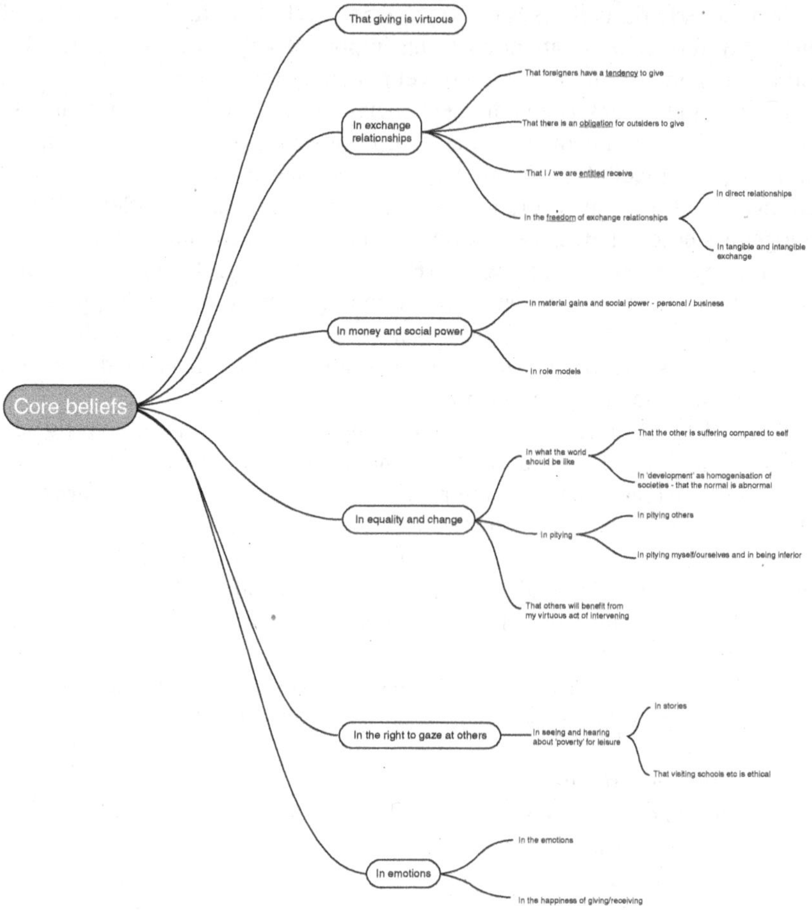

Figure 8.2 Core belief system.

By analysing beliefs as causes or consequences of behavioural norms for instance, it became obvious that a change in one belief affects the system as a whole (Maturana and Varela 1980). Eliminating one element (belief) from its structure would change the meaning of travel philanthropy.

Beliefs are therefore key leverage points in the system. This aligns with the Usó-Doménech and Nescolarde-Selva's (2016) argument that if the system is to be changed, or if the system changes organically, considerations are needed in merit to the cause and effect of different beliefs, the structure of norms linked to the belief system, the influences on actor behaviour, and thus system dynamics.

Figure 8.2 and Table 8.1 provide an overview of the travel philanthropy core belief system that has emerged from our study. These beliefs are mechanisms traced to the domain of the real (Bhaskar 1975).

Table 8.1 Core underlying beliefs of the travel philanthropy system

Core beliefs from which the system of travel philanthropy emerges: why the system exists			
Core beliefs		*Sub-beliefs*	*Indicative quotes*
That giving is virtuous	Gift-giving, donations, sponsorship, and other terms used for transfers and transactions in the travel philanthropy system are fundamentally based around the belief that 'giving is good' and that 'donating is a virtuous act'.	**n/a**	"It's a very beautiful thing" (U3_TB_16); "a beautiful human instinct" (Z1_NGO_21); "kindness" (EN_TB_70); "it's just from being good, from good heart, I have no objection about being good to help, they think it was for the right purpose…" (E2_NGO_57) "it's a good thing to help then not helping" (E2_B-COM_64)
In money and social power	Travel philanthropy is concerned, partly, with economic transfers and transactions, money as an object of exchange, as well as other tangible and intangible objects. A belief in money and in social power that are often associated with each other is very much a core element of the system. The belief in money as a solution to a goal underlies many decisions.	**In material gains and social power**	"people want money, there are so many needs that they have to meet, there are so many associations they have to join and they don't have money and they have to use this kind of donations, looking for funds for needy people to kind of enrich themselves" (U2_CBO-F_8)
		In role models	"they want to be rich… now people love instead of long way they take the shorter way to be, just because of the following of examples" (E2_TB_56)

Table 8.1 (Continued)

Core beliefs from which the system of travel philanthropy emerges: why the system exists				
Core beliefs		*Sub-beliefs*		*Indicative quotes*
In exchange relationships	There are four identified influential sub-beliefs that constitute the overall belief *in exchange relationships* for the travel philanthropy system specifically. In terms of, freedom of exchange relationships, bracketing whether the traveller is donating due to compassion or not, travel philanthropy is based on the belief that humans are free to exchange. If an individual would like to exchange with another individual, directly, or indirectly, if one would like to travel to a remote corner of the world and exchange with another human being in that place, one is free to do that. This is an informal system of exchange built upon these beliefs.	**That foreigners have a tendency to give**		"whenever they see me talking to the tourists they are seeing some light they know I am going to convince these people to do something for them" (Z2_1_B_INS_38) "People give them money! You know you shouldn't just be able to do nothing and receive something" (U1_TB_3)
		That there is an obligation for outsiders to give		"people here have the perception that people from outside [tourists] have just things, they have more than enough, they have more than they need and they should give those things to people…" (U2_INTER_12)
		That I/we are entitled to receive		"there is some kind of culture, anyway growing here that tourists or even foreigners, have to give something so it's like anyway people have that feeling that they should be given" (U2_CBO-F_8)
		In the freedom of exchange relationships	**In direct relationships**	"a tourist can easily come to school then sees the children walking barefoot and then he says I'm going to buy them shoes, I mean he really feels compassion for their wellbeing then

			In tangible and intangible exchange	he donates shoes or uniform or even a water tank if he sees them carrying water on their head" (U2_CBO-F_8) "some tourists give so they think every tourist can give so they are always expectant so its a culture" (U2_CBO-F_8)
In equality and necessary change	A belief in the existence of 'inequality' and in necessary change, that someone else's normal is abnormal. This is a deeply rooted belief based on 'othering', which is a significant element of the travel philanthropy system, and without this belief inequality would not be recognised. Beliefs which contribute to the structure of the system are therefore not only concerning intentions to give – the virtuous *act*, but also driven by what the outcome is intended to be, or the belief in a need for change. This tends to be in respect of subjective judgement of the Other.	**In what the world should be like**	**That the Other is suffering compared to self**	"she cried for maybe 3 minutes, she said wow, that people are really poor she cried" (U3_CBO-F_15)
			In 'development' as homogenisation of societies – the normal is abnormal	"they [tourists] feel that thing which I know is a normal thing for us" (UU2_INTER_9G)
		In pitying	**In pitying others**	"when they see this child is sitting on the floor it makes a huge difference" (Z2_2_TB_44)
			In pitying myself/ ourselves and in being inferior	"you find that like maybe you're begging or what, so to beg is to beg, because if you have money you can't beg, so you feel you have that inferiority complex in your heart" (U1_CBO-IN_4)
		That Others will benefit from my virtuous act of intervening	**n/a**	"we've chosen our place, we've chosen our guy or girl, we like their institute, whatever it may be, church, clinic, orphanage, school, whatever, we're looking at donating this much, what do you guys think is the best way to go about it?" (U1_TB_3)

(*Continued*)

Table 8.1 (Continued)

Core beliefs from which the system of travel philanthropy emerges: why the system exists				
Core beliefs		*Sub-beliefs*		*Indicative quotes*
In the right to gaze at Others	Exchanges tend to be based on and inspired by the gaze at 'poverty', 'inequality', or the lives of 'Others', and core to the system is a belief in the right to gaze at Others.	**In seeing and hearing about 'poverty' for leisure**	**In stories**	"sometimes people are interested to know how the wellbeing, how people live here" (U1_CBO-F_2)
			That visiting schools etc. is ethical	"in peak season you would found that the tourist is basically on the ground and almost every day coming to view our area going to school…" (Z2_1_B_INS_38)
In emotions	Emotions tend to be considered a sufficient decision-making mechanism for the behaviour of all actors in the system, whether donor, intermediary, or recipient for example. A belief in my emotions, whether happy, sad, or surprised for instance, are also believed to constitute ethical grounds for exchange, and decisions, and this contributes a unique aspect of this system. Whilst shock and pity are often tactics or reasons within exchange relationships in the system, fear has causes and effects, and positive emotions are arguably very influential in causing the existence of the system.	**In my emotions**	**n/a**	"it all depends on someone's heart, feeling of what's going on" (Z2_2_TB_41) "I don't know you know the children have a special approach to change the mentality of the visitors" (EE2_NGO_61LPI)
		In the happiness of giving/receiving		"it's good for them to bring and give to the people direct … people feel happy, when they see them giving them something, they feel happy" (U3_CBO-F_15) "I gave them to the right people so that man became happy, the children became happy, I also became happy" (U1_CBO-IN_4)

For Global North and South stakeholders, our study indicated that moral claims regarding the importance of donations continues (Stirrat and Henkel 1997; Singer 1972). Travel philanthropy steams ahead as a system encouraging donations in Africa, driven by the assumption that 'giving is good' and the power of the moral imperative to 'assist those in need' (Von Bergen and Bressler 2017:114). These are significant in driving this as a niche segment of contemporary philanthropy, with acts of giving hailed as 'symbols of virtuousness', remaining predominantly immune from critique. That giving is virtuous is indeed a core belief of travel philanthropy, dominating most participants' perspectives too, notwithstanding its entanglement within dilemmas. People rarely dare to question well-intentioned giving or bring attention to 'flawed practices' (Von Bergen and Bressler 2017:114). Moreover, if critiqued, people do not consider the idea of not giving at all a solution, aligning also with Berry's (2014) thinking.

In our empirical assessment, whilst participants debated consequences of giving, they rarely suggested that giving per se should stop. That is, if actors in the system did not believe in the virtue of donors giving tangible objects (e.g., money, books, clothes, buildings, etc.) to recipients, they would unlikely publicly participate as they normally do, for example, through ceremonial gift-giving handovers directly involving tourists (Honey 2011). If giving was believed to be immoral, or equally if such a belief was absent from the travel philanthropy belief system – thus supporting Mingers (2014) in the importance of using critical realism's ideas of 'absence' in systems thinking, the travel philanthropy system would not exist.

Similarly to Von Bergen and Bressler's (2017) approach, the aim of our study was not to state if giving is good or bad per se, but rather to understand the conditions through which helping may negatively impact others, an issue gone unrecognised in travel philanthropy scientific inquiry. If locals are brushing aside the negatives of philanthropy for fear of discouraging future donations, or enthusiastically facilitating donations without questioning possible negative outcomes, or if donors intend to be good persons by giving, then 'doing something' does not mean 'doing it well' by informing oneself of ways to do good effectively and ways to effectively avoid harm (Schmidt 2017). This dilemma was captured in an Ethiopian participant, who explained:

> *The money is very important for the local community, we are afraid, and the channel is difficult, but we have to do a better one.*
>
> (E2_NGO_61)

The I/we are entitled to receive belief is also the root cause of behaviour and exchanges within the travel philanthropy system. That 'Africans should receive' or that 'I personally should receive' emerged also as beliefs at the foundation of the systems' existence and as a norm amongst both recipients and much of the tourism industry:

> *There is some kind of culture, anyway growing here, that tourists or even foreigners, have to give something, so it's like anyway people have that feeling that they should be given.*
>
> (U2_CBO-F_8)

> *They should also benefit from tourism through donations through scholarships.*
>
> (U3_TB_17)

This 'entitlement' stance that pervades the system by recipients and donors is very much part of global norms, emerging from how the African continent is portrayed through the media (Ayittey 2015; Harrison 2013). This has been apparent throughout the researchers' extensive experience and observations during their involvement in academic, business, and consultancy practices in the continent over the last 20 years (see for example Novelli et al. 2015; Novelli 2015), and consistently on the rise as tourism in Africa grew.

The belief system underlying travel philanthropy affects decisions and types of development interventions, similar to how beliefs and assumptions by managers in development organisations influence a programme (Moore et al. 2019). It is important to acknowledge, as argued by Green (2016), that deep-rooted beliefs and norms can block change, so understanding this underlying structure is critical. The remainder of this chapter introduces non-core beliefs and the informal rules identified in our study that shape how the system functions.

8.3 Principles and norms framework – the unwritten rules enabling exchange in the travel philanthropy system

To understand the system, relevant beliefs, norms, and principles need to be captured, as much as the 'rules' and assumptions about the right way to act. Assumptions are 'rules of thumb' that influence choices and reflect deeply held beliefs, norms, values, and ideological perspectives (Vogel 2012). Our study looked at rules that actors may take for granted, not judging that they were correct, good, bad, right, or wrong, but that they were identified at the 'domain of the real' (Bhaskar 1975). Although socially constructed, they contain causal properties that are part of the dynamics of the system, with the ontology of these assumptions being important for our study.

There are regulating principles and norms, which have been previously identified as 'appropriate behaviour' (Cold-Ravnkilde et al. 2018) and are a set of informal rules that guide behaviour in travel philanthropy without the forces of law (Mahmoud et al. 2014). In order to transcend simple rationality (Taylor-Gooby 1998), identifying a normative framework could contribute to a better understanding of behaviour in the travel philanthropy market contexts, because this system does not normally implement laws easily (Baron 2018). Rather, informal rules are what shape the obligations and permissions that an individual has to this social system (Boella and Torre 2007). Principles represent values, beliefs, and attitudes about the way things are done appropriately (Manz et al. 2008), and norms are coordinated individual principles that refer to what ought to be done (Henderson 2002).

Table 8.2 provides a mapping of key unwritten rules that emerged from our study, highlighting also a clear overlap between principles, norms, and values (see Table 8.3). These have not been separated as it was unnecessary for the purpose of our study but are suggested as an area for future research.

The rules identified here are emergent properties of the system from social learning (Sen and Airiau 2007), with the informal rules of travel philanthropy reflecting mutual influence (Legros and Cislaghi 2020) embedded through processes of feedback. The more regular a certain behaviour has become – i.e., giving in Africa or getting a commission as an intermediary – the more individuals believe that there is a norm to comply with, making the behaviour widely accepted, normalised, and legitimised (Suchman 1995). However, moral legitimacy should be understood from multiple realities (Ehrnstrom-Fuentes 2016), and our study

Table 8.2 Informal rules enabling exchanges

Informal rules framework: Core principles and norms enabling exchanges	
Rule category	*The rule*
Interventionalism and private philanthropy	Entitlement to intervene
	Emotions can decide
	Us vs. Them – the norm of Othering
	Giving is better than not giving – but the meaning of the donated object and its pathway differs
	Responsibility is bounded
Mixing business and philanthropy	Business and helping
	Poverty sells (but indirect)
	That's good enough
	A proportion buy-in
Legitimacy – actors and roles	Anyone can give/not give, if they want to
	Anyone can be an intermediary
	Donor should decide
	Power can be used
Social learning	Imitating
	Othering

Table 8.3 Informal rules framework: core principles and norms enabling exchanges

Informal rules framework: core principles and norms enabling exchanges			
Rule category	*The rule*	*About the rule*	*Indicative quotes*
Interventionalism and private philanthropy	**Entitlement to intervene**	Any 'could-be donor' with the means to give is entitled to intervene in an African community. If one chooses, one has the right to decide whether to give, and to whom, what, where, when, how, and why. Sometimes formal restrictions are a constraint e.g., Government, traditional authority, but a rare barrier to the intervention entitlement. This principle guides the *act* of giving in Africa and should be judged in its own right, rather than on potential consequences of that act. The responsibility of the donor/ giver is generally bounded to giving, mainly because the act of giving itself is presumed 'virtuous' and 'good' in itself. Thus, the system is built upon the rule that private philanthropists are entitled to intervene in African communities without needing to understand the consequences of that act, due to the virtuous nature of giving in itself.	"if it is their call to do something" (EN_TB_68) "you can't say no because that's their pocket" (E2_TB_58) "it's not bad giving the money based on their interests" (E2_TE_59) "we can't say no, you cannot do this and so on, so we don't say so because after all we don't have any legal mandate to stop any client donating any given person" (EN_TB_65) "we had no right to say no and also happy that they have" (Z2_2_TB_41) "they [tourists] are going to see how the kids are taught and then afterwards they have gifts for the children, I don't know what the gifts are but they have gifts" (U3_TB_16)
	Emotions can decide	From compassion, to happiness, surprise or fear, when triggered, emotions are sufficient reasons to guide decisions about whether to give or not, as well as to whom, what, where, when, how, and why. The rule is: • Compassion can supersede planned strategies • Judging a cause can be personal and subjective • Evidence-based giving decisions not the norm.	"they didn't come with a purse of giving ... they were just doing community walks ... they were inspired by the situation and circumstance" (U2_CBO-F_8) "they are really like sad you know and then they wanted to help" (E1_B-COM_69) "people are touched, so that is better to give directly for the kids" (E2_TB_58)

	Us vs. Them–thenorm of Othering	The system is built upon an assumption that it is right to question and judge the lives of Others, often directly between two parties, and that to ask the needs of others, acting as a kind of hero or saviour, is a normal, moral, and an acceptable act. The principle is that there is an *Us*, and there is a *Them*, which sits at the heart of such interventions. If Othering was unacceptable the interventions and system would not exist.	"they feel touched and feel people are disadvantaged because of the way they have seen the culture" (U1_CBO-F_2) "they would love to know how is the Third World and how we do our things, are we still into our culture this and that" (Z2_1_B_INS_38) "visit that village and then they ask problems, why kids they do not go to school what is the problem and they ask problems from the village" (E1_TE_52)
	Giving is better than not giving – but the *meaning* of the donated object and its pathway differs	Giving is better than not giving, this is the norm, but different principles conflict when guiding what is a *good* or *bad gift* or *object to give*, what is considered a handout or not, who is an acceptable beneficiary or not, or the type of exchange. What is a 'good' object, and/or a 'good' pathway to give is a highly confusing and conflicting part of the system. Different actors have certain principles about what is right and/or wrong. This adds to the complexity of the system.	"there are so many ways tourists from different corners of the world interpret this idea of ours, we understand that, many tourists may not see it in a positive way, we expect if I brief them ... don't donate... they may relate it to some kind of racism issues like we are black and so on, we don't go to that deep level, so we are simply giving some advice if you want to donate then you can do it this way ..so we show them the positive side of donating to the school, not the negative side of donating to the individuals" (EN_TB_65)

(*Continued*)

Table 8.3 (Continued)

Informal rules framework: core principles and norms enabling exchanges			
Rule category	*The rule*	*About the rule*	*Indicative quotes*
	Responsibility is bounded	In these private interventions, due to the above principles, many actors are led by the principle that a donor does not need to take responsibility for the succession of events that take place after the object has been introduced because their responsibility is bounded to the act of giving as opposed to the consequences. This principle applies across actor groups for various reasons – see Chapter 10.	"it's just from being good, from good heart, I have no objection about being good to help, they think it was for right purpose, but it's the people who are using it for wrong purpose, it's not their fault" (E2_NGO_57)
Mixing business and philanthropy	**Business and helping**	In travel philanthropy business and philanthropic objectives are allowed to mix. One can be helping in a charitable manner whilst at the same time benefitting, financially. This is a rule which applies to all actors in the system, businesses and CSR, intermediaries, donors, and so forth. The understanding is "I can help you profit, if you help me profit too", or "I'll scratch your back if you scratch mine". These rules shape the exchange characteristics, and system perspectives. Generally, there are principles in the tourism industry, that high end tourism must go hand-in-hand with giving to communities, especially hotels due to their permanent presence in a destination. Reasons for this are varied from responsibility to local destinations, to reassurance of ethical practice, to keeping customers happy. On a community level, business and helping is increasingly popular as an economic alternative. The rule is intermediaries benefit from playing a part in the aid chain is the norm.	"the real motivation is a number of things... wanting to have the project in place will also be wanting to benefit" (U2_INTER_12) "it becomes almost a norm now and if I'm doing sales and marketing people say oh well what are your initiatives as well and yeah you have got to have them" (Z1_TB_29) "we just believed in doing it anyway, but as I am saying to you, each time you want to apply for a concession they are like, so what else are you doing for the community" (Z1_TB_28) "guests love that, they love knowing they are contributing and for a lot of people that is one of the reasons they chose to come to [lodge]" (U1_TB_3)

		Business and helping combined principles indicate different rules to traditional development interventions in Africa, different primary objectives, and moral principles. A principle of profits over poverty means that business can be the primary priority, and the philanthropic element secondary.	"I believe they [tourism businesses] should have philanthropic programmes and projects but they should channel it through a reputable ..." (Z2_1_NGO_37) "You don't just wake up because they have chased you from the job or because you have studied and cannot get a job and you say I am starting a CBO, I mean there should be some scrutiny of who should start a CBO and why it should be started" (U2_CBO-F_8)
	Poverty sells (but indirect)	The rule is that 'poverty' can be promoted to prospective donors using business principles, such as product development, sales, marketing, and it can be a theatrical display and/or a tourist experience. However, asking directly for donations face to face is rare – rather the rules say it is morally correct to tell 'the story' through different forms of communication, but the sale is 'indirect' that recipients cannot openly name the price and must take what they are given. The donation can become a voluntary payment for a tourist experience.	"we do advertisement, I mean we don't do advertisement, we do entertainment at the lodge and sometimes you may see somebody dancing and you know he entices your eyes and you say oh I can support that child" (U1_CBO-F_2)
	That's good enough	As long as tourism businesses are doing something it is acceptable, whatever that is. Evidence regarding outcomes or impacts tends to be unnecessary. There are no rules about reaching beyond good enough for any donor, intermediary, or recipient.	"they [businesses] don't really have a plan they are not following it, they just set up some token thing and throw some money at it and then forget about it" (Z1_NGO_21) "I would say it's good on the surface" (U1_TB_3)

(*Continued*)

Table 8.3 (Continued)

Informal rules framework: core principles and norms enabling exchanges			
Rule category	*The rule*	*About the rule*	*Indicative quotes*
	A proportion buy-in	It is a numbers game to get the donor audience, for fundraisers, intermediaries, and recipients. Having an audience to tell the story is considered the key to success. The principle is that because tourists are not obligated to give, but many do. By nurturing relationships between potential donors and recipients will mean that eventually someone will like the idea of giving.	"they keep trying different people, mentioning to them different projects... can push the thing they just go with that one. So people try different people" (U2_INTER_12)
Legitimacy – actors and roles	**Anyone can give/not give, if they want to**	The owner of the object has the choice about whether to give or not, it is not an obligation. An inquiry is not required regarding the legitimacy of a donor, they can be anyone from anywhere as long as he/she has an 'object' he/she is willing to give. Thus, a stranger can give to a stranger if they decide. To put this in perspective, if a Ugandan visitor arrived in a UK village and offered a gift to a child playing in the park, or offered to pay for someone's fees to go to college, without a parent's consent, very different principles would apply.	"no-one to need by like organisations, just the tourists want to need to sponsor individually, so if somebody is to sponsor by organisations it is possible, but most of the guests or the tourists want to help by individual" (E2_TE_60)
	Anyone can be an intermediary	Exchange legitimacy can be guided by personal trust, no formal agreements is required, no formal standards guiding legitimacy of donation or intermediaries. Formal intermediaries are not necessary. To be an intermediary, and to raise money from others to allocate it to beneficiaries, is considered a respectful role.	"Do you think it's a reputable person to be donating to, do you think they money will be spent wisely" (U1_TB_3) "run by the community and has been successful for 10 years, in theory" (U1_TB_3) "they have to give the money to the local community ...but it has to be legal and it has to be well organised" (ELPI)

			"we are an NGO and the only funding we have is what they [tourists] physically give us" (ZWSSG)
	Donor should decide	This rule links to *anyone can give/not give if they want to*. The individual donor is considered the legitimate decision-maker. This includes the right to allocate their own funds, or the funds of others (if a fundraiser or intermediary) in an aid chain, including being free to interpret needs, use of resource, and predicted impact, regardless of information and cognitive capabilities. The recipient is to appreciate the decision of the donor – the donor can be guided but it is ultimately their decision.	"we've chosen our guy or girl, we like their institute, whatever it may be, church, clinic, orphanage, school, whatever, we're looking at donating this much, what do you guys think is the best way to go about it?" (U1_TB_3) "giving is all about your pocket" (Z2_1_B_INS_38) "depended on the situation of the interest of the tourist, but it is not obligatory but it is based on willingness, willingness to do something" (E2_TE_59)
	Power can be used	Performing power – there is a principle enabling the moral right to ask the needs of Others and perform/display power in this way. To promote a socially and economically higher ranking to the visited. Performing powerlessness – playing with the perception of power, using power to perform powerlessness to make the Other (potential donor) feel powerful, with upper power – promoting what you *do not* have rather than what you have. The customer is king – this business principle influences these development interventions in Africa because for businesses keeping customer happy is the key to more business – tourism businesses/guides are avoiding negative emotions of their customers (tourists), or endeavouring to enhance positive emotions of their customers – to encourage further business.	"if you give me your email so I can practice my English" (E2_TB_62) "I sell our problems, I basically sell our problems that's what I do" (E2_TB_62) "someone doesn't have a project but I mean rents a house and then writes a very big poster and says orphans and vulnerable children, and then makes a website, makes a Facebook page then starts selling the project to the donors" (U2_CBO-F_8)

(*Continued*)

Table 8.3 (Continued)

Informal rules framework: core principles and norms enabling exchanges			
Rule category	*The rule*	*About the rule*	*Indicative quotes*
Social learning	**Imitating**	The rule says if they can do it, I can do it. Principles of mimicking cause the behaviour of different actors, whether businesses, local individuals, or tourists for example. Imitation of positive and negative behaviours are included but copying is a key principle – follow others. See Chapter 10.	"people have learnt to get money easily ... the same kind of story" (E2_B-COM_64) "They have seen the guides who did maybe the same ... they have seen they are well off, they have seen them leaving the poverty status, and they think when they do the same, they get their commissions and become a bit rich, so they want to do it after copying from others doing it" (U2_INTER_12)
	Othering	It is the norm to actively differentiate social groups, othering oneself or defining the Other by chosen social categories. Then guiding behaviour towards different actors based on this. This economy is built on the principle of social difference and exchanges are shaped by othering. For all actors. Principles include fundraising from outsiders are good, wise locals get money off foreigners, outsiders have enough should give to locals, you are different – stereotypes and you lack/we lack, etc. See Chapter 9.	"people here have the perception that people from outside have just things, they have more than enough, they have more than they need and they should like give those things to people so someone who took it off someone who is an outsider people don't see him as someone is very very guilty" (U2_INTER_12)

			"I think the international people are not happy to see children here in Uganda not studying, because they know what education means. I think with that it is what maybe comes in their mind, and they feel they really need to support these poor, poor families and these vulnerable kids" (U1_CBO-F_1) "it's embarrassing because when they go back they tell their people, that the people are suffering, and we are used of that way!" (U2_INTER_9) "M: most of, i told you before, only white people are good, black people only thinking about themselves they are not" (E2_TE_60)

attempted to look at how legitimacy in the context of travel philanthropy exchanges is understood in different settings.

8.4 Characteristics of exchange relationships and the informal rules of exchange

System patterns are characterised by a combination of proactive and reactive positions in the exchange. The proactive positions are lead actors of exchanges, activating a succession of transactions or transfers of some form, propelling the circulation of tangible and intangible objects in the system. The reactive positions respond either actively or passively.

> *I don't ask them to donate they are the ones who ask, how can we donate?*
>
> (U3_CBO-F_15)

For example, the 'spoken' rule (about asking for a donation or not) tends to suggest recipients are reactive – i.e., that they cannot thus do not ask a potential donor to give, rather the donor is proactive, and the recipient responds as a secondary act passively. However, our study also revealed that a common 'unspoken' rule is that local actors are not always subdued; they can also be highly proactive. A combination of active and reactive positions for different actors under different contexts exist. A proactive recipient may not be overt in asking for a gift, the request may not seem direct, thus there are also rules that steer a type of indirect or subtle persuasion. A key finding that emerged from our study is the core principles that recipients in travel philanthropy may not be passive recipients of aid, a view previously advanced by Silk (2004). Instead, they may be more influential in aid exchanges than one might assume, and sometimes more aligned with Retsikas (2016) theory of solicitation. For example, one participant stated:

> *People can anyway influence the tourist, they can easily guide them on how to donate their money on which organisations are genuine and which ones are not genuine and which individuals to give and who not to give, so that general category can guide the tourists.*
>
> (U2_CBO-F_8)

> *The white people, when we ask, some just they give.*
>
> (E1_TE_46)

Table 8.4 presents a summary of the overall framework of the most influential informal rules guiding 'exchange relationships' in travel philanthropy that have emerged from our study. It addresses how exchanges occur and those found to guide expectations of entering into a gift relationship as well as creating and maintaining relationships. Although unobservable, these rules are considered real, with causal impacts on the actors and their behaviour (Bhaskar and Callinicos 2003). The elaborated full description of these rules can be found in Tables 8.5–8.7.

Table 8.4 Informal rule categories guiding exchange behaviour

The unwritten rules guiding behaviour
Principles: How exchanges occur
Asking or persuading to give/receive
Offering or not offering a gift
Accepting or not accepting a gift
Reciprocity
Principles: Expectations of entering into a gift relationship
Promising
Obligation
Value
Principles: Creating and maintaining relationships
Trust
Integrity
Wilful blindness

8.5 The ontology of travel philanthropy stories

Belief systems are made up of the stories that assist in making sense of the world (Usó-Doménech and Nescolarde-Selva 2016), and this became very apparent when analysing the data, with stories and the emerging narrative framework forming a significant component for the travel philanthropy system to function. The most significant finding in this respect was that, firstly, stories are found to be passed from one actor to the next in some kind of automatic and legitimised circulation of exchange in the system, often involving actors telling other actors what they expect to hear, reinforcing the underlying belief system. Secondly, considering the causal properties of stories in the system, it was evident that stories were a cause and an effect of beliefs; whether based on facts or not, they influenced actors' emotions and behaviour and could legitimise action. Thirdly, the narrative framework and themes that emerged from across all studied cases was very similar, with the content of stories being particularised to the specific location, providing important new understandings about mechanisms that shape the system.

Table 8.8 emerged from a critical review of key thematic storylines within a set of emerging categories and themes as per Table 8.9:

1. Why exchange
2. How to exchange
3. What to exchange
4. Reassurance and evidence

Table 8.5 Informal rules – how exchanges occur

The unwritten rules guiding behaviour	*Comments*	*Indicative quotes*
Principles: How exchanges occur		
Asking or persuading to give/receive	The *spoken* rule guides recipients to be reactive, that they cannot ask to receive a donation, cannot ask a potential donor to give, rather *the donor is proactive*, and recipient responds as a secondary act. However, the common *unspoken* rule is that local actors do not always sit and wait, they can also be highly proactive. A combination of active and reactive positions for different actors under different contexts has been revealed. A *proactive recipient* may not be overt in asking for a gift, may not seem direct, thus there are also rules that steer a type of indirect or subtle persuasion. This is one of the key principles behind the finding that recipients in travel philanthropy are not really passive recipients of aid. If nudging tourist to donate, is it important to understand the difference between sales, charitable fundraising, and begging as the driver of an exchange. Are the way donations are generated relevant to that? If the results are the same does this mean from an efficacy and ethical standpoint? It is critical to understand the reality of exchanges if we are to shift thinking beyond idealistic notions of doing good, the one-way transfer of aid idea, and better understanding through candidness may lead to greater efficiency, for recipients and donors, and less harm too.	***Reactive recipient examples*** "we do not persuade them"; "we don't encourage them because… that would be like begging" (U1_CBO-F_2); "you don't force donations, you just wait when somebody likes" (U2_TE_6); "I don't ask them to donate they are the ones who ask, how can we donate?" (U3_CBO-F_15); "we don't force them, we tell them… they are not obliged to donate but if they figure out…we have seen kids walking here" (Z2_2_TB_41); "you do not ask from tourists, you do not solicit" (ZVWH); "better that the project is being sold by you being the headman otherwise if it is done by the management of [company name] it is more like soliciting" (Z2_1_B_INS_38) "not good to ask" (E1_TE_46); "when the tourist ask us to help someone we will explain" (E1_B-COM_50) ***Proactive recipient examples*** "the thing is to keep telling people what they can do, because you can tell one hundred people and you get like one person, one person to donate" (U3_TB_19); "why don't you actually give it to me and I go to school and finish my degree"(UF); "once they get their email addresses or whatever, it's email after email after email, and the donors suddenly don't want to donate anymore" (U1_TB_3); "people can anyway influence the tourist, they can easily guide them on how to donate their money on which organisations are genuine and which ones are not genuine and which individuals to give and who not to give, so that general category can guide the tourists" (U2_CBO-F_8)

		"I'm not actually pushing them to donate I'm more in generalising, not more in soliciting" (Z2_1_B_INS_38), "the whole idea is to try and stimulate them to support conservation work" (Z2_1_NGO_33), "you leave it to go naturally, you don't want to sound like the guy who wants money from people, so you leave it natural and it builds up from there" (Z2_1_NGO_33) "when the white people, when we ask, some just they give, they get before from the tourist" (E1_TE_46); "they say buy me book and faringi say ok, they go to the shop … give to owner of shop.. for them take … wow" (EHKU); "they have a system going … give me your email...there is one or more than one boy that writes the emails and he makes a fortune selling them" (E2_TB_56); "individuals will say they joined the university and they are going to graduate … still on the street begging rather than going to university" (EN_TB_66)
Offering or not offering a gift	Shaped partly by the belief that the donor has the freedom to decide if they want to intervene, the rule is that a donor can offer a gift freely, and it tends to be the *how* to go about it that is the question left open. The system is shaped by the principle that the donor should be in a proactive position of offering or not offering a gift. Giving per se, donor as proactive, is an accepted rule, but it is not always the case that actors are led by the principle that the donor can decide what or who to give. In Ethiopia some of the tourism operators are quite firm that they direct tourists to donate to institutions for example. In Zimbabwe the donor can take a proactive position in offering a gift, but how it is given is then led by other principles. In Uganda evidence suggests there is more freedom for the donor.	"someone is supposed to give when they feel like giving, you are not supposed to tell them who to give and not to give" (U3_TB_16) "someone just gives from their heart, they give what they want to give, they don't ask what to give" (U3_TB_16) "bring it, you're welcome to hand it, you do not give it to a child and you give it to a teacher in charge with all the school kids around and the headmaster or whatever, to adults that are in charge watching so they can see that" (Z2_1_NGO_37) "we will take you to the school or hospital you can donate there, that is the right place to donate" (EN_TB_65)

(*Continued*)

Table 8.5 (Continued)

The unwritten rules guiding behaviour	*Comments*	*Indicative quotes*
	The effects of the proactive donor principle are not always straightforward; it can have consequences for the beneficiaries. It raises ethical questions regarding responsibility, beyond the initiation of an exchange, and for how long that responsibility lasts. It is suggested that even if the act of the donor creates negative effects, then their responsibility is bounded, guided by the principle of the freedom to offer or not offer a gift. See Chapter10 for a further discussion on this.	
Accepting or not accepting a gift	One might assume that accepting a gift, being the recipient of a so-called kind act of a donor, follows the simple principle of appreciation, but it is more complex than that. It is important to understand the principles of accepting a gift, or not, from the local perspective. For some recipients, accepting a gift can appear simple, it can come with no debt to reciprocate. For others, when the recipient is to become a debtor in the exchange, if the benefit outweighs the loss, then the gift will be accepted, it is often an economic decision. In an exchange, with a succession of transactions or transfers, accepting a donation can also be based on an initial economic gain, a short-term view of reciprocity, yet down the line of the chain, different expectations in terms of transactions can follow. It is also evident that accepting a gift is to take what one is given.	"being given a gift is a god driven thing, it is god who brings someone to give a gift and I have always experienced that, it's when I get free things that I haven't worked for" (U2_B-COM_5) "the reason people accept that donation is that they are in poverty … people go for what is available they take up the material gain they don't look at anything else, so it is poverty that drives it. Some people cannot lose 1 million because of religion – he changes it!" (U2_INTER_12) "you may take it without knowing, it normally comes after" (U2_CBO-F_8) "because Ugandans are very poor… when you donate like 10,000 and when eventually the one who has been donated gets a 1,000 he will never mind. Instead of 10,000 he has got 1,000, it will be ok. Because that one is for nothing" (U2_TE_6)

	If one is aware that the intention of the donor is interfered by an intermediary, like a leak from the aid chain, because the gift is 'for nothing' then the recipient steps into a reactive position and accepts whatever given. Recipients accepting a philanthropic gift tend not to influence the object that is given, rather the principle of accepting a gift is to appreciate whatever it is, not to be impolite and question the imprudent nature of the object offered 'you cannot refuse'. The rationale behind the principle of accepting any gift can be related to building and maintaining future prospects or relationships, through this type of behaviour towards donors.	"… the toys and sweets, when she comes, she comes carrying bags, and you never know what is in the bags … when she's at school you can't say you just get away with your toys, it becomes rude" (U2_CBO-F_8) "when you are to be given a gift or a donation or what, you cannot say that I want the … you cannot decide in front of somebody that has donated you something and say I wanted this" (U2_TE_6) "it's not good to disagree a donation because if you disagree a donation it means that you don't want to be helped. So you agree, maybe this time someone can donate a spoon but next time can donate a plate!" (U3_TB_18)
Reciprocity	Not to reciprocate a gift, or to transfer rather than transact in certain travel philanthropy exchanges can have unintended implications. Principles regarding reciprocity can be inconsistent and certainly contextual. Once a gift is received, proactively or reactively, reciprocity can be activated. There are instances in which reciprocity is the rule, so to accept a donation means reciprocating something. This might be changing a religion, rewarding a social licence, etc. It has been argued that donors who do not activate the principle of reciprocity, they transfer rather than transact, can have negative implications. It may be that giving without obligation, in a disinterested manner, is considered altruistic than obligated transactions; however, it is argued that following principles of reciprocity can sometimes actually produce more positive outcomes.	"they are just carefree' (U2_CBO-F_8) … 'it's bad' (U2_CBO-F_8) "they are going to give their money and just walk away" (U1_TB_3)

Table 8.6 Informal rules – expectations of gift relationship

The unwritten rules guiding behaviour	*Comments*	*Indicative quotes*
Principles: Expectations of entering into a gift relationship		
Promising	When a tourist communicates an intention to donate to a local recipient it is assumed a promise, which is assumed a verbal commitment by many so whether the object promised is activated or not, it has implications in the system. That promise has effects such as expectations, anxiety, knock-on promises, etc., that can be long-lasting, when the tourist is back in their home country. Unfulfilled promises (even if not meant to sound like a promise) made by donors unfulfilled when home is one common occurrence as well as donors beginning giving and later stopping. This issue is also closely related to the fear of being involved in travel philanthropy for some actors because this contributes to the feeling of mistrust already existing amongst some communities and cultures. What is critical to note is that a promise alone can cause unintended consequences as important to consider as the effects of a tangible object. The informal nature of the exchanges is found to be creating some confusion. This issue of 'promises' is common in child education sponsorship i.e., without written contracts or clarity in the responsibility of different parties in the exchange, and if a donor withdraws after a certain period of time, such instances can leave a parent expectant of further assistance in funding the education of their child and create a burden for the intermediary and the parent.	"some of the tourists are liars, somebody may be excited, comes here, promises you, I'll give you this and you've not even asked and maybe I don't know if its emotional or what" (U1_CBO-F_2) "…they say we would like to help, let's reach in our country and see what we can do. Here we take it as a promise so we are expecting something a message a good message from them for helping…" (U2_INTER_9) "if a client says he will think and help for us, we take it as a true thing of helping. We don't go of saying he will think or she will think, which means sometimes it can come as a positive or a negative where the donor will go reach home, plan how you're going to spend the money, and you find there is not money to send, so there is not helping" (U2_INTER_9) "it's not a good thing because you raise the hopes of that person and at the end of the day nothing comes" (Z1_CBO-F_30) "like a tourist comes and promises I will do abc up to z and then after doing a, b then he withdraws, and then the other community will continue thinking that the client sent bcdfg … most of tourist always in the middle of the promise, before they fulfil the whole promise, they always withdraw… they will continue believing they are also sending this and for you, you are keeping it" (U3_TB_18)

	Once a promise is made, is it the responsibility of that donor to continue their entire education sponsorship? When a donor withdraws, it means the child withdraws from the educational establishment. So, although is it good that the donor provided the help they could within the capacity of their generosity, even if for a short time, it is a moral issue that needs debating. Does it become their responsibility to support the child till the education cycle is completed or should they not intervene at all? Why is the communication so confused between actors?	"because we accept here in Uganda we accept we think differently to a mzungu, because when a mzungu says he wants to help, he has his limit or sometimes is helping for one term or two terms and when here, when a parent hears that a child has got a sponsor he expects for a children from the primary 1 to the university and he finishes his education that's where the problem is" (U2_INTER_9)
Obligation	This leads to the principle of obligation, related to expectations, which suggests once you are involved, you are involved, but should you be obligated? The promises discussed above can cause an obligation for certain actors, especially intermediaries. If there is also an understanding that resources are not being allocated in the most effective way, or if there are signs of corruption for instance, there is confusion amongst actors as to a responsibility to raise this. It is evident that it is not a simple practice just to donate and walk away. Amongst these principles the rules suggest that if you become an intermediary for donations then you take on certain obligations to the donor, and the recipient. Donor expectations and the obligation of recipients to fulfil that, tied into obligations of reciprocity in order to sustain future donations. There is also evidence of the principle of obligation amongst recipients to receive certain objects from the donor. Gaining greater clarity on what role one takes on seems to be a paramount issue.	"but here we say oh I have got a mzungu and he is going to help me, so me that I have took the client in the field on a hike and I have got a burden because that parent will keep on coming, [name], how far how far" (U2_INTER_9) "her donation wasn't all that big donation because she would never ask me what she wanted to donate, she would always dictate what she wanted to donate" (U2_CBO-F_8) "just by association of us listening, once we're involved we're involved and that has the potential to sour everything, and no-one needs that" (U1_TB_3)

(*Continued*)

Table 8.6 (Continued)

The unwritten rules guiding behaviour	*Comments*	*Indicative quotes*
Value	Under the principle of expectations, the principle of value also emerged, suggesting different objects have different values thus different meanings. There are also rules about the right way to behave in relation to that. This has implication on what is exchanged in this system and between whom. The value of money can change depending on the nature of that exchange. Based on the perception that some gifts or donations are "free things given to poor persons" (UBB), this concept of 'free' has implications on the behaviour of the system. It is argued that if the recipient has not contributed anything to deserve that money, then it is taken for granted and the money is wasted. This was a theme in all three countries, yet many donations in travel philanthropy continue to have the characteristics of a free gift. This is further discussed in Ch. 11.	"I learnt that when people are given things, they take them to be just free things they have not contributed anything about those things they are getting, so they take them for granted. and soon every becomes history" (U2_INTER_12) "well the cause i would say is because they got the money easily, I mean if you get the money by working hard then you wouldn't use it for the wrong purpose" (E2_NGO_57E2_NGO_57) "if someone gives you money and on the way you lose the money you don't really feel the big pain but if somebody gives you money after working for it you will go back and look for the money" (Z2_2_NGO_45)

Table 8.7 Informal rules – creating/maintaining relationships

The unwritten rules guiding behaviour	*Comments*	*Indicative quotes*
Principles: Creating and maintaining relationships		
Trust	Trust is the key in travel philanthropy. In this system, creating and maintaining relationships of exchange are fundamentally built on principles of trust rather than formal agreements. Trust is effectively as an emergent property of building and maintaining relationships, it comes from the bottom up, and there are a range of scenarios in which such principles shape the patterns in the system. Trust is therefore not static. Trusting guides or tourism businesses is also particularly influential in this system. Lack of trust also shapes how the system functions. It can enable or constrain collaboration. Many tourism businesses, particularly in Zimbabwe, do not trust direct donations to local actors in communities. They argue that it is the tourism operators they need to trust their money with. Trust also shapes where exchanges take place because tourism businesses in Zimbabwe will not take guests to any village for example, therefore they pre-arrange the philanthropic experience and therefore the donations tend to be focussed in a specific geographical area.	"there are people who have a passion of helping but they may not follow too much because they think they trust but at the end you find the beneficiaries are misusing the donations of the help" (U2_B-COM_13) "when guests ask, oh we would love to donate, and we know how things can be here in Africa with donating, do you think this is a reputable source to be donating to or whatever project they are looking at" (U1_TB_3)

(*Continued*)

Table 8.7 (Continued)

The unwritten rules guiding behaviour	*Comments*	*Indicative quotes*
Integrity	To understand the concept of integrity in travel philanthropy a few insights will be highlighted to understand what this means in this context. To have or not have integrity is to be considered moral or immoral. As a system comprised of individual personalities, it is not possible to generalise such principles of integrity across actor groups, each individual personality is shaped by many different factors. However, there are arguably trends across cases. Integrity can have a different meaning depending on which actors are in the relationship. For example, if a local individual 'cheated' a foreign donor they are considered wise, and acceptable, but to do the same to an insider would be considered immoral. It is an understanding amongst recipients in Uganda that the level of honesty in accountability is never quite the whole truth. This is not necessarily considered immoral but just a norm. In Ethiopia and Uganda particularly the norm of lying has a significant effect on the exchanges in the system and the outcome of the donations.	"some of the people who are, who have good morals they spend all the money for the benefit of the poorest of the poor people, immoral people they may use the money for their own and they beg on behalf of other poor people" (E2_NGO_61) "they share a lot, it's never quiet and they are proud of it, because you find they are boasting over others, they met so and so, the rich guy, we did this deal, I got 5 million off him or her so they are boasting and sharing the stories and people consider them as wise… . they look at them and say this is another class of people, these are wise, if they did those tricks and got the money, the locals consider them to be wise" (U2_INTER_12) "there is a general one that is almost a general behaviour whereby if you talk of accountability there is always some accountability but it may not be real, yeah, they put accountability on the paper but it might not reflect what's on the ground" (U2_B-COM_13)

Wilful blindness	The recognition of wilful blindness (Heffernan 2012) as a principle guiding behaviour amongst actors in the travel philanthropy system has been revealed across actor groups and considered a significant principle that shapes the system. There are regular patterns emerging in the system caused by and an effect in which local individuals turning a blind eye towards inefficient, suspect, or immoral behaviour by another individual or organisation, donation recipients, to maintain some kind of stability in other relations. No one wants to be the whistle-blower; the rule is to turn a blind eye. Corruption in general is rarely questioned on this level in Africa, partly because creating 'enemies' has consequences and when there is a lack of formal consequences, such as a fully functioning police unit or court of law for example, the principle tends to be to stay quiet. To confront any immoral behaviour within a community is to gain nothing, to have no incentive, thus perceived the wrong thing to do. To confront an individual that has behaved immorally creates a risk to personal consequences such as risking the loss of status and popularity in the community, which most individuals strive for. Travel philanthropy is shaped by the principle of moral complicity, wilful ignorance, or blind eye. Details of this pattern can be found in Chapter 10.	"we are living together you will not say no don't give they will think you are jealous you know you will not say like this, tourists they want to give for them and you will not say no" (E2_TB_58) "that one is not advised because when you tell the tourist not to give them money, maybe they may think of, they may think maybe you want to cheat the pygmies, that's why the community leave them" (U3_TB_19) "good for the government if not they don't want to lose that resource so they hide their eyes and even if they know" (E2_NGO_61)

Table 8.8 Overview of travel philanthropy narrative framework – the current discourse and purpose of stories as mechanisms

The narrative theme
Why exchange: ***stories as mechanisms for giving***
What is lacking
What I need
Our vision, if … then
Emotion-based
A world different to yours
You can change the world
Fabricated stories
Positive Africa
My imported story
Exchange with me – I'm more trustworthy
What to exchange: ***stories as mechanisms for type of object exchanged***
Gift type
Local issues framework
Education
How to exchange: ***stories as mechanisms for giving pathway choice***
Our business is doing good
Responsibility – should
This is a genuine story
Development narrative
Reassurance and evidence: ***stories as mechanisms for legitimacy***
Accountability of resource use
Using pictures

Table 8.9 Story framework

The narrative theme	Indicative quote
What to exchange: ***stories as mechanisms for type of object exchanged***	
Gift type	"of course they are always briefed and told so many things about the school, yeah so they are so told so many things about the school, about sponsorship, about buildings, about helping who and who" (U2_CBO-F_8)
Local issues framework	"it is the poorest community, you see we are living by the national park. and living near the national park people don't have food, because being near the national park what they grow is being eaten by the animals, imagine someone plants his bananas and the gorillas come so" (U1_CBO-F_1) "actually saying focussing on us because they have seen we have got some conflicts caused by the elephants, lions coming to take our animals whereby they have got some interest in these animals when they want to bring their tourist to come and view the elephants and lions so in return they said we would rather help the community because we are benefitting for these animals by viewing them" (Z2_1_B_INS_38)

Table 8.9 (Continued)

The narrative theme	Indicative quote
Education	"they ask me, they first walk around and we go to the someone's hut inside the hut sitting down and then you explain them about the education and then Hamar are really it's like I told you it's really like backwardness and they are really like sad you know and then they wanted to help" (E1_B-COM_69) "mostly i would be saying that some students doesn't go to school and whatever, so he thought of an idea of helping, like sponsoring students" (U2_INTER_9)
How to exchange: *stories as mechanisms for giving pathway choice*	
Our business is doing good	"we went into the locals to find out exactly what they want us to do you know they gave us 101 things to do and as a priority we had to prioritise things for them so that can come up with something very conducive and a benefit to all which is a very good thing" (Z2_2_TB_41) "it's that ability to be a market leader in your projects that you are doing that drives you as well to say that if we do it" (Z1_TB_26)
Responsibility – should	"people these days that are travelling are very very aware of where they are travelling to and the fact that you are travelling to a place where there is also something else going on that has a purpose ... it becomes the reason for travelling that they can go to a place that is actually benefitting people, benefitting the conservation side but also on the community side" (Z2_1_TE_35)
This is a genuine story	"it's genuine, there is nobody begging for money, they are not being asked here to do this or do that, they literally go there and everyone is chuffed to see them, our communities know there are absolutely no payouts or anything like that but they can see the fruits of these funds" (Z2_1_NGO_37)
Development narrative	"we say it's better to show somebody how to fish than to give them fish so if you are to give once, something that will continue to make them do better so that you don't have to give the following way… maybe you give them something tangible they can do so the small things like the clothes they can get from what you have already given them" (U3_TB_18) "$10 is taken from each guest that comes through the camp specifically to fund for the admin work that we do as the foundation so much so that if a sponsor wants to sponsor a particular project they don't have to worry about it going to that admin part of it, it is going directly to that project that they wish to support…" (Z2_2_CBO-F_40) "we went into the locals to find out exactly what they want us to do you know they gave us 101 things to do and as a priority we had to prioritise things for them so that can come up with something very conducive and a benefit to all which is a very good thing" (Z2_2_TB_41)

(*Continued*)

Table 8.9 (Continued)

The narrative theme	Indicative quote
Reassurance and evidence: ***stories as mechanisms for legitimacy***	
Accountability of resource use	"that they have two people graduating university and these kids came from nothing and now they are here thanks to all the donations" (U1_TB_3) "being open that this has been funded to me, this has been funded to me… so whatever I always do is ok look someone donated me this I have managed to put a window, someone donated me this was able to put a toilet…" (U1_CBO-F_1)
Using pictures	"it's very easy to make up reports and to even fake pictures, take pictures of someone's kids, someone's buildings, and you send especially if you know your donor is not going to come so reports can also be faked" (U2_CBO-F_8) "they say ok now it is time for us to go and we go and we have pictures where we will issuing out everything and then the kids they get" (Z2_2_TB_41)

As stories are important in preserving culture and customs (Smaldino 2014), they are objects arguably with a similar role in the aid industry and travel philanthropy in preserving practices of how things are done and why. Stories have become myths about what is rational to do (Fastenrath et al. 2018), so reflecting on the growth of this phenomenon across Africa, with the content often mattering less than the number of story followers. For instance, the told 'what I am lacking' story, interconnected with 'I am entitled to receive' or "foreigners have a tendency to give" beliefs, intend to trigger emotions and justify the *why exchange* between donor and recipient. Occupying the recipient role appeared to have become the rational thing to do, as the norm evolving from 'giving donations to Africa' whilst on holiday or businesses had been normalised into a myth and a story followed by an increasing number of actors.

Stories are told by tourism business to tourists and international audiences through narrated corporate social responsibility (CSR) successes, confirming Gardner's (2015) previous findings and to local communities as an outcome of their desire to 'plough back' and importance of participation. Local guides reinforce these stories to tourists by using the 'plight of Africa' and need for sponsoring children's education. Local communities share them in the form of entrepreneurial 'business of poverty' narratives that encourage processes of emulation, as discussed by Liu et al. (2019). For the purpose of our study, the narrative framework of travel philanthropy was analysed to investigate stories as a key element in the system. These stories were found to be tightly interconnected with the belief system and mechanisms in behavioural patterns and their absence or a change in the narrative would arguably have an effect on the structure of the system, thus a potential leverage point for change. How and where stories are told also appeared to affect the system, through for example, the tourist–local village tour encounter or the increasing use of technology – i.e., email and social media affecting interventions. As these participants from Zimbabwe and Uganda highlighted:

> *It's becoming more visible and look you can't even discount the fact that social media plays a huge part in it.*
>
> (Z1_TB_26)

> *Blame it on social media, yeah, because now it's like people who run projects or individuals who are in touch with them, they are kind of seducing them, they are kind of trying to woo them to kind of confuse them.*
>
> (U2_CBO-F_8)

Philanthropic identities or identity storytelling motivating donors, proposed by Maclean et al. (2015), proved relevant in our study. It is a well-known concept amongst local actors that foreign donors, or tourism businesses, are craving stories, some believe this is for business purposes and some for personal purposes. Therefore, various locals actively assist donors in creating their philanthropic identities as part of their business of poverty. As one Ugandan participant said:

> *With my school she would really have a story to tell and I think she would attract many donations.*
>
> (U2_CBO-F_8)

Travel philanthropy combines global tourism (Salazar 2013) and aid campaigns through imaginative storytelling (Woods 2015). The stories do effectively create a simplified version of saving Africa for tourism consumption, but also maintain the system structure as different actors play their expected roles, often telling each other what they want to hear, or imitating the stories of others. As Salazar (2013) highlighted, in tourism local guides have become cultural brokers accepting imagined narratives about themselves with colonial tones. In travel philanthropy, the noticeable merging of travel narratives with colonial nostalgia (Scott and Selwyn 2010) and common aid industry narratives about how to develop Africa and address 'its suffering' create a refreshed role for local guides as development or poverty brokers.

People also tell stories because the audience is active – i.e., tourists are curious about what they see or hear and ask questions. This is common in travel philanthropy, unlike traditional aid. For example, according to these community participants from Uganda and Zimbabwe:

> *People have not maybe begged you or what and you [tourist] are the one to have brought the topic, asking how what what what.*
>
> (U1_CBO-F_2)

They [tourists] come to the village and they ask you what are your needs.

(Z2_1_B_INS_38)

Storytelling can generate *emotions* or represent *beliefs* that can generate certain behaviours, trust or mistrust, expectations, legitimacy, can refer to role models, or an individual-specific experience driving new norms. Stories travel, and as such they are borderless potentially spreading from a locally generated narrative into national and even global domains (i.e., through stories told by the tourists and/or through social media posting), with both potentially positive and negative effects. Interestingly these risks were recognised, as captured in one tourism business response:

You know sometimes you have to say, careful on the reality, because you don't want them coming back and you know blaspheming Zimbabwe and this is how they live and this is how the government looks after them because again that has a knock-on effect.

(Z1_TB_29)

8.6 The role of technology in the exchange

Although not a prominent aspect discussed in the extant literature, 'technology' was included as a key element of the travel philanthropy system because of its significant effects on the way the system works. Technology here refers generally to social media, email, mobile phones, online banking, or electronic money transfer facilities. This element was recognised to have an influence in enabling networking in relation to time and space of the travel philanthropy relationship, which emerged as a significant factor in the system.

Communication via generally accessible, low-cost electronic devices, social media and messaging facilities such as Facebook and WhatsApp have changed the way the travel philanthropy network and the donor–recipient relationship functions, enabling digital post-trip donations, and reconfiguring or creating new roles for different actors. This is captured in statements such as:

It is very easy to communicate to a person from anywhere else in the world.

(U2_B-COM_13)

I meet a donor here or a tourist going to track gorillas, then I ask him for his Facebook account when he goes back I take pictures of my school, send him the pictures using WhatsApp or Facebook then he responds, then we continue the process of getting in touch with one another, and then developing into partners for the donation.

(U2_CBO-F_8)

Tourists tend to spend limited time in each destination they visit, so to build a strong relationship with a tourist on-trip can be challenging. Therefore, technology enables a follow up to the short encounter in a destination to evolve into a relationship in the *post-trip period*, through digital stories too. This is a practice also employed by tourism businesses and their foundations, with fundraising from previous clients, and perceived as legitimised post-trip communication compared to local actors, as indicated in the following responses:

> *If they have shown an interest, we have a mailing list.*
>
> (Z2_1_NGO_37)

> *Nowadays you can have it [a link to donors] on your phone ... after that they will send you [previous tourist] hundreds of kinds of stories, messages and things like that to convince you to help, to help people.*
>
> (E2_B-COM_64)

Technology also aids *accountability*. For instance, sending photos to donors about a project appeared to have driven reassurance and trust for donors to channel donations through individual intermediaries or direct to recipients. According to a local participant in Ethiopia:

> *Most people are not come, they meet each other and then they promise them, just they meet each other by email, WhatsApp, Messenger, they not come here not see results, so that is one main point one main problem.*
>
> (E1_B-COM_69)

Technology arguably enables direct donations, and increased donation volumes, given the ability to communicate directly between donor–recipient. On the other hand, technology also constrains donations for others, as competition over scarce resources means *competition* amongst local actors, at times leading to unnecessary pressure on the donor, as corroborated by a Zimbabwean community-based organisation response:

> *You get clients that directly want to... they eventually know that it's not the right way because the community is now constantly getting in touch, asking for stuff, asking for, and then they can't handle the pressure.*
>
> (Z2_2_CBO-F_40)

The availability of mobile money facilities and *money transfer* companies such as Western Union and MoneyGram have made the direct transfer of financial donations from donor to recipient directly, or via a local intermediary commonplace, without too much questioning and assuming that the money is being used for the specific chosen cause. For example, one respondent warned of the potentially problematic consequences of such practices by stating:

> *The boy is young, you send 20k [ETB] to Western Union and he take this money and you think boy go to school?*
>
> (E1_B-COM_69)

Alternatively, local guides are often used as financial intermediaries to channel funds to intended beneficiaries via their bank account and in a trusted monitoring role.

> *They will send though to my account and I will take it to her.*
>
> (E1_B-COM_50)

Needless to say, this leads to questions about the effectiveness of the process and implications of the nature and trustworthiness of the actors involved.

8.7 The *gift* in travel philanthropy

In order to evaluate the role of the *gifts* in travel philanthropy, it is necessary to look beyond the lens of the givers' beliefs and more into the gift recipient perspective (Baskin et al. 2014). The travel philanthropy system includes transfers of tangible objects – i.e., either money or non-monetary items such as school materials, as well as intangible objects – i.e., emotions and services. For example:

> *They [donor 1] donated something that is tangible, something that one can see and is going to stand the test of time and help the children in the community, even the community at large, because the toys [from donor 2] are now finished and the sweets are eaten.*
>
> (U2_CBO-F_8)

Whilst potentially debatable from a local perspective and impacts being non-linear, the value of tangible objects given within travel philanthropy compared to other forms of giving within global philanthropy are relatively small. So, we might more accurately call this system a niche form of philanthropy or micro-philanthropy in Africa.

While no studies so far have measured the financial or in-kind value of tangible objects on an aggregate basis, nor was it the intention here, identifying the scale and type of donations became the focus of our study to uncover further meanings relevant to the system. Many participants were unsure about the significance of travel philanthropy, both in terms of scale and impact. Some perceptions of the size of travel philanthropy as a donation economy were variable. While some dismissed its value – 'numbers are minimal', "very small it only creates a begging culture", 'not significant', 'not that significant', 'it's just small amount', others went to the opposite end of the spectrum – 'it's a huge thing', 'many times', 'it's common', "one student a lot of sponsorship", "I don't think people realise how big it is". What this outlined is that there was very little clarity across all the destinations as to what was really going on:

> *When you go out with a group of tourists and $20 or $50 is handed out there.*
>
> (Z2_2_TB_44)

If you calculate this money it is big, individuals, every children if you go to internet house you can see the internet house every children has an email communication, why, because of the tourist, so they all this send money.

(E2_NGO_61)

Most participants suggested that approximately 20% of tourists that they interact with would donate something. What emerged though is that not only is the value of the object a significant consideration, but its meaning, power, as well as past, present, and future impacts are of relevance. For example, if a donation associated with school sponsorship suddenly stopped, this may cause termination of education for a child, as the expected sponsorship may be the only source to support that child. The absence of the donation would have substantial emotional, economic and social impacts on the beneficiary. This is a major social issue potentially on the horizon in Ethiopia if the dependency or non-monetary value on donations for some youth is underestimated. As a community participant in Ethiopia explained:

It's valueless, then when the donator or the sponsor stops that thing that person for sure he is not going to, he is going to do like stealing money from friends, neighbours, stealing things from people and becomes a valueless person that is problem like some problems what I saw.

(E1_TE_52)

Apart from the unknown scale of this noticeably fragmented philanthropy niche, the characteristics of the tangible objects given included the sporadic nature of such donations, from absolutely nothing one month, to something small or significant another. Giving, in this context, is very reliant upon tourist volumes and their capacity to provide sustainable funding. In another participant's words:

A donation from another source may be big and able to do something big on the ground but a donation from a tourist is small because most of them have travelled and money is water and he reaches on a project when he is completely bankrupt.

(U2_CBO-F_10)

Furthermore, given the seasonal nature of the destinations involved in our study, as for most others, the flow of gifts would follow such patterns.

While one-off larger value donations can occur (figures up to around $30k and $100k were quoted although very rare), the majority form the long tail of philanthropy, made up of much smaller value donations. Values quoted in Uganda in merit to direct gifts from tourists ranged from $5 to $500 and one participant referenced a tourism operator contributing approximately $30k per month as part of their bednight levy contract with a community-based organisation (CBO). Although this was a very significant amount for the location in question, the impact of that money remained debatable, thus confirming that it is not always relevant to use input value alone as a measure of success or a metric to communicate the whole picture, rather a more qualitative approach would provide a fuller picture.

In Ethiopia, the values quoted were slightly greater than Uganda. Individuals in a community received between $50 and $500 per month from an individual sponsor, with some reportedly having up to 40 sponsors each, aggregating significant salaries. Some gift values were undoubtedly greater than mere school sponsorship, with one individual having received a cheque for $1 million to build a hotel and another $10k to buy a minibus. In Zimbabwe, values ranged between $30 and $150 a year for child sponsorship, usually via an organisation rather than directly to a child, a few $1k–$5k donations towards boreholes, with occasional out-the-blue cash injection of up to $20k. The tourism corporate sector in Zimbabwe quoted higher figures than other cases and appeared to be growing over the three years that preceded the field assessment (2016–2019). Some of the largest companies spent $100k–$230k per year in support of communities' projects either fully funded by guests or budgeted partly by the business (i.e., as part of joint ventures or CSR) and supplemented by guests.

Donations ranged from the most common items for children – e.g., pens, books, clothes, etc., to school toilets, national park entry gates, water tanks, a new ward for a hospital, school buildings, funding for tribal ceremonies, t-shirts, hotels, boreholes, animal tracking devices, and so on. Although beyond the scope of our study, it is worth mentioning that determining the reach of those donations was also difficult. In terms of sponsorship in Ethiopia and Zimbabwe, reference was made to around 500 sponsored children in Lalibela (Ethiopia) only. While, across Ethiopia sponsorship was predominantly amongst young boys, in Zimbabwe it appeared less gender biased, but more aimed at younger children. In Zimbabwe, two of the larger tourism companies were sponsoring approximately 80 children at one time (2019), as an indicative figure. In Uganda, reported numbers seemed much lower in the research locations and of approximately 100 children in total across the three villages.

The significance of the donation depends on the recipient. If, for instance, a relatively small donation is given directly to a child, the value may be significant for them, and may potentially be life-changing either positively (i.e., an opportunity to attend school) or negatively (i.e., induction onto a begging culture or alcohol, etc.). While in Uganda, $30 can also be a significant contribution for a CBO, with an 'amazing' (U1_CBO-IN_4) effect of that for the children involved in the CBO, in Ethiopia, that may mean less to a guide used to receiving a salary of a few hundred dollars every month. One object could have non-linear effects and power within it that may not easily be visible from a surface event. Although evidence from our study would suggest that the scale of travel philanthropy is not financially staggering or towering over other forms of aid in Africa, this does not mean that it is not significant or unworthy of critically assessing and discussing.

Philanthropists have been criticised for projecting their personal values in judging what is good or bad onto distant places as part of moral geography (Burgold and Rolfes 2013). Within the travel philanthropy domain, findings indicated that such personal criteria also travel with the philanthropists when they are not at a distance, and such projections become relevant in the face of the 'suffering other' as well. The power relations that favour the subjective decision of donors with their preconceived notions about problems (Barkan 2013) were apparent, with tourists often deciding on the gifts before they travelled (i.e., school materials), or corporate philanthropy being influenced by the personal values of leaders facilitating the business' philanthropic activities, rather than an objective analysis of sustainable gift solutions. Our study could not find thorough ex-ante or ex-post (Arensman et al. 2017; Richards 2019) evaluation in terms of *the gift* in travel philanthropy, rather, beliefs, stories, and emotions appeared to inform the appropriate gift.

8.7.1 Emotions and services as intangible objects exchanged

While beliefs and stories have been discussed in the previous section of this chapter, here the focus is on emotions, which emerged as having a significant role in exchanges too as intangible objects of exchange – not physical objects but equally real (Bhaskar 1975), and treated as objects in exchanges. Gift-giving decisions are often emotionally led, and evidence of the 'warm glow' and the personal benefit derived from the act of giving overpowered levels of concern for the public good (Atkinson 2009; Null 2011). Sentiments not only shape the roles actors occupied in the system, but some emotions become objects of exchange, effectively transacted, perhaps as a means to an end. Donors (maybe moved by compassion), recipients, and intermediaries can in some contexts give and receive *happiness* in the exchange, often ignoring any possible consequences of their giving, as the following responses attested:

> *If a client gives something, that one who is donating feels happy, that he has given something, which is true, if I give something I feel happy.*
>
> (U2_INTER_9)

> *The kids were happy but they will be encouraged for begging.*
>
> (U2_TE_14)

> *Locals who are following the tourists are very smart, how to touch the hearts of the tourists.*
>
> (EN_TB_66)

> *It's more about getting attached, let me put it simply this way, attached to the little smiles that you see in the community.*
>
> (Z2_1_NGO_33)

It also emerged that tourists could exchange hope for promises, with *hope* as a positive emotion by the recipient, at risk of evolving into *sadness* (Z1_CBO-F_30), if not maintained. For recipients, for example, not only the tangible object, but also the feeling of respect that it provides them in their communities holds value (U2_INTER_12, E1_TE_52).

Concurrently, the web of exchanges in travel philanthropy are organised through a range of intangible exchanges of *services* between actors, again not physical objects, but part of exchanges in the travel philanthropy system. These identified services included delivering a tourist *experience*, providing *access* to donors or beneficiaries, *advising* on strategies about doing good, and presenting *accountability* or reporting, often not as paid jobs, but rather as intermediaries expecting something in return. Intermediaries for example most of the time tended not to assist for pure altruistic reasons:

> *They feel they should get something off everything they deliver, everything they do.*
>
> (U2_INTER_12)

In agreement with Pollock (2016), very little is said about the right of the destination community to say 'no thanks' to visitors coming, but they indirectly do get involved in directing visitors towards what they want or need in the case of travel philanthropy, as they sometimes shape the visitation, often offering services, with the hope of receiving donations. This research found saying no to donations per se, was unlikely. However, being part of the aid chain could be demanding as actors were expected to do everything for free, as a virtuous act. This has implications on covert behaviours, for instance, rather than resistance. One example of resistance emerged amongst teachers in Zimbabwe arguing against a lack of payment or tips for hosting clients of a tourism business, but their superiors shut them down because of the hope for philanthropic financial contributions:

> *I [teacher] don't want a pen and I am hungry ... so the teachers confronted me and said [head teacher name] you will host all the visitors who come through we are not going to do it, we are being ill-treated, this person wants give me a dollar and this guide is saying no that is not allowed, so when I confronted them [tourism business] to say I have a crisis, at the end they said [head teacher name] if that is the case we will go somewhere else because I knew what I meant in terms of the benefits that our children are getting, what the community are getting.*
>
> (Z2_1_B-INS_31)

Our study evidenced that the travel philanthropy system reinvigorates the symbolic domination of the West towards the 'developing' world discussed by Sahlins (1974) in terms of foreign aid relations and often disconnects with local realities. Aid is far smaller in value, but, arguably, its symbols are just as powerful on a local level. The gifts in travel philanthropy and the processes of giving represent a clear distinction between wealthy and poor (Kosansky 2002) and asymmetric relations (Korf 2007), so this form of philanthropy deepens and reinforces the belief that Africa is in need of the West and that recipients should be grateful, as previously discussed by Korf (2007).

8.7.2 The meaning and power in an object

What is the meaning of an object and the power within it? Does a modest pen have the power to affect how the system behaves? Arguably, such objects are loaded with meaning when considered. Our study supports Berry's (2014) sentiment that even if the community may be resource-poor, any gift is not necessarily a good gift regardless of quality, and the 'possible silences of harm' cannot be overlooked. Whether the gifts are interested or disinterested at the start of the aid chain, by the end of their journey, they become another resource for competing groups to manipulate and control (Stirrat and Henkel 1997). As a local Ethiopian participant said:

> *Some tourists come in and give pens and books but the children don't go to school but they don't know they just give it then they sell it because they are not a student and then they drink or whatever so this is not good.*
>
> (E1_TB_53)

Every object has certain powers, entangled with morality. A few t-shirts given for example can create a 'me-too' culture amongst a traditional tribe or generate cultural changes, as indicated here:

> *[Giving clothes] brings problems to the culture to the tradition ... they [females] take that t-shirt and run away from us, we don't control them they are out of us, even the big problem is tourists giving t-shirts to the girls.*
>
> (E1_TE_52)

Arguably, some of the gifts in travel philanthropy also lack mechanisms for real reciprocity. Gifts in travel philanthropy have been found to include both expressions of affection (Belk 1979) and expectations of reciprocity (Mauss 1925/2016), but the norm of reciprocity is very complex in this system because the feeling of obligation to reciprocate is inconsistent (Gouldner 1960). Whilst travel philanthropy may involve 'gifts', only considering the system as an alternative system of exchange of gift-relationships and social bonds leaves space for further reflection. Arguably, many of the gifts produced by this system do not represent Mauss' (1925/2016) 'Hau of the gift', because that suggests the recipient must return it and the donor has a hold over the recipient, but this was not always the case in the studied locations. For example, a Ugandan participant said:

> *I have learnt that when people are given things, they take them to be just free things they have not contributed anything about those things they are getting, so they take them for granted. And soon every becomes history.*
>
> (U2_INTER_12)

There were indeed some situations in which this theory could be argued, such as tourism companies' giving gifts with expectations of a social licence reward from a community, or a local aid broker gaining power (Levi-Strauss 1969), but in terms of tourist donors the context was significantly different. In terms of tourism business reciprocity, it is also worth reflecting on the quote below, which may imply that reciprocity may well be the simple gratitude for the act of giving:

> *Of course they [the tourism business] need us [the school], I think they need us, but sometimes let me tell you sometimes they say, sometimes there is that feeling for sort of a threat to say you guys if you feel you are not happy about this and this, then we will leave your school and go to other places.*
>
> (Z2_1_B-INS_31)

The findings indicated that a large proportion of the gifts in travel philanthropy could be considered conditional giving, which implies the giver has rights over the recipient (Boland and Griffin 2016). For tourists, there tends to be little 'force in the thing' in terms of reciprocity (Mauss 2016), which has implications on behaviour – see Chapters 8–10. However, the common gift-giving experience provided to tourists, formally or informally, does mean donations come with certain expectations (Abernethy 2011). These are often intangible, such as reinforcing the image of poverty as outlined in the literature review and reciprocating in gift exchange, perhaps with a 'warm glow' emotion for the tourist or performing as an entertainment service, as evident in the response by a Zimbabwean NGO:

> *They [children] would come out every time and see these older white people [tourists] all smiling at them, they know that they have got to smile and do a song and then they'll get stuff [a gift].*
>
> (Z1_NGO_21)

8.8 Beliefs and rules of the travel philanthropy gift

8.8.1 Beliefs influencing the gift

How and where the giving and receiving process occurs can assign a new meaning to the object exchanged according to the context. For example, the same amount of money given to a school has a different meaning to the same amount of money given directly to a child.

A range of beliefs influencing the objects exchanged emerged from our study. These are the following:

1. The object given judged by beliefs about *what the object is* – belief in the form of the object, in the outcome of the object, or in the moral of the object;
2. The object given judged by beliefs in *who decides what the object is* – belief in myself as expert or in you as the expert;
3. The object given judged by beliefs in *how the object is given/received* – beliefs in giving processes or in receiving processes.

Different beliefs determine different meanings of tangible objects, and what one considers a gift, donation, tip, or handout was unclear due to inconsistent beliefs amongst actors. As these participants put it:

> *I think donations and gifts are closely related, because they are free things given to poor persons. Then, but they are different. Donations is somebody donating you some money to raise something, to keep you remembering that this was donated by [name], so donations and gifts are also ... a gift is like a tip, like I am pleased to see gorillas you can get this tip... it is not for nothing, it is for your appreciation. You have appreciated that you have got a chance of seeing a pygmy. Now this chance I have to give him something. If you are not kind you will not. But if you are kind you will.*
>
> (U2_TE_6)

> *Tips are allowed but then you cannot have a tip of a whole house being built for you.*
>
> (Z1_TB_28)

A dominant belief is that it is better to give things than to give money (unless given to something or someone approved or legitimised to do so – i.e., a school), that giving clothes directly to a handful of people or pens to children has positive effects, whereas giving money takes on a different meaning. In all our study locations, most tour operators interviewed believed in donating non-monetary objects – i.e., school materials and clothes. Tourism businesses particularly supported this in Zimbabwe and Uganda. In some contexts, however, certain objects such as pens had negative connotations compared to monetary gifts:

> *We are trying to stop people handing out pens in the countryside so making that a big no no ... because you create a begging culture, kids come up and say give me a pen... it's all over.*
>
> (EN_TB_67)

> *Sometimes someone comes in the house and someone gets sick and they ask why are you not in the clinic and they say I don't have money and the tourist can say this is not nice and helps them giving some money.*
>
> (E1_B-COM_51)

Additionally, if the object donated had rules attached – i.e., some expected reciprocity, the meaning of the object given might be affected. One example came from a recipient perspective linking a foreign religious tourist to travel philanthropy in Uganda:

> *He's given houses to people ... it's an exchange, the reason why I see it as an exchange number 1, he has helped me because I have gone to his church... there*

> *He is not helping me, because what if it was help, he wouldn't force me to his church.*
>
> (U2_TE_14)

As previously mentioned, one of the most common transactions in travel philanthropy is sponsorship of children in education, with donors rarely questioning the belief in the rightness of the formal education system:

> *People are always interested in children getting an education so I think that is a soft button.*
>
> (Z2_1_NGO_37)

In regard to this, beliefs range from primary education being a highly valued area of intervention for local people (Z2_2_NGO_45, Z2_1_NGO_33, U2_TE_6) and a sustainable investment (Z1_TB_28), to parents in Ethiopia, for example, believing that their child being sponsored in school is negative:

> *Hamar people don't want education they always thinking bad tourists want to help us even with the money.*
>
> (E1_B-COM_54)

Actors also judge the object transacted based on its *outcomes* or its purpose. This can include beliefs about the meanings of development or sustainability, or the impact of money or other objects after they are given. Locals raised the issue of equality, associating the solution with the preference for donations to institutions rather than individuals (E2_B-COM_64, EN_TB_65). For example:

> *What we suggest is, tourists is good to give donations, but let's do it to institutions, schools, hospitals or local village leaders so that one everyone will benefit.*
>
> (EN_TB_65)

What is believed as a successful outcome may also vary considerably. Outcome-based beliefs are often about the actual object given rather than the impact that it has, with success being judged by the amount of money donated rather than what that money actually did (U1_TB_3, U1_CBO-F_2). While some believed that short-term assistance is a valuable exchange (Z2_2_TB_43), others felt that this was not the case:

> *Someone can be compassionate about something that is negative in the community also.*
>
> (U2_B-COM_13)

Beliefs in who decides was also identified as relevant in the system as what is gifted is often driven by tourists' interests, passions, or values or their beliefs in the tour operator as the expert, as long as they have provided them with a personal experience of the cause:

> *Someone just gives from their heart, they give what they want to give, they don't ask what to give.*
>
> (U3_TB_16)

> *We get our clients to donate through us then we can ensure that the money is going to what it is supposed to be going to i.e. helping with school fees, uniforms, books, water pumps.*
>
> (Z2_2_TB_43)

Beliefs in how the object is given/received can define the meaning of the object itself and influence what is donated – i.e., giving pens/clothes directly to children in a school or a lodge may be considered good vs. giving pens and clothes directly to children on the road may be considered bad. Beliefs about this issue were so diverse amongst respondents that it emphasised the inconsistencies in the system, with the giving process or pathway often defining what is an acceptable and/or sustainable handout or not. For example:

> *I've seen so many tourists like even in their own vehicle like throwing sweets to kids to passersby, they throw sweets, they throw pens, books to whoever they pass by.*
>
> (U2_CBO-F_8)

> *We don't allow them to give out to the kids because will create a culture of begging, we don't allow them, but if a guest want to specifically give to a child, they will have to consult the teacher.*
>
> (Z2_1_B-INS_31)

The level of follow up by the donor after gift-giving was also believed to be important in defining the meaning of the donated object. The donor might be giving an object they believe is morally good or of a certain value, but this might be either changed by the recipient or affected by transaction leakages associated with the number of individuals involved in the chain, providing a lower value to the intended beneficiary. This determined widespread mistrust in the travel philanthropy system and in the giving process occurs – *how you give*:

> *I don't take it to the teachers, no, because sometimes the kids don't get what they are supposed to get.*
>
> (U3_TB_16)

> *If it goes directly to them you can't manage where that money is going it could be going to buy the chiefs next cow or the loaf of bread or something like that.*
>
> (Z2_2_TB_43)

From a recipient point of view, contrasting views emerged about acknowledging donors through naming them at their request vs. doing it in recognition of their contribution to a cause. Some believed that naming a donor at their request devalued the object:

> *If you give you give you don't need to kind of blow out that you have given and then your name is all over, so I think you just give in a humble way and let go.*
>
> (U2_CBO-F_8)

While others felt that showing appreciation for receiving a gift is believed to be the right thing to do, even if unrequested (U2_B-COM_5).

8.8.2 Informal rules (principles) influencing the gift

The informal rules, intended also as principles, guiding what is exchanged have an influence on the behaviour of the system and have been identified under the themes of *authenticity*, *overt*, and *covert* principles.

Many similarities about what is an *authentic* object to give emerged across case study settings. School materials were the most commonly donated object, dominated by the belief and principle that a genuine donation creates change (U2_INTER_12). Authenticity was also associated with the rule that the donor should decide (U1_TB_3), or the beneficiary, whether an individual or an institution, defined it (EN_TB_65, E2_B-COM_64, EN_TB_66, U2_INTER_9).

What also emerged is that there is a combination of *overt* and *covert* principles guiding donations in travel philanthropy. Some exchanges within the system were somewhat hidden for various reasons – i.e., a local may not want to risk being challenged as a legitimate beneficiary by other locals, which makes it challenging to measure and monitor the donation and its relevance to the travel philanthropy system. Other local actors were open about their involvement because they might receive some benefits – e.g., raise their status in their community (U2_CBO-F_8). Tourism businesses tended to be more overt towards social licence (U3_TB_17) purposes amongst the community and provide transparency to their clients that they had linked to CBOs, as a responsibility towards the donor.

In Uganda *overt* engagements were more of a voluntary principle, compared to Zimbabwe, where the rules appeared to be that they have to report donations to the government:

> *It is paid from our trust account into a school account that is monitored by the ministry of education and they cannot take out without their [knowledge], they are then answerable for it.*
>
> (Z2_1_NGO_37)

Various mechanisms also enabled *covert* exchanges within the system. While in Zimbabwe, contrary to the above, evidence suggested that some schools avoided transparency so as not to discourage other donors to support (Z1_NGO_21), in

Uganda there was a dominant covert rule that "you take what you can get" (U2_TE_6). In Ethiopia, the principle of soliciting donations for personal gains and keeping it under the radar was significant (EN_TB_66, E2_TE_60).

Tourism businesses also indicated that they could not possibly know the whole story, with everything being "all so behind the scenes, under the floorboards… it's so hidden…" (U1_TB_3).

The informal rules are further unpacked in the following chapters as the emergent travel philanthropy system is further explained.

Chapter 8 has presented the elements of the travel philanthropy system, helping to inform a better understanding of its underlying mechanisms. Both tangible and intangible objects, the observable and unobservable, were found to be critical in shaping further understanding of the travel philanthropy phenomenon. These findings have challenged the way travel philanthropy is conceptualised – a system as an epistemological construct rather than ontological. The analysis in this chapter began identifying core beliefs (part of its the belief system) necessary for the system to exist, from which the travel philanthropy system emerged. Further beliefs and principles which affected how the system functioned followed. Subsequent analysis explained the influence of informal rules, or principles and norms, the narrative framework, the influence of emotions and other intangible elements that steered the behaviour of actors, such as becoming decision-making mechanisms, or defining the meaning and power of the gift. Technology was also identified as a key element new to enter the travel philanthropy debate. Chapters 9 and 10 will introduce travel philanthropy as a system, identifying the *emergent* patterns that form the bigger picture, rooted in the underlying mechanisms discussed throughout this chapter.

9 The travel philanthropy system and the business of poverty

This chapter introduces the travel philanthropy system, which emerges from the foundational elements discussed in Chapter 8. To simplify the complexity of this reality, a system map is presented, highlighting the key events and mechanisms that best explain the causal themes identified in our research. The analysis delves into different sections of the system map, where details are examined through a range of interdisciplinary theories. This chapter focusses on the local consumption of donations, including the business of poverty.

9.1 The emergence of the travel philanthropy system and its drivers as an exchange economy

To better understand the impact of travel philanthropy, the non-linear emergent patterns within the travel philanthropy system must be considered, as well as the characteristics of micro-exchanges between actors that cause the higher-level patterns of exchanges. These provide the bigger picture of the travel philanthropy system rooted in the underlying mechanisms discussed in Chapters 7, 8, and 10. At first glance, the travel philanthropy system as cause–effect relationships looks rather chaotic, with Figure 5.2 offering a visual representation of the emergent system, and showing part of the analysis process used for our research, which also informed the model elaborated throughout Chapters 10–11.

Whilst not ignoring the complexity of reality, a system map can be more useful as a communication tool by interpreting and highlighting the *key* events and mechanisms to illustrate the structure and dynamics of the system (Forrester 1994). As an abstraction of reality and a simplification of 'the territory' (Korzybski 1931; Bateson 1971), in a world in constant flux, such maps are inevitably incomplete and temporal, and characterised by some level of bias by the observer/cartographer. Nevertheless, this chapter intends to discuss the travel philanthropy system both visually and in a narrative form, highlighting parallel themes that emerged from our empirical research conducted across Uganda, Ethiopia, and Zimbabwe.

Figure 9.1 illustrates the *travel philanthropy system* that emerged from our study, which was informed by more intricate findings in Figure 5.2. This chapter and Chapter 10 investigate causes, effects, and systemic emergent patterns identified

DOI: 10.4324/9781003497929-9

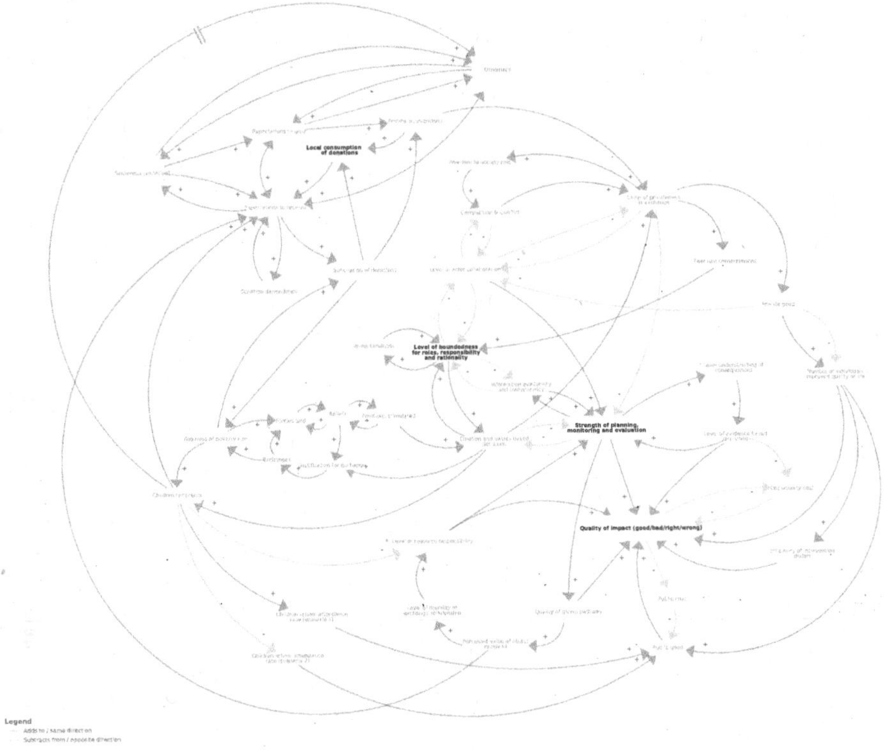

Figure 9.1 The travel philanthropy system map.

and illustrate how a change in one element tends to affect others, generating the dynamics of the system. Here, different system elements will be unpacked and interconnections of events understood better, providing insights into the travel philanthropy bigger picture.

The exploration of the system begins with the elements found to drive the *local consumption of donations*, a pivotal effect of this informal exchange economy. The area of the system map discussed in this section is highlighted in Figure 9.2 (within the yellow dotted line).

Figure 9.3 illustrates a section of the zoomed in system map, to indicate what drives the local consumption of donations in African tourism destinations. A growth in the local consumption of donations is often associated with the altruistic philanthropic behaviour of donors and their gift-giving (Honey 2011), but it is more complex than that. The orange text (vs. blue) merely highlights this key event, local consumption of donations, which is the focus of the discussion.

What follows also provides an analysis of system interrelationships (arrows), events and mechanisms, and influential contexts of the system, by analysing *otherness*, *sameness*, *expectations* and the *pull/push* of donations in tourism destinations.

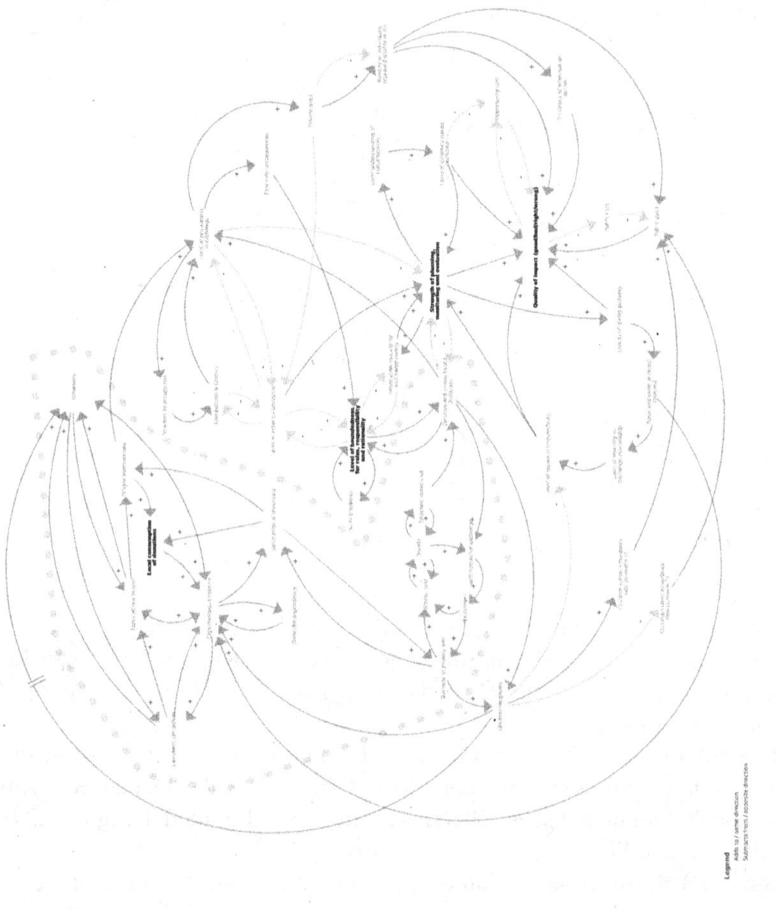

Figure 9.2 Area of system map: the local consumption of donations.

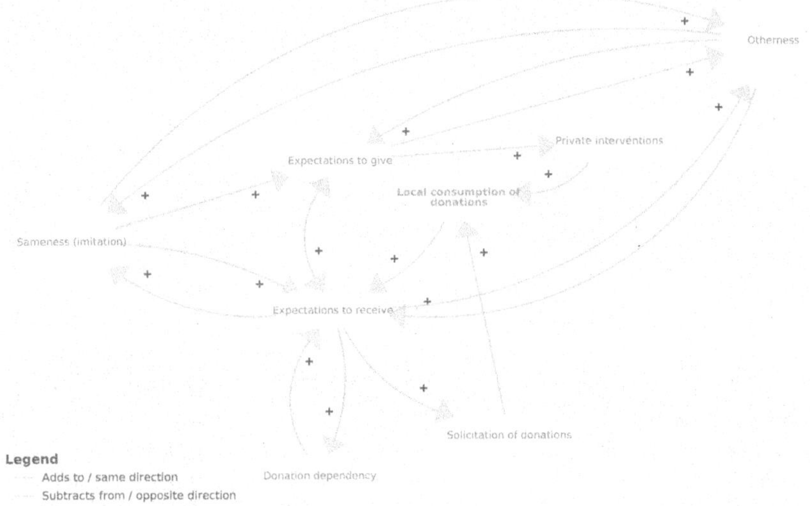

Figure 9.3 Local consumption of donations: causes and effects.

9.1.1 The ontology of otherness *and* sameness

Otherness and *sameness* were found to be powerful elements in the travel philanthropy exchange economy (Figure 9.4), not only deeply rooted historically but also evolving in the present practices; thus, what we may see on the surface comes from somewhere at a very lower level of 'the iceberg' (Vandenbroeck 2015; Mingers 2014). If we are to change the structure of the system for improved outcomes, we need to understand what the system is, how it functions, and why. Otherness and sameness are significant findings in this respect.

Our research highlighted that an absence of *otherness* and *sameness* in the travel philanthropy system would cause a significant change to its structure and nature, and the purposes of the system outlined would cease to exist. Both concepts rely on complex underlying sub-systems of beliefs, principles, and norms that are systemic in nature – from which these higher-level system structures emerge. They are causal mechanisms triggering other dynamics in the system, such as expectations to give/receive and the local consumption of donations. *Otherness* and *sameness* are thus both a cause and effect in this system – see Figure 9.4.

9.1.1.1 Otherness

Otherness is comprised of socially constructed identities shaped by beliefs, experience, and reason informed by the past and reshaped in the present, as well as being connected to forward looking expectations. Yet understanding the ontological properties of otherness is fundamental to grasping the systemic nature of

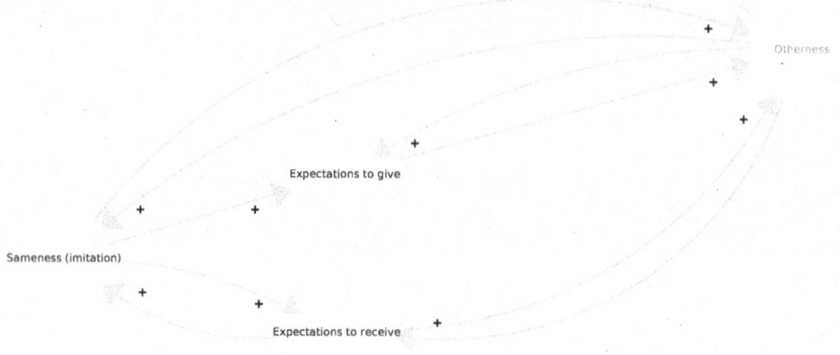

Figure 9.4 Otherness and sameness as events and mechanisms in the system.

Explanation for Figures 9.3 and 9.4

Interconnections in this segment of the system map are 'same' or positive – e.g., if otherness increases, expectations to give increase. If otherness decreases then expectations to give decrease, etc. The findings indicated that otherness drives the travel philanthropy system, and caused or reinforced more otherness. Ideas of sameness, such as identity and imitation, also shaped, sustained, and expanded the system. Othering tended to be a key cause of expectations to give and receive, and expectations to give/receive could reinforce otherness. Sameness such as imitation or copying caused expectations to receive and give. A donor may imitate another donor, a local person may imitate a local recipient, etc.…

travel philanthropy. Travel philanthropy exhibits processes of ontological separateness based on socially formed categorical boundaries, which shape actors' ideas about who they think they are (Kovacs 2017) and the expectations one has of another (Ehrnstrom-Fuentes 2016). This is particularly relevant because tourism is, for obvious reasons, powerful in generating a higher level of cultural mobility for both international tourists and the local visited population, as they both physically and mentally encounter differences in a specific geographic space (Sharma 2019). Thus, both visitor and visited are confronted with alterity and strangeness (Levinas 1999; Kovacs 2017), and so Cooley's 'looking glass self' (1902) is recognised to have a significant influence on how actors in the system navigate and reflect this ontological separateness in their decisions. As two Ugandan participants stated:

> *I see people getting more connected to tourists who come to these ends. I see the relationship of the tourists and people who come in, I see it strengthening a bit and they end up forming some things that leaves money behind and hence people do things with the money.*
>
> (U2_INTER_12)

> *The way we are, we think in other places people are well off, like when I go from here to Kampala I see communities which are better than here so coming here and you say you are a bit poor it is not your problem it is not my problem because I was born here and I found me here, the area is like that.*
>
> (U3_TB_19)

Referring to the underlying belief system, causes of otherness include beliefs in what the world should be like, in the right to gaze at others, and closely linked to the sentiment of compassion, which would not exist without an *other*. As an emergent entity in the system, *otherness* is also caused by an interpretation of the *other* stemming from cultural mobility enabled by tourism. This combined with current or past events leads to an association of the *other* with the act of giving or receiving, and a perceived difference between *self* and *other*. Experience and historical events, such as those found in scholarship on the lingering impacts of colonisation and the aid industry also have an effect (Ndlovu-Gatsheni 2015). The business of poverty and direct philanthropy perspectives are built upon otherness, arguably presenting the other or self as an object of charity and built on the self-perpetuating imagination of otherness, whether the giving or receiving *other*:

> *I have learnt that there are some people who really feel sorry for the very poor people here, there are some who honestly feel sorry because I have seen some of them in tears, like when we give them the story, I have seen some in tears crying and really someone wouldn't pretend and end up crying I don't think so, so I have learnt some people really feel sorry and wants to help.*
>
> (U2_INTER_12)

Emerging from the structure of the travel philanthropy system is a vivid extension of Ayittey's (2015) curated African image defined by its woes. As organisations and individuals attempt to change the world according to their beliefs (McDonald and Scaife 2011), they lock in the same narratives (Dunning 2012). Traditional subjects of poverty in Africa have also adopted the narratives ingrained across the rest of the globe. They join the production of this image in an attempt to change their world in preconceived ways, with stories seemingly normalised and influenced by colonialism, despite this opportunity of greater agency in storytelling about Africa (Ndlovu-Gatsheni 2015).

Otherness in the system reinforces or forms new beliefs – i.e., that all foreigners are rich, foreigners have a tendency to give, that giving is virtuous, and so on. Moreover, *otherness* shapes norms of pitying others, interpreting the normal as abnormal, and providing grounds to act superior. As these participants explained:

> *There are tourists they visit that village and then they ask problems, why kids they do not go to school, what is the problem, and they ask problems from the village and then the village the elders tell the problem.*
>
> (E1_TE_52)

> *Mostly when you take them to the community and they are seeing an old house or they see like a women carrying heavy stuff on their head, they feel that thing, which I know is a normal thing for us... but for them they are like ohhh it's a big thing because they have never experienced that, that's how sometimes they do want to help.*
>
> (U2_INTER_9)

Said's (1978) *Orientalism* informs 'Us', as donor from the North, and 'Them' as recipient in the suffering South. The promotion of the colonial gaze is very much alive in travel philanthropy and identifying with binaries of rich and poor, white and black, foreign and local, play out considerably. While Said associated such binaries with Western thinking, binary thinking occurred across stakeholder groups. While fewer researchers study the 'Gaze' towards the West (Jedlowski and Thomas 2017), our study observed that the Gaze operates in many directions across the system.

Travel philanthropy appeared to be reproducing myths, categories of cultural difference and common stereotypes about *otherness* (Salazar 2013). However, our study showed that the 'voyeuristic gaze' identified in the tourism literature (Sharp 1999), in which one objectifies and trivialises racial identity, occurred both ways, with *otherness* being equally present from a local actors' perspective – i.e., using 'foreigner' to categorise tourists – with a consequent effect on a range of behaviours and decisions made by both. How one acts towards an outsider/foreigner can be very different to what is appropriate and acceptable towards an insider/local (Sahlins 1968; Kragh 2016), as evidenced in the following statements:

> *When they reach here of course they [tourists] see the community is very poor, that's why I think they donate, they see people here very poor, and they think the communities they should also develop.*
>
> (U3_TB_19)

> *Someone who took it off someone who is an outsider, people don't see him as someone is very very guilty, they think someone from outside has things to give.*
>
> (U2_INTER_12)

> *They expect, all tourists are the same, all white is white for them.*
>
> (E1_TE_52)

It is also important to distinguish *otherness* from the 'white saviour concept' (Zane 2016), as local actors were found to be othering themselves, objectifying and homogenising themselves as objects of development. This reflected Sharma's (2019) suggestion that local people retain agency and present a simplified version of themselves that confirms the Western idea of the *other* through negotiated power (Zevallos 2020), thus, *otherness* becoming a strategic mechanism to inspire the act of giving or receiving.

The combined efforts of all actors in travel philanthropy produces yet another example of Africa being an object of pity and reinforcing stereotypes of African poverty (Jones 2017). Our study also identified narratives of deservedness, as argued by Rudrum et al. (2016), with claims about targeting the most vulnerable without really understanding the meaning of poverty in specific contexts. However, in the travel philanthropy system, not only do we witness African people positioned as victims and subjects of development by outsiders (Lahiri-Dutt 2017), but also the subjects of development explicitly position themselves as the *other*. For example, in our study, it became evident how local people would position themselves as victims of poverty, othering themselves from the visiting other:

> *They donate us because we are not like them, we are poor, you will find we have poor families which cannot support themselves, they see us as people who are in need and they maybe the way they see us, the way they hear about us, so you find that maybe they pick us, because someone told me for us in our country we have got poor people but they are not like the Africans.*
>
> (U1_CBO-IN_4)

Also prevalent was a process of *othering* of the foreign white donor by local people, reducing cultures to bounded totalities (Deifelt 2007), as a cause and an effect of travel philanthropy. Embracing well-traversed global African development narratives, the tourism industry increasingly positions travellers not so much as exploring neo-anthropologists, but as development professionals with the right to save African *others* from their plight (Burgold and Rolfes 2013). Today the souvenir is arguably a personal story of saving the *other*, often designed and sold through staged ideas of *otherness* (Staszak 2009). Novelli et al. (2015) argued this as a vivid example for the democratisation of charity. Travellers are now philanthropists (Zane 2016) and new authorities intervening in the development scene. Some tourists arrive with preconceptions based on global narrative, while others develop beliefs during their visits. However, the majority of donors may not represent an organic surge of altruism, but the effect of imaginations strategically shaped in powerful ways (Jedlowski and Thomas 2017). This reflects what Staszak's (2009) discussed as 'bland otherness', staged and transformed into merchandise and human zoos, as evidenced in the following response:

> *I called him and told him to organise because these people want to find children in class, they want to see how the children are taught, so I told organise and make sure the teachers are there, we are going to come with our guests they are going to see how the kids are taught and then afterwards they have gifts for the children, I don't know what the gifts are but they have gifts.*
>
> (U3_TB_16)

Local actors can perform individualism or collectivism (Allik and Realo 2004; Tomasello 2020), aligning with Tomasello's (2020) views that local actors can wear both hats. For instance, in Ethiopia part of society has shifted from a collectivist sharing society to an individualistic one:

> *In the former time we used to share what we had with especially neighbours, now for instance I have no neighbours I have my own fence... we used to eat together, we used to drink, we used to have feasts together, and also with our neighbours but these days nobody is having shared life with neighbours.*
>
> (E2_NGO_61)

In Ethiopia and Uganda particularly, a significant shift to individualism has occurred due to the system of travel philanthropy, as exemplified by the following responses:

> *[There is] competition for development [and] envy.*
>
> (U2_CBO-F_10)

> *Money is now god, in some parts you know, so they are blind and they are looking for money, to get money you have to step over others, and even on their dead bodies.*
>
> (EN_TB_66)

On the other hand, there are also collectivist traits that affect resource allocation, focussing on in-group responsibility and shared value, soliciting whatever resources one can get from an outsider group for sharing with the insider group. In Ethiopia for example:

> *They beg off tourists, but I don't think it's good or bad for the economic activity of the country... it has its own advantages because the income flows in the country... giving money has its advantages only if the money has flowing in the country because the aim is growing up of the country yeah, one country one peoples one growing up and good.*
>
> (E2_TE_59)

Within a community, the business of poverty enables a more granular 'us vs. them' thinking. In a Ugandan village, which to an outsider may appear culturally homogenous, power and class divides indicated that the Bakiga population tends to objectify those of Batwa origin as a saleable product in their business of poverty:

> *We live around, but like we lost the track, we lost the way, we lost ourselves, so there is a way we feel inferior and we are not proud of ourselves... we are not freely participating in the community as like others, so there is somewhere we reach and the inferior limits us to reach.*
>
> (U2_CBO-IN_7)

Similarly, in Ethiopia, the national tour guides objectify Hamar tribe residents as poverty products. In Zimbabwe, historical inequalities and rivalry exists between Ndebele and Shona tribes and this affects the way business is conducted too.

Impacts are never as simple as outsider vs. insider, or foreigner vs. local. To understand the reality of exchanges in systems such as travel philanthropy and pathways of aid, it is also important to understand the ontology of *sameness*.

9.1.1.2 Sameness

Three perspectives of *sameness* are relevant to our study. Firstly, sameness intended as 'identity' (Klein 2014; Ricoeur 1981; Olson 2004; Staber 2010) becomes useful to analyse different roles of actors, hybrid identities, and decision-making. Secondly, reflecting on Smaldino's (2014) concept of cultural traits, we must draw on some of the socially learned norms and behaviours that emerge within the travel philanthropy system. Thirdly, and mostly relevant in this analysis are ideas such as mimetic desire (Girard 1977), imitation, emulation (Whiten et al. 2009; Schlag 2012; Maness and Cirillob 2016; Staber 2010; Babutsidze and Chai 2018; Richerson and Boyd 2005), and role models (Bandura 1977; Arjoon et al. 2018; Fastenrath et al. 2018; Koumakhov 2009; Liu et al. 2019).

Interconnected with *otherness* as an element in the system, sameness is understood as personal or collective identity (Staber 2010), so a reflection onto the *other*. In the context of travel philanthropy, cultural mobility through tourism in Africa amplifies differences between Us and Them. Identities were also hybrid. For instance, a local actor may regard their personal identity as the educated elite in a community when reflecting on the *other* in their own community, yet he/she may take on a personal or collective identity as a member of a disadvantaged local community in relation to the tourist or donor *other*. Similar trends appeared across all stakeholder groups. An ordinary tourist in Africa may assume membership within an 'elite', 'altruistic' community in relation to local people, in turn reinforcing collective identities of poverty and inferiority amongst the visited people. The influence of personal and collective identities affects cultural traits and feedback into the idea of role models too.

Research findings showed that it could take just one individual to initiate change in the system, as the power of sameness can lead to exponential growth (Mingers 2014) and positive change. Exposure to the *other*, a new behaviour, idea, belief, and so on can create role models. However, this may also play out negatively too, as during our research, some local people expressed cynicism towards international donors and a widespread belief that any actor occupying the role of intermediary is always taking a personal cut. These perceptions emerged into behaviours and have become the norm, inspiring imitation behaviour, sameness, and reinforcing the cycle of mistrust of others, which can cause lying, and inter-community rivalry, arguably significantly influencing exchange systems in the localities under investigation and in Africa as a whole.

Similar to Mueller's (2020) thinking, actors affect other actors and adapt to what other people do by imitation, as opposed to market forces. Found to a certain extent across all stakeholder groups, Girard's *mimetic desire* in travel philanthropy is perhaps most strongly interconnected with expectations to receive donations, particularly amongst local actors driven by a desire to become the intermediary in the 'business of poverty' and prosper financially. Mimetic desire also existed amongst some tourism businesses, particularly in Zimbabwe, where a greater proportion competed for responsible tourism status and a social licence partly through travel philanthropy involvement. Travellers also arguably demonstrated mimetic desire in seeking the warm glow (Dickert et al. 2016) associated with giving.

Other significant characteristics of the travel philanthropy system were *imitation and emulation* (Whiten et al. 2009; Schlag 2012; Maness and Cirillob 2016; Staber 2010; Babutsidze and Chai 2018; Staber 2010; Richerson and Boyd 2005) of the *other*, including expectations to give or receive. In Uganda, the aspiration to be a local intermediary was typically an example of an imitation and emulation process aimed at generating covert 'commission' from selling the ideas of a 'project' to a prospective donor, being the middle-man of donations, using community-based organisations (CBOs) for personal gains, at times leading to rivalry between recipients, and worse the involvements of minors in the transactions. For example:

> *They kind of tap the donors ... they use lies to bring someone down who is in fact using the money well so that they can take the advantage ... it started in Buhoma and now it has come it has come here.*
>
> (U2_CBO-F_8)

> *They have seen what has happened in other places, they have heard what has happened in other places and they want the same to happen.*
>
> (U2_INTER_12)

In Ethiopia, similar patterns amongst intermediaries and recipients emerged, with young children receiving the donations:

> *There was a 5 year old teaching a 2 year old the language of hassling – "hello how are you no I don't want any money could I get your email address so that I can practice my English" – and then all the emails start coming... now one guy writes all the emails and sells them to the other hasslers who send them out.*
>
> (E2_TB_62)

Imitation and emulation also concerned the behaviour of some tourism businesses. While most Ethiopian businesses imitated non-involvement, in Zimbabwe imitation of involvement was most prominent. Lodges, tour operators, and hybrid models copy each other, even in the types of development interventions for the 'fear of missing out' (FOMO):

> *There's a great thing of FOMO here like you know you don't want to miss out and if a company sees oh wells so and so is now involved in that, why are we not on that programme, so I think it's just a part of that, don't get left behind or don't be seen to be not doing anything.*
>
> (Z1_NGO_22)

Geographical and social context affect the density of the social network (Staber 2010). In Zimbabwe, ideas, practices, and expectations circulated at a faster pace than in Uganda. National-local guide imitation, on the other hand, grew in Uganda and Ethiopia through greater mobility and communication technology.

Frequency-based imitation (Banerjee 1992) and trait-based imitation (Richerson and Boyd 2005) produced expectations to give or receive. This happened through copying beliefs and doing what others were doing, or imitating other businesses. Tourists also imitated each other across all studied destinations, with giving being partly 'pushed' and partly 'pulled' by peer-pressure, belief, and emotion, projects that had received donations previously (U1_CBO-F_1) or simply imitating the trend or the norm of giving in Africa, or the desire of being identified as a travelling philanthropist in Africa. As one CBO said:

> *By showing them things that is already done...this has been funded to me... look someone donated me this I have managed to put a window, someone donated me this I was able to put a toilet.*
>
> (U1_CBO-F_1)

Whilst imitation patterns can be local–foreigner or Africa/South–North, imitation patterns in travel philanthropy appeared to more commonly occur between actor groups' peers – i.e., traveller-vs.-traveller, business-vs.-business, guide-vs.-guide, child-vs.-child, and/or peers of the same age group (Babutsidze and Chai 2018). The role model of the West, portrayed through tourism and TV, also drove 'Africa' to copy the West, striving for example, for comparatively improved wealth and luxury lifestyles:

> *The technology, there were a few radios to hear news, now there are TVs in each house so I mean there is no wonder people are asking [for donations]... lots of demand now ... living demand, you need a better life, a better infrastructure, a better house, a better car.*
>
> (E2_NGO_57)

> *We're copying each other and as Africans we are all of us copying western.*
>
> (U2_CBO-IN_7)

Role models motivated by social and economic gains (Sharma 2019) or by the fear of social losses (Fastenrath et al. 2018) emerged across all our study locations, with only one company or individual needed to spur imitation. Positive role models did not necessarily create positive, ethical, or responsible behaviours, as they merely provided a model to imitate. Examples of what was portrayed as 'positive role model engagements' included the following: Ethiopian children receiving money from a portfolio of sponsors via Western Union each month; Zimbabwe's tourism businesses implementing copycat corporate social responsibility (CSR) activities; and Ugandan guides becoming donation intermediaries. However, these does not necessarily translate into positive impacts.

Sameness highly influenced behaviours that emerged amongst recipients and intermediaries. The Ethiopian child who directly receives monthly sponsorship from an eventual portfolio of foreign tourists may gain a new identity that offers new freedoms and luxuries. That child may become a 'positive role model' for others, but not necessarily have a positive impact, as him/her may grow up with an idea that somebody will provide in perpetuity, ultimately affecting their motivation and ability to look for entrepreneurial ways to get out of the 'giving and receiving cycle' in the future. In Ethiopia, other youths taught children the tricks of the trade and initiated them into the 'business of poverty', who thus gained respect from community leaders for becoming recipients of easy money. This spreading culture of easy gain allowed those without education to profit from the business of poverty as new, successful roles models, a situation creating great concern in Ethiopia:

> *We will not get leaders from the community if individuals who are not educated become rich people all people prefer this way so we will not get… . they will have the power even still they have the power some of the, you know you can compare me and I may have second degree masters the other one has a hotel but complete grade 10 or 12 now who is an influential, me or him, for most of the people he may have an influence he may have a power, when he talks something they may say 'oh he's a' even nobody wants to be friends to me, if he come he can have many friends you know that is power, money has the power you know, so the town will be led by illiterates in the long run.*
>
> (E2_NGO_61)

Role models are potentially one of the most critical mechanisms in shaping the system's behaviour (Morgenroth et al. 2015; Liu et al. 2019). For example:

> *Many other previous tourists have donated to other people in the past, that means everyone wants to take another option like ok if my older brother gets some sponsorship ok it's money from somebody I look up to him and then try to get something.*
>
> (EN_TB_68)

Negative role models on the other hand refer to the *avoidance* of similar distresses. Avoiding what the role model did can also influence those in the travel philanthropy system, going somewhat against Koumakhov (2009). In our study, examples included actors not 'touching the money' from tourist donations or occupying the intermediary role, thus avoiding some known risks emerging from being involved, at times leading to accusations of cheating and being embroiled in local politics:

> *They [the community] believe that once the tourist says I will give you a flask and a cup and a tea bag and a spoon, they don't know that one time, the tourist, after giving the flask may withdraw or hold to give the rest, for them [community] they will continue believing they [the tourist] are also sending this [to Uganda], and for you [me], you are keeping it… . you find the one who can trust you tomorrow cannot trust you for things you didn't do.*
>
> (U3_TB_18)

In addition to this, in Zimbabwe, there is a growing donor fatigue and dependency from development intervention models – i.e., learning what not to do in development, emerged as characterising both positive and negative role models and shaping the system. While imitation and role models were identified as key mechanisms for the system growth and function and valuable leverage points to generate changes, increased tourist and business donations caused by role models copying the *other* (whether virtuous or not), these practices created or fuelled higher level emergent behaviour in the system. From a negative perspective, the emergence of the new powerful role models appeared to be leading to dependency and further expectations for donations and greater donations solicitation:

> *They think when they do the same, they get their commissions and become a bit rich, so they want to do it after copying from others doing it.*
>
> (U2_INTER_12)

In Uganda, the desire to become recognisable intermediaries to attract donors and benefit financially pushed the establishment of CBOs and/or private schools, at times through evidently fabricated organisations and stories. The style of relationships with foreign donors were also the result of copycat practices, with sameness drawing individuals into the system as participants and shaping ad hoc goals and behaviours, and the same models emerging across tourism destinations in Uganda and Ethiopia.

On the other hand, some positive–positive role models also emerged, in that role models used donations for furthering their education or starting sustainable businesses. For instance, whilst one might perceive the fact that in Ethiopia some individuals had received donations to establish a large hotel business, purchase vehicles, and support education with money invested, rather than merely fund a life of alcoholism, drugs, and begging (an unfortunate consequence of philanthropy discovered in some instances) as a positive outcome, some may raise the moral dilemma and systemic paradox of travel philanthropy as fuelling younger generations' wealth gained through skipping education and by begging on the street.

9.2 Expectations and different moral economies

Although *expectations* are not necessarily always observable, when triggered, they generate behaviours in the system with significant causal powers. This research found *expectations* to give or to receive to be key systemic aspects of causality in actors' economic decision-making in travel philanthropy (Manski 2004), a cause behind peoples' behaviour, and an effect on other mechanisms. Expectations are generally based on imagination about what can happen (Shackle 1961), and therefore not necessarily rational, meaning that actors may not be able to use all information to make informed predictions regarding future outcomes (Fastenrath et al. 2018). This research further highlighted that actors can overestimate the probability of good outcomes basing decisions on expectations rather than facts. Actor expectations, as shown in *Figure 9.5*, occurred as a key mechanism underlying the local consumption of donations, both to give and to receive, and as a cause and an effect of *otherness* and *sameness*.

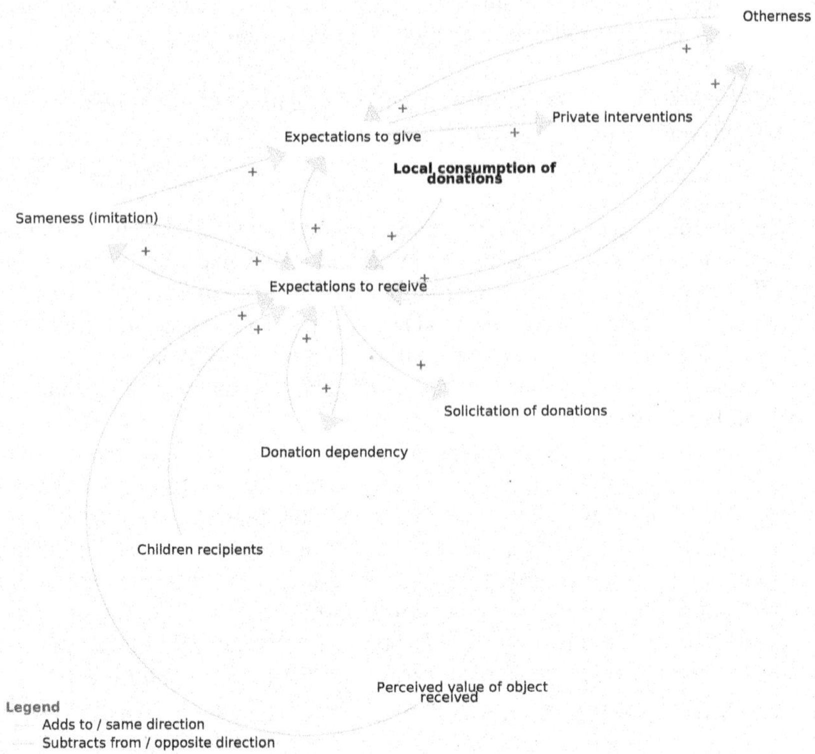

Figure 9.5 The ontology of an expectation.

Explanation for Figure 9.5

An increase in expectations to give, or to receive, if occurring within an enabling context (e.g., an exchange pathway was possible) tended to result in an increase in the local consumption of donations. Conversely, lower expectations cause less giving/receiving. The set of mechanisms underlying expectations to receive includes beliefs, principles, norms, emotions, stories, and experiences.

The greater the expectations to receive, the greater the propensity to depend on donations, and the greater the dependency on donations, the greater the expectancy to receive. The perceived value of an object received, if high drives expectations. This is a slight anomaly because a low-perceived value can also increase expectations (but not shown in this map – see Section 10.4 *and Chapter 11). Children particularly, but not only, by receiving gifts, have increased expectations.*

9.2.1 Expectations to give

Based on earlier discussions about the assumption and core belief associated with giving as a virtuous undertaking, a non-consequentialist perspective driving the expectation to give often exists. Tourists' expectations to give are triggered by different mechanisms depending on the stage of their trip (pre-trip, on-trip, post-trip), amongst other conditions, and expectations to give tend not to focus on *whether* to give, but which pathway to choose.

Aligning with Burgold and Rolfes (2013), our study contributes to the idea of 'new moral tourists' as aid workers. Expectations of moral integrity in travel philanthropy stem from beliefs in the freedom of exchange relationships, the existence of an underlying expectation and/or obligation to give, and aspirations to change the world, facilitated through stories (Usó-Doménech and Nescolarde-Selva 2016). They shift decisions from economic to social and lead to unquestioned myths that this is the way to do things (Fastenrath et al. 2018), reinforced through imitation (Meyer and Rowan 1977), and influenced by the number of story followers (Fastenrath et al. 2018). In this regard, tourism businesses investigated here demonstrated mixed philosophies in terms of expectations to give. In Zimbabwe, giving has become a competitive marketing tool and prerequisite for a social licence, and expectations to give from the tourism sector are greater than in Uganda and Ethiopia. Ethiopia appeared to be at the other end of the spectrum, with tourism businesses showing particularly low expectations to give and less social pressure. This low expectation to give by tourism businesses did not reflect a reduction in the expectations to give by tourists, however. The absence of tourism businesses' expectations did not affect the level of local consumption of donations, but rather raised the question of who should give and how.

Expectations to give in some contexts could stem from a desire to enrich the donors' lives rather than the beneficiaries' (Bertini et al. 2014). While exchanges in the travel philanthropy system are not necessarily an act of freewill or free from obligation, the belief that giving is virtuous may not always drive expectations to give. Evidently, other underlying mechanisms exist, such as expectations to give as a form of reciprocity found in gift-giving (Mauss 2016) for an enjoyable tourist experience, or driven by exchanges for a social licence. The causal properties behind expectations to give are complex and contextual and might stem from obligations.

Few mandatory requirements exist for tourism businesses (or tourists) to provide financial support to communities and/or conservation projects through a proportion of earnings (Wolff 2011; McLachlan and Binns 2014). In Zimbabwe and Uganda, formal arrangements existed to grant concessions or award benefits as a social and unwritten expectation, rather than a mandatory obligation, still influencing expectations to give.

Lastly, as tourists experience novel, unexpected, uncertain situations (Stutzer 2013) when faced with alterity during travel (Levinas 1999), moral intuitions often come into play (Woods 2015) based on beliefs fostered by *personal emotions*. Emotions theoretically have a role in human rationality (Dornschneider 2019; Dunning 2012; Koumakhov 2009; Matthiesen and Klitmøller 2019), but actors were found both to imagine future situations based on their emotions, which affected their present decisions (Hanoch 2002), and to drive the System 1, gut-level responses as a reaction to stimuli (Woods 2015; Smaldino 2014):

> *They come here for 3 days maybe and they don't think about sponsoring and at the end of 3 days they have the idea.*
>
> (E2_TE_63)

Furthermore, donors' expectations to give may increase due to fear of saying no, driven by feeling pressure to meet the expectation or perceived obligation to give, but also as a consequence of positive emotions, or by being horrified, shocked, and sad by how they view the world around them.

9.2.2 Expectations to receive

The beliefs that foreigners have a tendency to give, that there is an obligation to give and that I/we are entitled to receive have causal powers and were worldviews embedded within a high proportion of local observed practices. One could argue that the story has gained myth status and become a societal expectation and a norm (Fastenrath et al. 2018). According to a local guide in Ethiopia:

> *They [locals] see one thing once and they are always looking for that thing, if farangi [foreigner] is distributing t-shirts or candy, biscuits in the village, always they are looking when tourist cars are coming to the village they are always looking for these things, they are expecting.*
>
> (E1_TE_52)

An increase in the local consumption of donations caused an increase in expectations to receive, whilst an increase in expectations to receive inspired soliciting donations and an increase in the local consumption of donations as a feedback loop. Causes underlying such expectations stem from historical and recent belief systems and norms, such as the persistent social labelling of Africa by the Global North and the rise to meet the expectations of others (Maxwell and Dornan 1997) or the arguably locked-in expectation to receive as an unequal African (Dunning 2012).

Beliefs in money and social power also drive expectations as part of the wider social system of travel philanthropy. What emerged was an increasing desire for material gain (Stern 1991) amongst local actors across our study locations. The broad findings were that local people increasingly required and desired more money than previously, influenced by a context of limited economic opportunities or lack of knowledge/ideas to create alternative opportunities. Local actors therefore increasingly turned to donations to increase their personal wealth, with money deriving from donations perceived as 'easy money' and an opportunity to "change your life very short [quickly]" (E2_TE_59):

> *They want to be rich... now people love [prefer] instead of [taking the] long way they take the shorter way to be [rich], just because of the following of examples.*
>
> (E2_TB_56)

The example mentioned above refers to others in the country that have received donations directly from tourists visiting Ethiopia. Similarly in Uganda:

> *People want money, there are so many needs that they have to meet, there are so many associations they have to join and they don't have money and they have to use this kind of donations, looking for funds for needy people to kind of enrich themselves.*
>
> (U2_CBO-F_8)

Gaining money and social power through travel philanthropy had become an increasing expectation amongst locals. The belief in money combined with a belief in social power suggested that gaining money leads to status in the community:

> *When it comes to decision making, when you are poor ... they also think your thoughts are poor ... they have to first look at someone who is a bit well off, they think that one has a sound mind, they think it is someone who can reason, so there is a way that person is respected, so the poor person faces a lot of constraints, even if you are a person with better thought.*
>
> (U2_INTER_12)

> *That one has that social status to kind of get from the community, anyway which is unfortunate because you are supposed to make decisions for the community not for your own selfish needs.*
>
> (U2_CBO-F_8)

The travel philanthropy system has partly become a tool for fulfilling societal expectations of status and a cause of the expectation to receive donations. Many CBOs appeared to emerge to fulfil expectations of growing financial resources available from the travel philanthropy system, but also by personal beliefs in money and social status:

> *They want money, most people who start up these projects and who bring the ideas to tourists and other people that they do projects, they are targeting some benefits, some money.*
>
> (U2_INTER_12)

S*ameness*, including ideas of social learning, imitation, emulation, or role models, also drove expectations to receive (Richerson and Boyd 2005; Staber 2010). Within the context of holding a belief in money and social power, the behaviour of peers appeared to be very influential:

> *They [guides] are seen as well off, they have seen them leaving the poverty status, and they think when they do the same, they get their commissions and become a bit rich, so they want to do it after copying from others doing it.*
>
> (U2_INTER_12)

They know somebody somewhere have got money from tourists and build hotels or have received from a generous guy and bought a car and that kind of thing so that's actually keeping kids on the street because they are looking for that guy who is going to do the same kind of thing and they believed one day instead of going to school they will find that guy.

(E2_B-COM_64)

Furthermore, some of the findings of our study align with Gardner's (2015) paradox of moral economies. Amongst some local recipients in Zimbabwe, the perspective was that a tourism company 'should do more' and "owes them benefits from their business that exploits their resources" (Z2_2_B-INS_42), evidencing a sense of entitlement. Promises made by companies to do good things in the future appeared to have raised expectations (Gardner 2015), as much as tourists promising donations while on-trip, but often not delivering. Promises of this kind have social, economic, and ontological implications as they cause certain emotions and expectations:

They [tourist] say we would like to help, let's reach in our country and see what we can do. Here we take it as a promise, so we are expecting something.

(U2_INTER_9)

It's not a good thing because you raise the hopes of that person and at the end of the day nothing comes.

(Z1_CBO-F_30)

Particular contexts that enable the solicitation and/or local consumption of donation tend to trigger expectations to receive. An actor may have an *expectation* to receive a donation for example, but it may not produce that event – i.e., the consumption of a donation. Various contextual factors, such as the type of face-to-face interaction with tourists, corporate giving practices in the destination, etc., are necessary to interact with this mechanism (Pawson and Tilley 1997). Donors are also not homogenous, so expectations to receive might arise differently depending on the expectations of the donor. For example, they may donate 'carefree' as they called it in Uganda or with 'strings attached' (U2_B-COM_13), which affects the level and type of expectation. To change *expectations*, and perhaps attempt to adjust system outcomes, changes to *otherness* and *sameness*, as well as the underlying belief system also need to change.

9.3 Local consumption of donations: push and pull factors

The local consumption of donations is a key element in this informal exchange economy (Figure 9.6). Travel-related giving in the studied destinations is growing, supporting geographically broader claims that travel philanthropy is a growing phenomenon in Africa (Phocuswright 2015; Goodwin et al. 2009; Novelli 2015). Although not investigated quantitatively, growth of tourism in a community tends

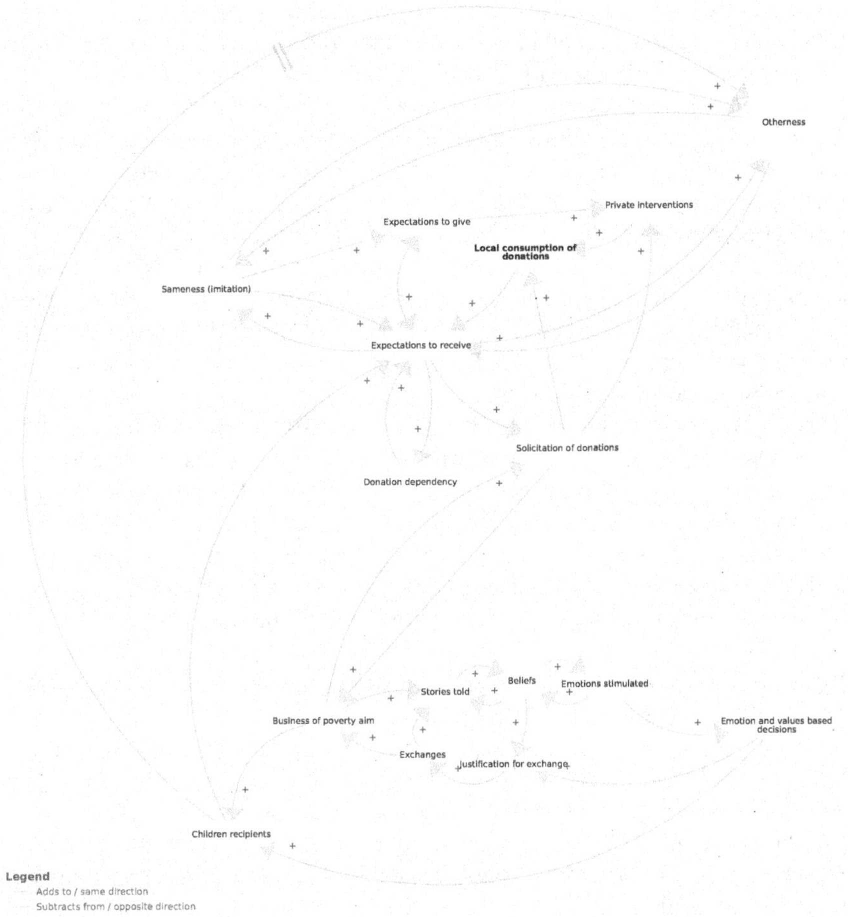

Figure 9.6 Local consumption of donations – cause/effect.

What is driving the local consumption of donations?

Explanation for Figure 9.6

Push and pull factors increased or decreased the local consumption of donations and were interrelated. The pull factors included the business of poverty and the solicitation of donations. The push factors arise from expectations to give and private [philanthropic] interventions.

This diagram illustrates that all of these connections are same/positive relationships – i.e., if the solicitation of donations increases, the local consumption of donations increases, and vice versa. If there is a growth in the business of poverty, there is a growth in the solicitation of donations, which increases the location consumption of donations. If the business of poverty increases there tends to be an increase in emotion and values-based decisions, which can increase exchanges, the solicitation of donations and the local consumption of donations. These are general patterns so the analysis below provides explanation that is more specific.

to correlate with growth of donations (Honey 2011), with mechanisms in rural areas triggering more giving than urban destinations due to the type of interactions. Philanthropy manifests differently in the travel and tourism context, although it mimics some behaviours found in other philanthropy systems (Ardoin et al. 2016). Travel philanthropy may be part of the long tail of private philanthropy, but the large numbers of micro-donations compared to traditional foreign aid can also have significant socio-economic implications. Therefore, this research supports Atkinson's (2009) call for more attention to the growth of this phenomenon as a perfect example of a shift in the Global North's compassion from the armchair to personal encounters with distant suffering *others* in Africa (Silk 2004; Korf 2007). Our study also contributes new findings on travel philanthropy as an exchange economy and a system shaped by all actors and argues that the often-presumed *passive* beneficiaries of donations in rural African communities are in reality not passive recipients of aid (Silk 2004). Rather, they often *actively* seek donations not as charity, but as business, for profit.

Various mechanisms within specific tourism destinations appeared to influence the scale and type of donations. The location of the tourist accommodation influenced the giving pathways. For instance, the closer the proximity of tourist accommodation to a community, the greater the giving of tourists. The same appeared for businesses. For example, a hotel business with multiple sites in Uganda explained that those properties embedded within a community were more likely to experience pressure to give compared to those out of sight:

> *We're lucky to be in a national park in the other ones, so there's not as much to give to.*
>
> (U1_TB_3)

This does not mean that businesses in national parks did not donate to communities. The vast majority of relationships were primarily with the nearest village, accessible to guests, as the proximity enabled enhanced levels of direct relationships between tourists and local people.

The supply of new 'responsible tourism' products meant greater access to potential donations and greater propensity to consume donations. This emerged as the drivers of push and pull factors. *Push factors* include the local consumption of donations including systems perspectives of direct philanthropy, social licence, and tourist product. Pull factors include capital for conservation and development, with travel philanthropy being a strategic undertaking to raise funds for various interventions as a resource platform, and the business of poverty. The social licence objective also existed across the two.

Digging deeper into the philanthropic practices of tourism undermines its potential as a *private sector for development* (McEwan et al. 2017) per se. Whilst tourism businesses were thought to be in the driving seat of travel philanthropy (e.g., Honey 2011), this does not represent a particularly accurate representation of the system. A growing proportion of tourism businesses in Africa are indeed shifting towards mixing business and philanthropy, essentially through CSR, but this is not a ubiquitous norm across the industry. The majority of tourism businesses in Uganda and Ethiopia actively avoided any involvement in travel philanthropy because philanthropy made treacherous territory in terms of local politics (EN_TB_68, U3_TB_18), whereas in Zimbabwe a greater proportion of businesses were more eager to be involved. Although similar system behaviours existed across cases and countries, hence the use of one systems map, actors driving the local consumption of donations differed.

Although the tourism industry is not alone in providing financial resources or infrastructural developments within the physical spaces in which they operate, with similar trends found in the mining sector for example (Gardner 2015, Z2_2_B-INS_42), what makes travel philanthropy particularly unique is the direct involvement of individual tourists (the customers) in the system. Unlike mining, for tourism businesses their product's success, business stability and client satisfaction directly depend on the good sentiments of local communities. Therefore, the incentives to engage in giving are arguably somewhat different. The obligations of the tourism sector in either maintaining, compensating, or repairing the natural, social, or cultural spaces they operate within generate a push factor for the local consumption of donations, as described by an environmental non-governmental organisation (NGO) in Zimbabwe:

> *It's always a double-edged sword, you might have the situation where tourism is impacting negatively on the environment in certain ways and you're the organisation that's trying to deal with that negative impact so you're trying to derive an income from a tourism-based organisation through philanthropy but dealing with an impact that they created so it is a cycle.*
>
> (Z1_NGO_27)

From the demand-side, new travel experiences, including a growing interest beyond wildlife into people as historical and cultural tourism products, as well as a growing trend to donate and have a 'giving' experience, combined with a growth in the mass segment of adventure travel in Africa, were identified across all three country cases as push factors. Partly enabled by recent years' exponential growth of tourism suppliers, both local and international, as well as expanded national infrastructure such as air and road that have enhanced the speed and ease of travel and opened new rural destinations, a greater volume of tourists were evidently reaching all the remote areas of our study.

A community representative from Zimbabwe explained that, in peak season, they receive tourists in their school every day with most giving donations (Z2_1_B_INS_38), and that interactions with local people were through either a formalised tour or an informal encounter, triggering gift-giving spurring from physical and emotional interaction:

> *Whenever these tourists are coming ... we are having them at school, they are there at the school, they talk to the children and from there they come to the village and they ask you, what are your needs and this.*
>
> (Z2_1_B_INS_38)

The mechanism of seeing, and believing, and the personal connection to the perceived cause, has become a powerful giving mechanism, which arguably even surpasses the emotional arousal used by NGOs for faraway causes leaning on moral intuition for marketing campaigns (Woods 2015):

> *They see this child sitting on the floor it makes a huge difference.*
>
> (Z2_2_TB_44)

Tourist itineraries that appeared to leave little room for additional activities to 'meet the people', restricting formal or informal interactions, most importantly limited decision-making time concerning donations, as further discussed in Section 10.3 on boundedness.

For local individuals looking to capitalise on resources available within their community, travel philanthropy has grown into an appealing opportunity for making money with low capital investment, in which the product is essentially 'poverty'. The pull factor is in the opportunity to generate a personal income from turning 'poverty' into an economic resource, or amongst tourism businesses, it is about striving for a social licence or business marketing, using their business as a platform. For example:

> *They don't channel their profits to here, they are only like a vehicle to help us get help.*
>
> (Z2_1_B-INS_31)

The structure of the business of poverty, and most of the details, appeared the same across all cases. If tourism businesses had not designed a specific tourist experience around poverty, locals stepped in with alternative ways to capitalise on it. In Ethiopia, the local children learnt how to stage the 'productive interaction' (Honey 2011:3) for tourists as much as a tourism business may have designed an experiential product. Locals deliver a service and an economic activity of a comparable nature, whether delivered by the children of Lalibela (Ethiopia), the local guides of Bwindi (Uganda), or a tourist hotel in Victoria Falls (Zimbabwe). Therefore, travel becomes less about the idealistic notion of the diversity of humanity coming together and sharing their wealth. Instead, we see the commoditisation of tourists and their interactions with locals staged for social and economic gains:

> *Because of what tourists give ... kids on the streets waiting for tourist and this is what tourism has caused.*
>
> (E2_B-COM_64)

The findings from this research therefore do not consistently align with familiar debates proposing that aid interventions and the gift-chain are designed by outsiders for passive observers (Korf 2007), or the construction of a Northern vision for people in the South (McNamara and Morse 2004). Although one could argue that effects of colonialism (Ndlovu-Gatsheni 2015) are a force, perhaps the North indirectly designs interventions, and there are interventions based on Western assumptions of development theories. However, the recipients in our study align more closely with Retsikas' (2016) theory of solicitation and taking, because aid beneficiaries in travel philanthropy very often shape the gift-chain and interventions, seemingly more significantly than in the bulk of debates around other forms of development aid. These actors tend to *pull* donations into their communities.

> *They said it is better that the project is being sold by you being the headman otherwise if it is done by the management of [tourism business] it is more like soliciting.*
>
> (Z2_1_B_INS_38)

9.3.1 The solicitation of donations and the business of poverty

As indicated in the system map, the solicitation of donations plays a core role in travel philanthropy and deserves attention as an important pull factor, also linked to the business of poverty (see Figure 9.7). Unique to travel philanthropy, recipients have direct access to donors, and these donors tend to be comparably inexperienced, driven by emotion, and acting informally. This is of course potentially true for some development aid, but in travel philanthropy, these emerge as distinctive mechanisms. Outsiders may control Africa's natural resources, but perceptions of poverty and pity for Africa are not a scarce resource. Soliciting donations has become a means of capitalising on that. This includes solicitation for private and public purposes:

> *My people know that the headman is doing a good job whenever they see me talking to the tourists they are seeing some light, they know I am going to convince these people to do something for them yes.*
>
> (Z2_1_B_INS_38)

This research, as previously mentioned, adds to Retsikas' (2016) argument about "the other side of the gift". Understanding the anthropology of the gift requires grasping the perspective of the taker, bearing in mind that taking follows soliciting, which can involve "more covert, secret and stealthy ways of acting" (2016:6). Many of the 'transfers of wealth' that occur in this system are caused by acts of soliciting, by locals particularly, and in some cases tourism businesses. Therefore, our study contributes evidence by recasting the so-called gift receiver from a passive to active role. It makes a further case for conceiving the gift in Retsikas' (2016:5) suggestion, "not as the adjunct of giving but as the accomplishment of soliciting" where giving is 'a by-product of taking'. Soliciting remains hidden from gift theory (Retsikas 2016), but demands attention as a space for new ethics, with this research contributing new insights.

To shift questions beyond the altruistic motivations of the donor (Honey 2011), we need to both question the motivations of the taker or the actor soliciting and consider this new space for ethics and moral responsibility. Analysing the solicitation and consumption of donations requires a deeper look at the business of poverty, in which actors (intermediaries or recipients) effectively commercialise the concept or perception of 'poverty' or inequality for private socio-economic gains. For example, many CBOs were established and numerous individuals acted informally as the face of a cause to tap into the donations market:

> *They are seeing them as businesses, they can get much more money from community organisations than setting up a shop.*
>
> (U2_CBO-F_8)

As Rudrum et al. (2016:1330) suggested in their Ugandan study discussing humanitarian aid, by linking theories with Mauss and Marx, there are "a set of specialised producers with professional interest in encouraging gift giving and gift taking" and whilst the stakeholders may be different in our study, the findings in the research echo the above. The findings also align with Swarmy's (2017) argument that humanitarian gifts are bound to structures and practices associated with the formal economy such as commodities, markets, and exchange relationships. The business of poverty (Figure 9.7) associated with travel philanthropy is less directly entwined in global politics and macroeconomics compared to the humanitarian sector, but it is tied in with the commercial industry of tourism.

While the common discussion in the literature focus on either demand-led or supply-led travel philanthropy (e.g., Ham 2011), our research adds an argument that one also needs to consider the recipient-led 'enticing' of philanthropic impulses (Goodwin 2011; Burgold and Rolfes 2013) pulling donations. Recipients understand that 'seeing is believing' (Bertini et al. 2014) and are very much part of the tourism product development to generate a greater local consumption of donations, through the 'productive interactions' the industry talks about and the manufacturing of gift relationships (Honey 2011:3). The strategically designed experiences to entice visitors (Spalding 2011) become a business. Not only tour operators capitalise on the vulnerability of the destinations as a commercial experience (Dawson et al. 2011), but the residents in the destinations play an active role in reproducing this pattern.

Tourism businesses and local stakeholders have made great efforts in ensuring poverty and deprivation have real faces (Crouch 2011). The marketisation of human suffering and mobilisation of empathy is part of the travel philanthropy system (Dundar 2016), and whether this occurs to satisfy the voyeuristic needs of the traveller only is arguably up for debate, as patterns of the business of poverty suggest otherwise. Travel philanthropy recipients have also made efforts to tap into the new moral tourists (Butcher 2003), by at times engaging even in questionable practices, as indicated in the statement:

> *Someone doesn't have a project but I mean rents a house and then writes a very big poster and says orphans and vulnerable children.*
>
> (U2_CBO-F_8)

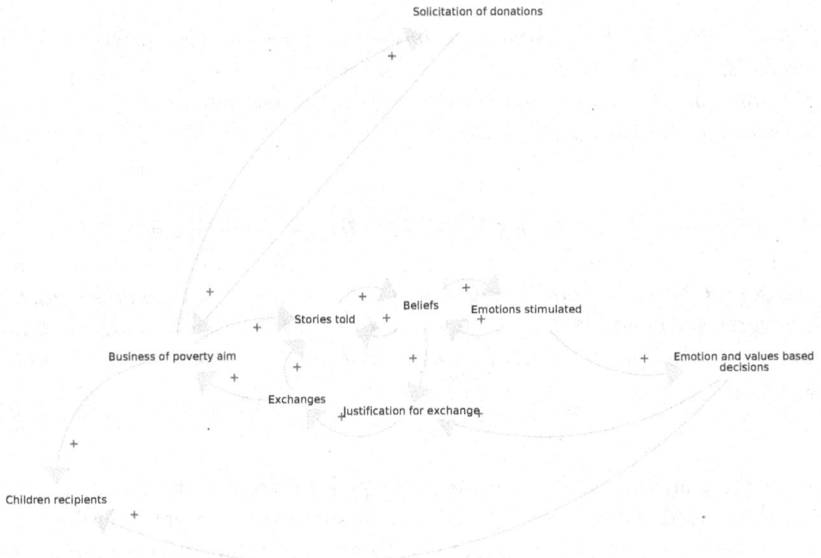

Figure 9.7 Events and mechanisms within the business of poverty.

Explanation for Figure 9.7

Findings indicated that as business of poverty intentions increased for actors', this tended to trigger more storytelling with related themes, generated new beliefs or reinforced existing ones, which tended to trigger emotions, have an effect on decisions, justify exchanges, drive giving/receiving priorities, and loop back to drive the business of poverty, as an actors' choice.

The business of poverty had a positive relationship with the solicitation of donations, thus increased business of poverty activity, and produced increased donations. This tended to result in a greater number of children occupying the role of recipient, as a core product 'marketised'.

The business of poverty was found interconnected with other elements of the system, and further mechanisms may exist; however, these were identified the most prominent mechanisms. The key focus for this section of analysis concerns the sub-system of the business of poverty nested in the wider system of travel philanthropy as a whole.

CBOs design a 'shop front' or brand to legitimise their cause/product. Certain keywords have become the norm such as vulnerable, needy, orphans, and children. The product is effectively a belief brought alive through storytelling, predominantly based around a negative plot, triggering emotions and sentiments that trigger the act of giving and receiving. The product can also be an idea or vision such as a proposed CBO, 'group' or solution to solve 'poverty' to legitimise transactions. Children tend to be the most common character in the stories based around the theme of education and schools. They can also be the storytellers and direct recipients. Health issues are also a saleable product. The product tends to unfold around a manufactured scene, particularly in schools, aligning with Staszak's staged 'bland otherness' (2009) in which local people have learnt to represent themselves. For example:

> *Some people don't have established business like us, but they think about starting projects, when they meet a visitor they say, we would like to start a project in this this this, and the visitor goes with an open mind and then the visitor goes home he sends money for the donation.*
>
> (U2_CBO-F_10)

> *You know tourists come not again to Lalibela ... always new tourists come and they have information but, when come to the ground, when come to see the actual, and when children talk to them, they impressed and they start, they change their mind even, they may be informed [about not giving to children] but when they get the actual communication they forget that one, and they start.*
>
> (E2_NGO_61)

In Zimbabwe and Uganda, the business of poverty tends to be about 'them', the '*other*' that needs help – i.e., a guide or tourism business promoting the woes of someone else. In Ethiopia, it tends to be about 'me'. This difference in approach affected giving pathways and the same input into the system (e.g., money) can have very different effects (e.g., donation impact). See Section 10.4 for a deeper look at this. The business of poverty has, however, over all the cases in our study become a story itself, from the leaders to the followers of the myth (Fastenrath et al. 2018), a norm evidently diffused from destination to destination, and actor to actor. The good, bad, and ugly were somewhat blurred in a blend of honest and dishonest practices amongst recipients of aid in this system, and donors were perceived as 'stupid' and 'naive' as well as 'kind' and virtuous.

Scott-Villiers' (2011) 'theatre of political participation' is an excellent way to describe the behaviour of actors in this system, not only prospective local beneficiaries in building alliances and defining themselves as oppressed people, but all actors including tourism businesses, their non-profit arm, CBOs, and so on. They are the brokers and benefactors as they suggest.

The business of poverty is arguably very much aligned with playing with moral intuitions through sympathetic education and perhaps an even more powerful presentation of 'victims' and the promotion of the single child in Africa compared to humanitarian campaigns of distant individuals. Tourists are directly face to face with the single victim, driving moral intuitions, in which the smaller the suffering the greater the intuitive responses (Woods 2015). The biases generated by 'cute faces over statistics' influence the way emotions enter decisions to donate, and how those resources are allocated. For example:

> *When visitors come ... they twist their mind to supporting and donating I think the way they see the world here is a bit different.*
>
> (U1_CBO-F_2)

In Ethiopia mostly on the streets you see poverty ... and even these locals who are following the tourists are very smart, how to touch the hearts of the tourists.

(EN_TB_66)

Our research supports Dornschneider's (2019) theory that emotions or affect interlink with beliefs and influence decision-making as well as Homer-Dixona et al. (2013), and Cold-Ravnkilde et al.'s (2018) thinking, as emotions, especially through processes in the business of poverty, produce and revise beliefs, which in travel philanthropy are often generated on-trip:

The children have a special approach to change the mentality of the visitors, even they know most of the visitors come and even guidebooks, it is written in the guidebooks ... they know the situation but when they get to the ground from their hotel to the restaurant from their hotel to the church in the meantime the children within a short period they change their mind, they have the capacity to change their mind.

(E2_NGO_61)

Photographing poverty has become a socially acceptable norm in Africa. Recently, the ethics of such practices has captured the attention of the media as Comic Relief stated they would "stop using images of starving people or critically ill children to portray the continent" (BBC 2020: online). When people travel to Africa, however, this is a practice employed by locals and tourism businesses in the solicitation of donations. Entertainment in the form of songs and dances has also become a marketing strategy with donations acting as a kind of payment for the 'product'.

Equally, whilst the practice of visiting schools is extremely common, participants questioned the ethics of such practices:

The song and dance think I just think is awful and they would come out every time and see these older white people all smiling at them they know that they have got to smile and do a song and then they'll get stuff, I think it's a horrible, a really horrible practice.

(Z1_NGO_21)

The creation of a tourists' own story is also a common product, influential to shape the giving pathway around enriching donors' life (Bertini et al. 2014):

Scholarships, that's why I think it's important we keep students in areas that are accessible, you know, so that's why we look at schools in that sort of area, because they will want to come and see them.

(Z2_1_NGO_37)

Driven by evidence of local cynicism towards donors and the international development community, particularly in Uganda, local actors were more than willing to assist prospective donors in creating their stories for marketing, as long as there is an exchange:

> *Some tourists also want to make things here to be able to get some balances.*
>
> (U2_INTER_12)

Through social learning and embracing unspoken rules, the business of poverty has become quite an enterprise in rural African tourism destinations. The characteristics of the processes within the business varied across cases but ultimately followed the same theme. The private nature of the business of poverty, driven by memetic desire, has created competition for donations as a scarce resource, which can generate conflict as a result. For example:

> *If you look at the characteristics of these people who start the CBOs ... the characteristics is that they are not employed, they don't have jobs, or they have lost jobs and they anyway cannot get good jobs or they cannot get any job so ... taking it as employment... they also rely on them for their livelihoods ... then they have to anyway buy land or buy buildings from the money... we would say they are also run as businesses instead of being run as charities or community organisations.*
>
> (U2_CBO-F_8)

Travel philanthropy provides an example that Silk (2004) says is rare, of those at the end of the aid chain having real power, with our research arguing that the power in the aid chain is less top down than one might expect. The findings also respond to Scott-Villiers (2011) argument that aid recipient communities rarely dispute the goodness of aid. However, although the framing of their position as "undeveloped, poor, ungoverned and uneducated" may go unquestioned, this framing has indeed become a lucrative enterprise.

As donations are involved, we might describe travel philanthropy as a form of 'professional beggary' using innovative strategies that tap aid resources to further individual interests (Hailu 2015:832). The unregulated social assistance coming from NGOs that caused this trend in Hailu's study of Ethiopia also appears in our research, albeit with different characteristics. Our research might also align more with Thomassen (2015), who considered the concept of 'begging' a micro-economy, an informal economy at the margins, but within the structures of a market economy, as well as a social relationship and cultural exchange. Whether it is begging, taking, or selling, many of the strategies employed are no different from what have emerged in typical corporate entities.

Not only locals generate the business of poverty system; some tourism businesses mirror the behaviour in much the same way. Negative storylines framed the poverty product as commonly as positive stories about businesses saving the local population from poverty. Print and digital literature often appear at various touch points of the tourist journey from airport lounges, to hotel website and their rooms, to the public areas in lodges and attractions, as well as through experiences designed or promoted to clients. As explained by a school head teacher in Zimbabwe, there is an awareness that the tourism operator they collaborate with sends tourists to their school to promote giving, and that their donations in turn help them to 'tout for tourists' in the marketing of their business (Z2_1_B-INS_31).

One operator in Uganda stated that a 'bare minimum' in efficacy is sufficient regarding the CBO receiving their funds, because whilst tokenistic, as long as this story sustains, it grows their business (U1_TB_3). Tourism businesses are not development organisations and their approach to interventions is similar to the approach of local actors in soliciting donations. The difference is within some of the purposes. The decisions of both tourism businesses and local actors in the business of poverty are important to understand. For example:

> *Like when they are coming to school they would love to see donations going through them so that whenever they are selling their product abroad especially whenever they are doing the marketing it helps them to tout a lot of tourists that we have connected a lot to the people and we have done this and that... it is a selling idea, they are selling their product like getting the tourist, they are connecting the tourists so whenever they are selling, they are selling this village as well, so it's actually helping to get more tourists.*
>
> (Z2_1_B_INS_38)

Our research does not necessarily condemn the business of poverty because it uses 'business' principles in a system supposedly based on altruism. Rather the findings suggest that this phenomenon must be critically understood as an approach to development that does not differentiate from other traditional development practices (i.e., INGOs), as many of the practices appear to mimic those on-going in the development field, with interventions led by an INGO or tourism-related actor often undifferentiated by local actors. In travel philanthropy, the end goal of 'good' often stops at 'donations' and this is enabled as a guide of morality in the business of poverty, rather than a consideration of the entire pathway. Therefore, actors rarely questioned the inadvertent consequences of using children and schools, for example, as a marketable product or platform to raise donations. Instead, they celebrated the *result* of generating an input of gift resources into the system. This finding does arguably attach different sets of moral questions, and the ways resources move along the pathway emerges as a contentious moral issue (Sinervo 2015).

The bigger picture of travel philanthropy as a system (and a map) has now been introduced, highlighting key events and mechanisms, exploring causes and effects, and illustrating the system structure and dynamics. This system map has contributed a better understanding of what the overall system is, how it functions and why, identified as important knowledge if one is to change its behaviour. The focus of this chapter explained elements found to drive the local consumption of donations across all case studies. This included findings of *otherness* and *sameness*, the causal powers of expectations (to give and to receive), and the system's significant push and pull factors, such as the solicitation of donations and the business of poverty. This chapter will explore the remaining interrelationships in the system, looking at the pathway beyond the local consumption of donations.

10 The travel philanthropy system

Resource pathways

This chapter looks beyond the gift exchange, into the resource pathways and the interconnected questions about impact evaluation. This includes how resources are distributed and behavioural tendencies associated with the travel philanthropy gift and key concepts illustrated in the area of the system marked with the yellow dotted line in *Figure 10.1*.

Here, we delve into how the informal nature of the travel philanthropy economy in terms of *roles* and *responsibilities*, the private nature of exchanges, and decision-making patterns such as bounded responsibility influences the types of giving and receiving in travel philanthropy, which produce three key resource pathways. We also explained that these resource pathways sit within a highly fragmented landscape of micro-donations which generate feedback patterns in the system. Through multiple and detailed visual illustrations of the travel philanthropy pathways, relevant events and mechanisms that represent the structure and dynamics of this part of the system are discussed.

10.1 The private nature of exchange relationships

Travel philanthropy is an informal economy in terms of *roles* and *responsibilities*, situated in a highly fragmented landscape of micro-donations that feed back to create this pattern. Private goals tend to drive individuals and businesses: they are not purely altruistic or public, but privately centred. Hence, an incentive to compete and covert action. *Figure 10.2* identified the key causes-effects only, which the wider system influences.

10.2 Giving–receiving pathways

The private nature of the travel philanthropy system influences the types of giving and receiving into three key pathways as shown in Figure 10.3.

Our study contributes to the gap identified in the literature about the typology of travel philanthropy and the need to shift our understanding beyond the idea that the system comprises homogenous schemes, products, activities, and effects. Each

DOI: 10.4324/9781003497929-10

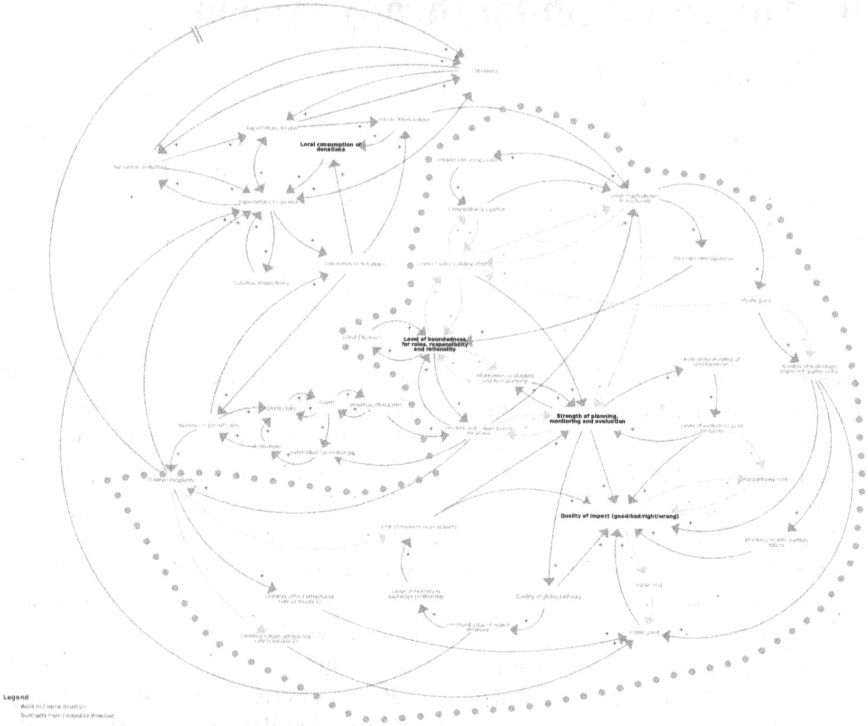

Figure 10.1 The travel philanthropy system: resource pathway (dotted line).

model does not indicate a neat pathway, but represents certain characteristics identified in our research.

Beyond loose change (Honey 2011), and pens, books, clothes, etc., the majority of financial donations occur post-trip, regardless of pathway:

> *When they are here they give what they have, the little that they have, when they go back home they donate more.*
>
> (Z1_CBO-F_30)

> *It's fishing so they don't come with 'I need this straight away', they will exchange letters, develop a relationship, get a couple of small gifts sent to you and then go for the bigger one and then you might get a minibus, so people have asked, can you support me to buy a minibus, thousands of dollars.*
>
> (EN_TB_67)

Figure 10.2 Privateness in exchange.

Explanation for Figure 10.2

What is the structure, underlying mechanisms, and interrelationships producing the privateness in exchange?

Map explanation:

The private nature emerges from underlying perspectives and system of beliefs, principles, and norms identified in Chapters 7 & 8 – e.g., freedom to fill a role and freedom of exchange relationships – as examples of unobservable mechanisms found to produce travel philanthropy's private characteristics.

The system pushes the local consumption of donations through private interventions and pulls through the solicitation of donations, forming an environment for high levels of privateness in exchanges. This characteristic led to a profusion of micro transactions within an uncoordinated and fragmented landscape of giving and receiving. Causes included competition and conflict over donated resources, and as actors competed for donations the level of privateness tended to increase, in turn creating feedback to reduce the level of collaboration.

Privateness also reflects covert transactions, or those hidden from the view of others, except parties directly in the exchange. Decisions to initiate or enter into an exchange tended to originate from private beliefs, interests, and individual personality, rather than some sort of organisational strategy or public sector rules/laws/regulations. Many of the exchange relationships between actors were private connections between two individual actors only, enabling private access to donations for recipients, and private access to recipients for donors. The system produces a very individualistic economy.

The resulting 'private good' as a resource outcome can be economic or social: economic in that a recipient or intermediary receives a financial transfer from a donor; social in gaining social status, power and influence.

Private interventions for private good in a highly fragmented landscape of donations with low levels of collaboration tend to affect planning, monitoring, and evaluation levels.

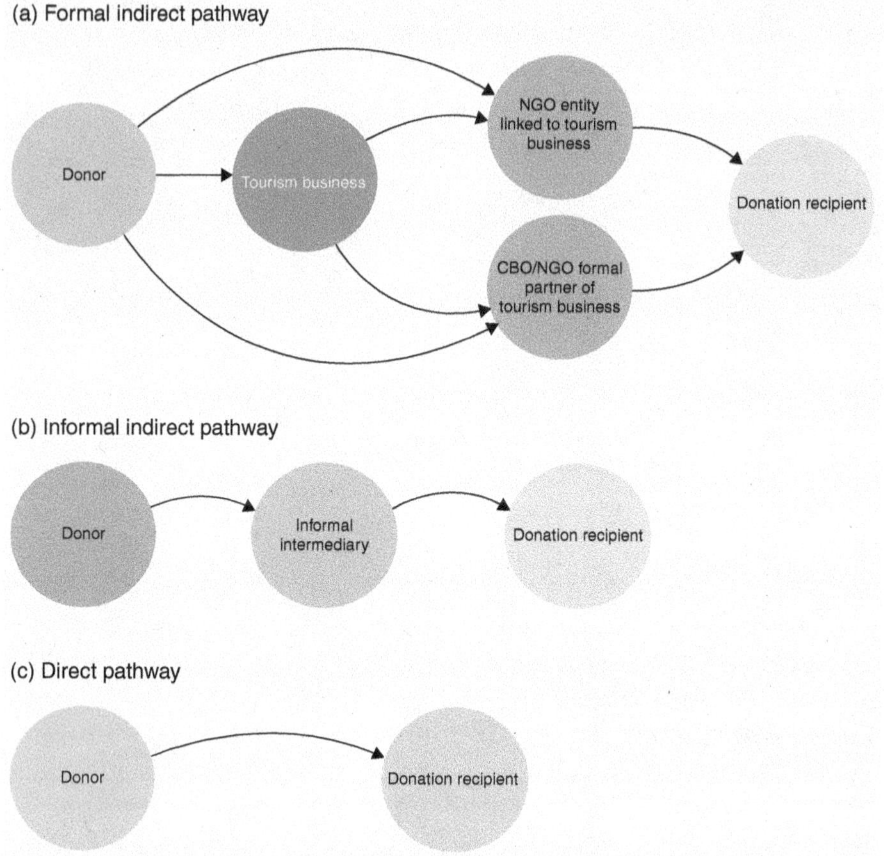

Figure 10.3 Giving–receiving pathways: (a) formal indirect pathway; (b) informal indirect pathway; (c) direct pathway.

> *100% of our scholarship programmes come from [the website].*
>
> (Z2_1_NGO_37)

All plant the seed of giving on-trip through an experience and emotional connection to the 'victim' and convert the donation post-trip.

10.2.1 Hybrid models – formal indirect pathway

Travel philanthropy is not an example of the traditional symbiotic relationship between a tourism company and charity (Turner et al. 2001). Rather, the tourism companies, through their formal programmes or hybrid models, acquire a role in the non-profit sector (Strachwitz 2015). Even though many of the non-profit arms of companies claim to be separate entities, if the tourism business ceased to exist so would most charitable arms. Evidence from our study showed that some tourism companies nevertheless increase the flow of donations (Bill 2011; Phocuswright 2015), and many now develop formal hybrid pathways (see Figure 10.3).

The drivers behind many of the hybrid models identified include:

- The desire for a social licence to operate amongst the local community, the travel industry, and in some instances the government or protected area management
- A belief in the business case for corporate philanthropy
- Individual business culture or management influence

Other mechanisms identified for tourism companies to invest in a hybrid model include attitude, the willingness to have an open relationship with a community-based organization (CBO), be a larger company, or have the resources to deal with the administration required to manage sponsorships (U1_TB_3, Z2_2_CBO-F_40). Financial incentives, such as that tax does not "come out of the owners pockets" (EN_TB_67), also drive companies. In other contexts, businesses approach communities offering gifts saying, "we want to help the community" (Z2_2_TB_41), or tourist motivations spur involvement. One of the large operators in Zimbabwe explained, "It started with one or two children saying they [tourists] would like to sponsor students" (Z2_1_NGO_37). After this, they decided the giving pathway required a structure in order to protect recipients from direct contact with tourists, as well as protect tourists from begging from sponsored students (Z2_1_NGO_37).

The hybrid model was more widely adopted in Zimbabwe, with various drivers including imitation, and a higher proportion of Western-led operators and pressure from international agents for a 'responsibility' discourse. This was not the case in Ethiopia. Operators suggested that this might be partly due to the tourism industry being young, margins being tight, and attitudes amongst businesses that doing business simply meant doing 'good' for destination communities. Hence, travel philanthropy did not incorporate sales tools or social licences. In Uganda, the trend ranged greatly, from zero desire due to local politics tied to donations, to more formal arrangements.

Some businesses provided the vast proportion of donations, with only around 10%–15% from guests (Z1_TB_28, U1_TB_3); others relied on 100% guest donations or other external funding sources for projects (Z2_1_NGO_37). Some hybrid models donated admin costs from the company and operating costs came from guest donations or external sources (Z2_1_NGO_37, Z2_2_CBO-F_40). The type of donor behaviour for businesses varied, from "I will just supply the funds and then they go and do that..." (Z1_TB_28) to "they don't control the purse strings" (Z2_1_NGO_37). Most tourism businesses were fundraisers rather than original donors, with money coming directly fundraising, or indirectly through the promotion of a village tour. Where the hybrid model existed, especially in Zimbabwe, the power was with the tourism operator to control the donor and the pathway:

> *They are quickly reminded by the guide that you are not allowed to donate on the spot you have to go back to the lodge and you will redirected where to put your money and then that money is put there, then after 3 months or so.*
>
> (Z2_1_B-INS_31)

From a community perspective, the hybrid model appeared in respondents' views such as "they are right here and our daily cry they attend to that one" (Z2_1_B_INS_38), but "it's us to tell them what should be done on the ground" (Z2_1_B_INS_38). A fine line was emphasised by tourism businesses regarding being too involved with the community and the importance of "keeping objective … because everyone has a sob story" (Z2_1_NGO_37). As suggested in the literature (Lindkvist 2011; Spalding 2011; Stirrat 1997), the intervening company only occasionally assumed the role of a partnership (ZWSMX, ZWSSH). For example, the recipients at an associated school suggested, "it's not a partnership but it's them helping us" (Z2_1_B-INS_31) or "they have to dance to my tune" (Z2_1_B_INS_38). This indicated a conflicting aspect within moral economies, including disconnection, as discussed by Gardner (2015).

Our study also questions the definition of travel philanthropy as a partnership with grassroots organisations (Honey 2011). What may sound like a partnership may not necessarily look the same from the receiving end. In Uganda, for example, some celebrated a ten-year partnership where a bed levy was allocated to a local community, but disrespect and mistrust from both sides posed questions about what partnership really meant:

> *If you don't keep an eye, they can get like maybe 300 bednights and they just tell us there is 150, so now every day we have someone who monitors and we also get information from the butlers, so we can know how many people are sleeping there.*
>
> (U1_CBO-F_2)

Tourism may have the potential to come close to the *shared value* model concept, with socio-economic or environmental gain purposefully embedded within the mission of the company as opposed to a side-line initiative (Porter and Kramer 2011; Smith 1994). However, travel philanthropy models almost entirely focused on giving gifts rather than restructuring business models to solve real issues. They shared more with traditional non-governmental organisation (NGO) models than corporate innovation (Kourula and Halme 2008) – some in Zimbabwe considered themselves no different from NGOs (Z2_1_NGO_36), with the tourism sector remaining in a donations approach rather than an alternative approach to integrating business practices (Ashley and Haysom 2006). Furthermore, hybrid models do not represent altruistic corporate social responsibility (CSR) (Lantos 2001) or cause-related marketing (Kotler and Lee 2005) *per se*, with the role of the corporate sector in travel philanthropy aligning most closely with strategic CSR (Baron 2001), as 'doing good' drives competitiveness.

Tourism companies cannot easily shift practices and incorporate new models of philanthropy (Shah et al. 2017), nor may they wish to do so, despite Aberenthy's (2011) argument that they have the moral obligation to identify legitimate local projects and reduce the spontaneity of giving by tourists. Indeed, when a tourism company did not occupy the role of donor or intermediary, the informal indirect and direct giving pathways become the most dominant.

10.2.2 Informal indirect pathway

Some donors choose an informal indirect pathway (or had no alternative) with donations often solicited by local individuals with English language and other knowledge and communication skills (Figure 10.3).

An individual guide or other local actor can facilitate the informal indirect pathway. A tourism business not adopting the hybrid model or appearing as a CBO can also gain legitimacy or simply be a vehicle to access donations:

> *I think most of these CBOs that you see around here or have emerged from tourism and they have gotten into contacts with people who come here for tourism.*
>
> (U2_CBO-F_8)

For donors to channel donations through individuals, 'friendship' emerged as a triggering factor (U2_TE_14), highlighting Bertini et al.'s (2014) relevance of trust and emphasis placed on the subjective nature of giving pathways that are informal and indirect. Individuals with the ability and personality to converse with outsiders appeared able to shape this pathway:

> *The educated I mean can easily manipulate the tourists or even donors who are not tourists because they can easily use the internet, use Facebook, WhatsApp to kind of convince donors to give them much more compared to those who are less educated.*
>
> (U2_CBO-F_8)

In Uganda, those promoting giving followed the themed quest 'for others', with individual gain being more covert. However, in Ethiopia the majority of giving tended to be driven by the less educated who were more overt in soliciting donations – i.e., 'for myself' as direct recipients, discussed further in the next section:

Get a sponsor only to think about yourself ... not think for any persons, any.

(E2_TE_60)

Concurrently, similar patterns amongst tour guides occurred in Ethiopia and Uganda. In Ethiopia, national guides occupied informal intermediary roles. This involved both honest and dishonest transactions, with several respondents mentioning a 50% share taken by the intermediary and/or the lack of direct contact between donors and beneficiaries:

The bad thing is when the money is transferred by the driver or by the guide they don't give properly like for example a person sends maybe £100 per month and then the person who bring that money maybe bring £50 so that such kind of problem we face before.

(E1_TE_52)

You give a guide some money [for someone they met] ... now I want you to take that money to him. Now, do you think that will reach me? Even though he is a guide he will eat [use] that money.

(U2_TE_6)

In Zimbabwe, critiques of both tourism and non-tourism related giving emerged, with specific reference to NGOs' ability to engage within the local context – "[they] come into this country they don't know the right people" (Z1_TE_23). Hence, donation pathways are highly dependent on the intermediary. It emerged that often the 'elite' took over this role and were highly influential in the allocation of resources intended for the 'community', with individual subjectivity and private gain often driving the giving pathway, over objective and public interests. For example, in Uganda:

How decisions are made is at times also personal and at times even on a selfish end.

(U2_CBO-F_8)

In Uganda, the growth of the travel philanthropy economy appeared to have led to a simultaneous growth of CBOs as a legitimate badge for local intermediaries and source of economic and social power for those spearheading them:

[New CBOs believe] you easily access good money from donors which is not the case because some of them have started them and never even get a single tourist visiting them... so they get frustrated ... they think maybe if they started then they would do what, they would get money.

(U2_CBO-F_8)

As previously mentioned this was not the case in Ethiopia due to stricter regulations for setting up CBOs or NGOs. Those that did, such as the estimated 15 children homes in Lalibela, were informal and covert (E2_NGO_61) and generally supported through an informal indirect giving pathway, with some relative risks associated with operating outside the law and often depending on a sporadic travel philanthropy pathway.

The informal nature of this pathway, its subjectivity, and individual interests add to the complexity of the system. The system enables leakages from the aid chain to the intermediary. The *informal direct pathway* is far more significant than recognised.

10.2.3 Direct pathway

Western (2011) and Bertini et al. (2014) touched on the subject of direct pathways, suggesting that direct giving is characterised by small amounts, can involve negative outputs such as boosting power of corrupt leaders, and is mainly outdated. However, it is actually one of the *most* critical issues, and the most dominant travel philanthropy pathway (Figure 10.3), demanding further critical elaboration.

The informal nature and freedom amongst roles appeared to influence direct pathways, as a fertile ground fostering direct relationships between individuals in the system and the chance to have a direct connection with the suffering other (Korf 2007). Other factors driving direct giving included a disbelief in formal intermediaries and lack of awareness of intermediaries or availability of intermediaries to channel donations. The underlying belief system, emotions, and storytelling together with otherness and the level of understanding of consequences affected the choice of this giving pathway. It is increasing in Africa, with most prominent cases in Ethiopia:

> *Increased yeah because they [ex]change email [with] each other, and then they promise to help ... no-one take responsibility around, the tourist is not advice anyone, so he take email from women or from man yeah, for phone, and then he's direct!*
>
> (E1_B-COM_69)

> *[There is] no well-organised NGO or local government offices where they can donate.*
>
> (EN_TB_66)

EN_TB_66 also suggested that much suspicion and mistrust existed in so-called development organisations and other intermediaries, such as USAID, who were providing funds to Ethiopia that were "supposed to go to the needy", yet were denounced to end up in the hands of the elite or government officials in Addis. Hence, the preference for direct giving:

> *Directly for children better... even of shoes, if not clothes, directly he gives for children it's good because when the money they can give for the teachers maybe he's used in his pocket.*
>
> (EHAG)

> *If they [tourists] meet some boy or some women from the village they will say I want to send her some money please can you give to her and they will send though to my account and I will take it to her, that person.*
>
> (E1_B-COM_50)

Handouts are very much a critical issue in direct pathways. These include small items such as pens, books, sweets, and clothes, which may be given to 'projects' or individuals on-trip, or sent post-trip, when it would generally include 'mobile phones, clothes, shoes etc' (E1_B-COM_50). This shows that Spalding's (2011) claims that handouts are outdated was not at all the case. Actually, defining a handout proved challenging due to the fine line between tips, handouts, gifts, and development interventions was blurred, but important.

Bertini et al. (2014) observed that tourists will give back whatever happened across cases. In Zimbabwe, for example, the tourists' freedom to give during village tours was emphasised as something difficult to manage:

> *$20 or $50 is handout out there... it happens when you take people and you can't stop it from happening.*
>
> (Z2_2_TB_44)

> *They were just giving to the children ... they assemble the children together and then they started distributing themselves, the guys, to the children, and the children were very very happy.*
>
> (ZVMT)

Such practice emerged as encouraged by tourism companies, with one participant from a hotel adopting a hybrid model explaining that:

> *We go and we have pictures where we will be issuing out everything and then the kids they get.*
>
> (Z2_2_TB_41)

> *They [girls handmade sanitary products as a donation] are there in the shop and cost $20 so guests purchase that ... if they reach up to $40 or so we then take it to the school and then we give it to the school kids.*
>
> (Z2_2_TB_41)

A local NGO emphasised problems with conflicting intervention philosophies occurring in the communities, criticising some of the tourism companies for their practices:

> *I really am not at all positive about the idea of just handing out things, I think that's a problem there and there are big risks when tourism gets involved with community development work or anything like that.*
>
> (Z1_NGO_21)

The vast majority of local actors did not regard handouts as a good practice as they generated expectations, caused inequality, and conflict, suggesting that the recipients would be better off with nothing:

> *We try to stop... the rest of the children fight each other... because you don't have clothes for everyone in the village.*
>
> (E1_B-COM_50)

> *They had one ball, and there were 2 teams at the playing field so they did like this, they remained in the car but threw the ball into the playing field so people ran for the ball, I saw people were like, these ones pulling this side, these ones pulling this side, fighting for the ball and they ended in blows. I remember someone was bleeding the nose and I did not like it.*
>
> (U2_INTER_12)

In Ethiopia, a major issue raised was the symbolic impact of t-shirts handed out to women by tourists. This may appear unremarkable, but it is a classic example of a non-linear effect in the system that has great impact on a culture that has not changed much for thousands of years:

> *They take that t-shirt and run away from us... the big problem is tourists giving t-shirts to the girls... they don't dress properly like the others, they dress their skin [traditional clothing] and then they put clothes under the skin and that is also not good ... that clothes is [gifted] from tourists.*
>
> (E1_TE_52)

Bornstein (2009) argued that there is a place for impulsive giving as well as more formalised forms of giving. However, most locals viewed impulsive giving negatively. Whilst impulsive giving may be deemed good if a circumstance warrants immediate, fleeting attention to save someone (Singer 1972), there is very little benefit from handouts. While donors' impulsive, emotional decisions may produce immediate positive gains for some recipients, these may be short-lived and problematic in the long-term. A more structured form of giving is better.

10.3 Boundedness

The level of boundedness for roles, responsibility, and rationality lies at the heart of the system (see Figure 10.4). Overall, our study identified a high level of boundedness in its subjects.

Boundedness is a key challenge within the system currently not addressed in the extant literature, but requiring critical consideration. As one respondent put it:

> *It's so interesting, and it's all so behind the scenes, under the floorboards.*
> (U1_TB_3)

Without a more candid discussion about the types of boundedness in the system and greater understanding about why this occurs, endeavours to claim improved impacts from travel philanthropy in Africa are likely to be limited. Boundedness is

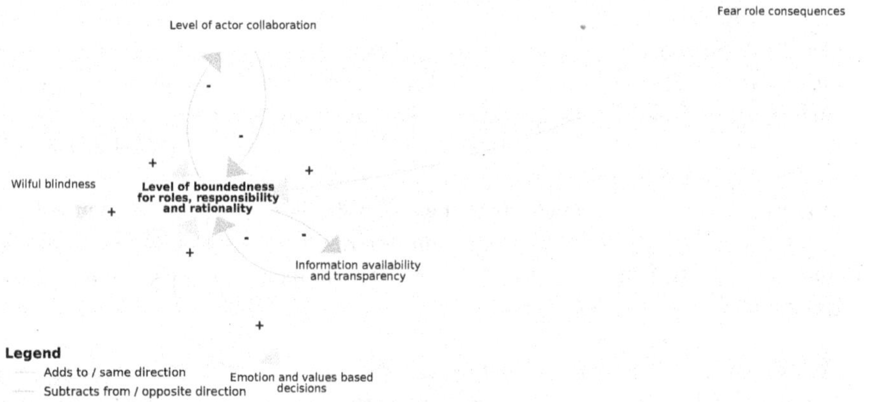

Figure 10.4 Boundedness of roles, responsibility, and rationality.

Explanation for Figure 10.4

Lower levels of collaboration cause higher levels of boundedness, and higher boundedness, lower collaboration.

Low information availability also caused high boundedness, and high boundedness caused low information availability.

Wilful blindness, emotion, and values-based decisions caused an increase in boundedness. For instance, increased emotional decisions led to high levels of bounded rationality. A high level of wilful blindness caused a high level of bounded responsibility.

The greater fear of role consequences, rooted in the private nature of the system, also caused a greater boundedness in terms of roles, meaning people feared to expand the boundaries of their roles and responsibilities in the system to avoid the personal/private consequences.

a fundamental flaw in the system, and a deeper understanding of choice processes across the entire system warrants further research. As a respondent said:

> *There's no responsibility, almost, because we just have to do what we are required to do, and I'd rather not have to start up projects that I don't feel passionate about, and I'd rather not have to get involved in the locals lives day to day if I don't have to... of course I think it may annoy the people that we are not doing enough.*
>
> (U1_TB_3)

Our study revealed substantial evidence across all case studies to suggest both internal and external influences on boundary decisions. A subjective decision landscape emerged, and decisions took place within a terrain of significant uncertainty from a practical, aspirational, and ethical viewpoint amongst all stakeholders.

10.3.1 Bounded rationality

To understand the emergence of boundedness in this system, and to unpack its causes and consequences, the theory of bounded rationality (Simon 1987, 1979, 1991, 1994, 2000) offers a valid framework. The choices actors make shape the system and so we must make sense of choice processes, including constraint mechanisms that influence decision-making (Koumakhov 2009; Manzini and Mariotti 2014). Both involuntary constraints, such as difficulties accessing information, and voluntary constraints, such as ignoring accessible information, are aspects of bounded rationality.

While travel philanthropy may represent an emergent exchange economy, decision processes are not solely economic, but also social, cultural, and deeply moral. Some decisions stem from self-interest – i.e., generating profit or gaining social licence – and others from a sense of moral responsibility towards others (Arjoon et al. 2018). Therefore, to assume actors are fully rational and that a sum of total preferences creates economic equilibrium is problematic for many reasons (Aumann 1997).

Not only do goals between actors conflict but also goals for an individual actor. A tourism business may make optimal decisions to maximise the profit for their business, whilst also trying to fulfil community development obligations, or create a satisfied tourist customer and ethical pathways of giving. Goals often contradict, and deciding which one to attend contributes to bounded rationality (Hanoch 2002). *Tourists* occupying the donor role tend to lack knowledge of what occurs beyond the act of giving. Their spatial distance from the location of the recipient, combined with limited knowledge and experience can accentuate this, supporting Burger and Owen's (2008) claim about information asymmetries found in the NGO sector. *Local actors* occupying the recipient role, if via an intermediary, tend to have limited access to the details of the donation they receive. There may be constraints to information because of the behaviour of other actors in the system. Privateness of exchange may drive down levels of transparency, and lack of knowledge, as Roehrich (2014) argues, contributes to bounded rationality.

Overall, information asymmetry and incompleteness cause more access problems than information overload (Schlaile et al. 2018; Dubbink et al. 2008). For example, available information may weigh towards the good in giving rather than possible negative consequences. Giving and receiving continue even with unknown outcomes (Hanoch's 2002), caused partly by actors' limited access to information. Donors to Ethiopia might discover after some time that the recipient has been using the resources for socially damaging purposes. Alternatively, the information regarding donor expectations and obligations comes later. An example from Uganda shows a decision to occupy the recipient role with limited information:

> *Such expectations shouldn't be there if you are given a donation... . you [recipient] may take it without knowing, it [obligations] normally comes after.*
>
> (U2_CBO-F_8)

Many information asymmetries arise because the information used are heavily weighted towards beliefs. Our research findings support Koumakhov (2009) because the belief system underlying travel philanthropy influences decisions and behavioural norms linked to roles. Our study echoed Langer's (1989) 'mindlessness' and rational fixation driven by actors seeing what they want to see, using beliefs to inform decisions, which contributes to bounded rationality:

> *People are always interested in children getting an education, so I think that is a soft button first and foremost.*
>
> (Z2_1_NGO_37)

Confirmation bias heavily influences giving behaviour. The most obvious example is that a large proportion of travellers to Africa decide pre-trip that they will donate to local people, predominantly children. For example, a tourism business in Zimbabwe explained:

> *When guests come, they always bring the stationary, the pens and pencils... .*
>
> (Z2_1_NGO_37)

Whilst many of the beliefs stretch across actor groups, Ehernstrom-Fuentes's (2016) points about the influence of culture on what is rational and Smaldino's (2014) theory of cultural constraints in perceiving situations are also relevant. Biases generated from one's own cultural framework, may not match with culturally influenced values and norms of the other person. Thus, this interaction may affect shared agency and trust in exchanges between a tourist and local recipient (Woods 2015; Matthiesen and Klitmøller 2019):

> *Well you know they [tourists] trust you, people are really I would say they have good heart, they trust, what we tell them they trust, the problem is we [local actors] don't keep our words, the foreigners keep their words, but all of us are not keeping our word because you know we ask the money to do business but we do it for different reasons, so if we do that for different purpose that means we don't keep our word, otherwise if they have the money they send you the money.*
>
> (E2_NGO_61)

Psychological or social consequences of gifts for recipients can be misaligned with donor reasoning because of information constraints. Many assume gratitude from recipients for any gifts, further supporting Berry's (2014) claims. Bounded by beliefs that giving is virtuous, donors may lack information to consider alternatives:

> *They feel insulated and for you, you know you have done something great.*
>
> (U2_INTER_9)

The challenge of achieving shared agency in travel philanthropy exchanges emerged from our study as a key issue stemming partly from cultural asymmetry. Furthermore, narratives in travel philanthropy become established myths over time, and such stories and/or myths are key elements of bounded rationality. As Fastenrath (2018) argued, the number of story followers can hold more power than the property of a story. As an example:

> *Those schools are surviving by the virtue that the children who have got people paying for them, if you stop then those children may go and sit home and their parents could not manage, so education would drop.*
>
> (U1_CBO-F_2)

Social learning and sameness or imitation have significant effects in the system as a source of information. These mechanisms contribute to decision-making of actors, such as new behaviours regarding the solicitation of donations or CSR and intervention practices amongst the tourism industry (Hanoch 2002; Simon 1956). As stated in the previous chapter, donors also imitate other donors, which creates a reinforcing loop of a certain type of giving pathway and allocation of resources. In a similar vein, the concept of innovator and imitating firms with choices biased by story following affect resource allocation and types of interventions (Fastenrath et al. 2018). For example, an NGO in Zimbabwe criticised the community intervention decisions by tourism businesses to be simply mimicking each other rather than designing aid around other types of information:

> *The flavour of the month at the moment seems to be sanitary ware, reusable, everybody is suddenly doing sanitary projects.*
>
> (Z1_NGO_22)

The ability to know all possible alternatives (Levi 1997) is constrained by the issue of limited and imperfect access to information, but we must also consider cognitive capabilities (Simon 1978). Even with incoming information that enables the ability to generate alternatives, actors may encounter difficulty generating facts and drawing inferences (Hanoch 2002), and default to *satisficing* (Simon 1965). Satisficing (Simon 1979) occurs amongst tourism businesses accepting a 'good enough' result – "I would say it's good on the surface" (U1_TB_3).

The boundedness of actors in this system from a cognitive capability standpoint also stems from a lack of understanding about intended beneficiaries, meaning a high risk of interventions failing (Lambe et al. 2020). With rationality bound by experience and values (Matthiesen and Klitmøller 2019), tourists intervening in development lack experience in the complexity of poverty alleviation. Time available to process information and evaluate a decision to donate is very limited, with most visiting several destinations in a short period:

> *When an American comes over to Africa for the first time they want to try and sort of save somebody ... the amount of people that are befriended and they start paying for the kids schools fees, that's a really big thing... I think people want to feel they are doing something with their lives.*
>
> (Z1_TB_29)

Even good intentions following similar giving pathways frequently produce inconsistent outcomes. Giving money to kids in Ethiopia who do not attend school or giving non-monetary items also creates similar effects because of inabilities to predict the response of recipients and know how they think, decide, and act:

> *Some tourists come in and give pens and books but the children don't go to school, but they don't know, they just give it, then they sell it because they are not a student and then they drink or whatever so this is not good.*
>
> (EHK)

Respondents did not accuse donors of malicious intent, but did raise concerns, particularly in Ethiopia:

> *That's how they see it, they just don't know.*
>
> (E2_B-COM_64)

> *That's the damage that's what I am feeling the gangsters it will be happen maybe later on after 10 years or 15 years ... and the fault is the tourists.*
>
> (E2_TB_62/ E2_TB_56)

Tourism businesses may be less constrained by time and distance from recipients in some respects, but limited cognitive capabilities in terms of choice processes emerged, conflicting with Lindkvist's (2011) argument. Capabilities did not arise due to merely being located in a destination. Rather, those in charge were primarily hospitality professionals, lacking expertise in sustainable development. They based decisions on individual beliefs and business models, and tended not to have the relevant resources nor dedicate time for rigorous searches to find alternative options, such as choosing optimal resources and pathways (Hanoch 2002):

> *The clothes they were given directly to each person, the book they were given directly to each person, the pens they will give directly to each child.*
>
> (U3_TB_18)

Cognitive limitations also occur amongst intermediaries and recipients, such as CBOs. These actors may be highly knowledgeable about their community and more easily access information about local development needs, but often lack experience, exposure, knowledge, and ideas to be able to interpret that information in a broader sustainable development context and plan. Cognitive capacities bound many donation recipients:

> *Someone gets money but finds he has no knowledge of using the money or what, say I've been donated $1k, what can I use it?*
>
> (U2_CBO-F_10)

> *We have not been educated to even visualise those situations.*
>
> (U2_CBO-IN_7)

> *There was no plan, strategic plan.*
>
> (U1_CBO-F_2)

Potential benefits exist to strengthen agency at the grassroots level in African development (Escobar 1992) and to bypass widely criticised inefficiencies in delivering aid down the global chain (Sinervo 2015). Whilst travel philanthropy donors can encourage recipients and intermediaries to have far greater agency, this may be overly idealistic and certainly contextual.

Aumann's (1997) constrained problem-solving abilities and bounded willpower occurred in respect to choices made against an actor's long-term interest (Sen 1999). Decision-making, especially in Ethiopia, lies in the hands of the youth with rationality bounded by various factors, such as age, experience, and willpower. Recipients' ages ranged from four to late 20s, thus favouring short-term outlook and lower investment in long-term goals and donations resulted in relatively lavish lifestyles rather than education. For example:

> *One boy is 6 or 7 years old ... this boy is not used this money and not use it properly ... he forget already the school the boy is young you send ETB 20k to Western Union and he take this money and you think boy go to school? ... with that money good clothes, good shoes and then where, maybe he go to bar maybe he looking girl you get me?*
>
> (E1_B-COM_69)

We see the darker side of travel philanthropy's cognitive limitations amongst the recipients in Lalibela (Ethiopia), where numerous individuals with resulting mental health issues and addictions were the continuing recipients of considerable donations:

> *He was getting so much sponsorship on the street... he was drunk out his mind and he just stopped going to school.*
>
> (E2_TB_56)

The issue of mental illness amongst these recipients was argued to be getting 'bigger and bigger' (E2_B-COM_64), with travel philanthropy worsening the situation as the money received was not being earned, creating cognitive limitations and lack of willpower in choice processes:

> *These people get a lot of money, these kids get a lot of money, and quite early, if you get money as a teenager a lot of money which you haven't really worked for then why would you care, you can do whatever things you want to do ... after working would you have been more cautious and more careful, they go they start alcohol early they get commercial sex early they can get to drugs to khat very early so you will see in some places that I often go I see a lot of young people with mental illness that I have never seen it is very frustrating.*
>
> (E2_B-COM_64)

However, it may not be appropriate to pinpoint backwards blame or forwards obligations (Scheffer 2001; Schlaile et al. 2018) to a single actor, such as a donor or recipient.

10.3.2 Emotions and bounded rationality

Emotions influence choices and play a key role in decision-making (see Figure 10.5). Many decision-making processes in travel philanthropy pursued an urgent objective shaped by emotions, risking mistakes (Hanoch 2002), and using fast and frugal decision-making heuristics (Smaldino 2014). Many decisions by donors, recipients, and intermediaries, as well as storytellers and audiences, reflected the gut-level responses driven by emotions, thus shaping 'System 1' (mental) processes, which make non-utilitarian choices (Smaldino 2014), rather than 'System 2', which tend towards consequentialism (Woods 2015).

Time allowed for decisions partly caused this pattern but also the novelty and unexpectedness of events increased affective reactions (Stutzer 2013). An unexpected 'windfall' for a recipient, for instance, increased affective reactions in decision-making around that resource:

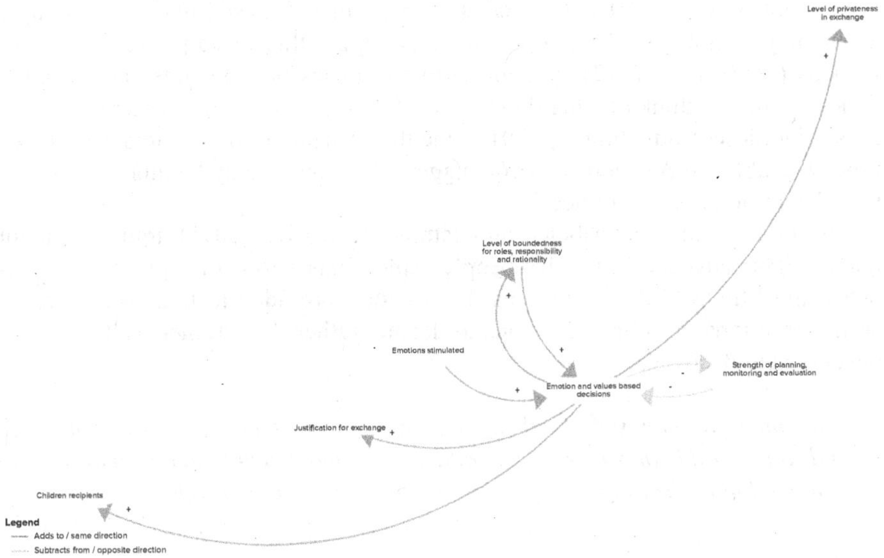

Figure 10.5 Emotions in decision-making.

Explanation for Figure 10.5

Emotion and value-based decisions interrelate with levels of boundedness and justification for exchange.

Greater levels of emotional decisions tended to cause to greater bounded rationality.

High emotion and values-based decisions caused lower levels of planning – i.e., an opposite/negative relationship.

> *Because it is unexpected, you end up misusing the donation, because maybe you never had plan of the donation you get excited.*
>
> (U2_CBO-F_10)

This did arguably provide another example of suboptimal judgements and decisions because of gut-level responses (Woods 2015). The novelty for tourists of seeing perceived inequality triggered affective reactions to donate, causing a bias, as emotions drive intuitive actions about how to act and take over objective reasoning (Hart et al. 2018):

> *When I took her around the school when she saw how the building was looking like she cried for maybe 3 minutes, she said wow, that people are really poor she cried and then she said he can go back home to Canada and then can send money for the dormitory.*
>
> (U3_CBO-F_15)

Spontaneous action caused by emotions inspired many travel philanthropy-giving events, not only with these two stakeholder groups (Picard 1997). The example above may well have been an optimal donation, but this was a prime example of Locke's (1690: II. xxi. 12) "passion [which] hurries our thoughts" constraining donors' ability to think of other things. Empathy has shortcomings as a moral compass (Matthiesen and Klitmøller 2019) and the structure of this system as a choice process enables it. As Bloom (2016) suggested, empathy may be more to do with the alleviation of pain for oneself.

The findings also contributed an example of Wood's (2015) argument about human rights advocacy, in which people exploit biases to get people to feel sympathy and blindly follow intuitions. Information provided to tourists intends to arouse emotions to inform decisions to donate, rather than trigger well-informed development decisions:

> *Sometimes you may see somebody dancing and you know he entices your eyes and you say oh I can support that child, so you find you are interested to support that one but in the long run, you will find that one was supported.*
>
> (U1_CBO-F_2)

> *It becomes meaningless if you help those children who would go to school, of the rich.*
>
> (U2_INTER_12)

Trust, rather than formal agreements tended to inform decisions. Trust, as discussed by Dunning (2012), is not a rational economic decision, because emotions determined how people evaluate the potential outcomes of their actions. Whilst outcomes might matter (Dunning 2012), the immediate payoff for performing the action of trust was significant.

The business of poverty aligns with critiques such as Woods' (2015) playing on moral intuitions, or Rorty's (1993) 'sympathetic education'. The reliance on emotional arousal and moral intuition caused a deliberate distribution of resources driven by the reaction of audiences, which can be suboptimal. This may be one reason why the majority of resources generated in travel philanthropy go to young children and schools. Practices of Wood's (2015) *moral outrage* appeared in strategies to drive moral indignation, rather than deliberate consequentialist thinking. Ample evidence suggested bias through 'cute faces over statistics', which encourage shallow thinking (Woods 2015):

> *They might think this pretty little girl or sweet little boy, but.*
>
> (Z2_1_NGO_37)

Assumptions about what appealed to these moral intuitions, namely education and children, informed those soliciting donations, and those donating assumed the recipients and intermediaries knew best. Dickert et al.'s (2016) warm glow feelings also influenced decisions, which confirmed the absence of Woods' (2015) advised education of donors to evaluate interventions. Bounded rationality narrowed the view of broader systemic development issues and the potential use of resources available through tourism.

Other emotions influencing decisions included envy, amongst local actors particularly, caused partly by donations. Both malicious and benign envy were present (van de Ven et al. 2009; Elster 1991), as was jealousy in protecting relationships with donors (Siltaoja and Lähdesmäki 2015). Resentment of wealth, status, power, and success as desirable objects or qualities of a person (Siltaoja and Lähdesmäki 2015) inform malicious envy (van de Ven et al. 2009; Elster 1991):

> *Now they are like bringing you down, so they kind of tap the donors so they use propaganda, they use lies to being someone down who is in fact using the money well so that they can take the advantage so, that one is so common here, and I think even in Buhoma, it started in Buhoma and now it has come it has come here... 2 years back, because that is when people started in fact getting into network stuff and Facebook and WhatsApp.*
>
> (U2_CBO-F_8)

Benign envy relates to issues of sameness, imitation, and emulation incentivising self-improvement (Siltaoja and Lähdesmäki 2015). Whilst predominantly found between local actors in shaping behaviour, it emerged as an emotion relevant from Southern actor to Northern actor, in terms of emulation, rooted in feelings of inferiority (Goel and Thakor 2005). The desire for admiration drove people's ambitions to be rich, rather than aspiring to wealth (Adam Smith in Brown 2018). The type of information available and cognitive abilities were evidently bounded, and, reflecting on Simons (1978) willingness to search for information as another factor, became useful to understand wilful blindness (Heffernan 2012).

An actors' willingness to search for information (Simon 1978) is useful to understand wilful blindness (Heffernan 2012).

> *I don't want to know, let it just be there because at the end of the day, they will say I was given the money, I ate the money or something of the sort.*
>
> (U3_TB_18)

Cognitively inaccessible information due to information blocks leads to epistemological ignorance. However, knowledge was also refused, ignored, or deliberately rejected (Croissant 2014; McGoey 2012; Robichaud 2017; Mallard and McGoey 2018; Heffernan 2012). Proctor's (2008) agnotology as a theory of absence provides a relevant explanation for what emerged from our research. Equally, strategic ignorance (Wieland 2017), and wilful blindness (Heffernan 2012) explained this systemic issue.

Whilst global development aid scrambles around the politics of knowledge and argues for an advancement in monitoring and evaluation (M&E) and more knowledge (Arensman et al. 2017), our study agrees with McGoey (2012) that we must place attention on the politics of ignorance. Knowing the least possible amount became a convenient tool for managing risk and avoiding blame. The refusal of knowledge enabled by this system affects accountability (Jungkunz and White 2013), amongst other issues. Wilful blindness is arguably one of the most important silences of the entire system:

> *It is frustrating knowing it could potentially go in the right place... I mean I would be very surprised if there was any money going back.*
>
> (U1_TB_3)

Across all cases, local actors' greatest concerns emerged about the unintended damages caused by tourists and donations, and by the behaviour of some recipients. However, knowledge was principally an ontological, rather than epistemological, ignorance. Especially in Uganda and Ethiopia, evidence suggests parents disregarded children skipping school to make money from tourist donations:

> *Instead of encouraging keep silent they don't encourage but they keep silent you know some of the kids they bought a car ... this is a silent killing.*
>
> (E2_TB_58)

In Ethiopia, wilful blindness escalated growth of expectations, from children to parents, and drove growth as a feedback mechanism:

> *Not directly encourage but they prefer to be silent, when you get some money, thank you he said, where do you get, now one tourist give me some money, oh don't forget your education for the first time, the second time, then after that, how much money you get, he also totally changes, this is what I have seen.*
>
> (E2_NGO_61)

In Uganda, rumours about a CBO receiving donations over a ten year period, but having nothing to show for it was the subject of an exchange on examples of wilful blindness:

> *Because we don't want to disrupt our contract and we are making a hell of a lot of money off this contract and if we go and we start asking questions then it might create a bit of a sour relationship and then it benefits us not to ask questions.*

In Ethiopia, the government also played a part in the culture of wilful blindness:

> *The government has a strict policy regulation but they didn't apply here in Lalibela, you know why? ... they want the resource in any form, if it is legal that's good, if not the money is coming to Lalibela to Ethiopia, so they, you know, their commitment to stop this informal associations are very little and limited.*
>
> (E2_NGO_61)

Our study further supports Heyer's (2015) Freudian concept that individuals know the truth but turn away from it, *deliberately avoiding* undesirable realities. This also adds to Lynch's (2016) concept of the wilful embrace of ignorance and evasion of unwanted truths, as well as Wieland's (2017) understanding of ignorance as driven by one's own will, rather than external barriers.

Suspicions are multi-directional, not only from donor to recipient. Turning a blind eye is a prevalent emergent behaviour across the system, evidenced by a common narrative that donations were in reality causing much damage in Africa. Yet giving persisted through a wilful embrace of ignorance:

> *People give them money! You know you shouldn't just be able to do nothing and receive something, that's not how it works, and a lot of tourism and charity work is [swear word] up a lot of areas. It's quite sad actually. It can cause a lot of damage I think.*
>
> (U1_TB_3)

> *Fuelling the anxiety and expectations from the communities and actually escalating the dependency... they have money in the pocket but they wouldn't want to pay school fees or whatever.*
>
> (Z2_2_NGO_45)

> *The growing bad side and you cannot really imagine where it goes, I mean these kids are growing up to be uneducated... the crime rate we have never seen may rise and it is rising so, of course in the long run if you look at it the problem is not going to last within just the people ... who are just school dropouts now, but it will be reflected on the community.*
>
> (E2_B-COM_64)

The travel philanthropy system produces little pressure for actors to loosen their mental rigidity, tending to see *what they want to see* (Langer 1989; Niehaus 2014; Heffenan 2012; Von Bergen and Bressler 2017). Some tourists insisted on preconceived notions about 'giving to Africa' on their trip, including ignoring local advice about appropriate gifts:

> *After they say sorry 10 times, but they are doing [giving handouts].*
>
> (E1_B-COM_69)

> *She [donor] would never take me seriously, and never agree with my proposals, she would only kind of ask me to take what she has brought or the ideas that she would be thinking of.*
>
> (U2_CBO-F_8)

Whilst actors may be aware about negative consequences of giving in their communities, many actors were concerned about drawing attention to the negative side of donations, caused by a fear of these funds stopping altogether (Von Bergen and

Bressler 2017). Even though actors know about socially harmful impacts, known benefits overpower decisions:

> *Because it circulates here in Lalibela, they may drink local beer here in Lalibela from one poor women, so she benefits because of that money.*
>
> (E2_NGO_61)

> *You've got the ones who it means so much to… 1 positive story must outweigh 99 negative stories.*
>
> (E2_TB_62)

> *I mean still there is a big advantage, even if some people died and some people are out of control.*
>
> (E2_TB_56)

A great level of confusion also existed amongst research participants about moral responsibilities of drawing public attention to problems, adding to Taylor's (2018) argument about the complexity of whistleblowing, Marginson's (2011) 'muted criticism', and Bouville's (2008) 'messengers of the truth'. Deliberate social silences, the known unknowns, warnings signs that are actively ignored or concealed and ignorance used as an excuse have all been identified in our study and appeared as mechanisms in the travel philanthropy system (Robichaud 2017; Davies and McGoey 2012). These norms of silence also contributed to *the fear of role consequences* because of personal vulnerability caused by threatening beliefs systems through exposing certain behaviours (Francis et al. 2015; Wettstein 2012; Bouville 2008):

> *They have their brothers, sisters, uncles, they are all clans there, so when you go to the street they may crash the car and even they might go to the extent of confronting our local guides even the tourists, everybody will be affected.*
>
> (EN_TB_66)

The general feeling was that it was better to maintain the norm of silence rather than trigger accusations of disloyalty, harassment and the like, as found by Bok (1981) and Munro (2016). In the wider socio-economic system, norms of silence were common in sustaining 'corruption', also found to have implications on the behaviour of local actors towards travel philanthropy, and affecting blowing the whistle on negative outcomes:

> *That [misusing donations] seems to be the norm now because no-one comes… to challenge it… it's a new culture yeah, we don't have to deny what is happening… . and also it is not only in donations… it is in almost every sector, corruption.*
>
> (UF & JA)

Our research led to the discovery of a disconnect between the socially acceptable actions carried out by individuals, and the unintentional collective harm that donations may cause (Williams 2019; Senge 1990), adding to the complexity of ignorances, especially in terms of moral responsibility (Robichaud 2017):

> *But they [donors] don't realise the damage they are doing, you have got all these youngsters who are getting no education ... it will be happen maybe later on after 10 years or 15 years... and the fault is the tourists.*
>
> (E2_TB_62/ E2_TB_56)

Our study furthers D'Souza and Adams' (2014) thinking on the unenlightened altruist, adding evidence that other stakeholders act out of ignorance, subject-ivity, and myopia regardless of their intention. Travel philanthropy arguably adds to the 'development speak' highlighted by Cornwall (2010), which enables feel-good words that admit no negatives. However, these ignorances can reinforce the ongoing discourse that represents the 'Third World' "as incapable of pursuing 'developments' on its own", in need of any donations (Chakraborty et al. 2017:39).

Overall, *wilful blindness* and the *ontology of ignorance* are critical components of the travel philanthropy system. Most actors are more than willing *not* to ask questions (Wieland 2017), and choose *not* to say the truth (Bouville 2008), sustaining glossy narratives by turning a blind eye. They thus enable certain patterns of behaviour and reinforce ideologies. The lack of a bigger picture per-spective and its fragmented structure contribute to this.

10.3.3 Bounded roles and responsibility

There were low entry barriers to occupying most roles and *actual* responsibility associated with a role was undefined and highly subjective, with few formal rules/ regulations. Perceptions of responsibility assigned to roles were inconsistent, and self-boundedness drove belief systems, individual capabilities, and personal interest. Furthermore, responsibility for actors was often personal rather than organisational; a broader set of stakeholders emerged with regards to responsibility, compared to earlier discussions in this chapter, such as those *affected by recipients of resources*:

> *You don't just wake up because they have chased you from the job or because you have studied and cannot get a job and you say I am starting a CBO, I mean there should be some scrutiny of who should start a CBO.*
>
> (U2_CBO-F_8)

The informal structure of the travel philanthropy system generated high levels of *freedom to occupy a role*. Freedom to occupy a role was a key element in the system map – e.g., any individual person can become a donor or intermediary, whether a tourist, a local guide or CBO, because 'responsibility' linked to roles is largely informal and unmanaged:

> *There are no structures that would follow up such kind of fake beneficiary. First of all, the way the money is sent there are no agreements, and also he gives*

> *Accountability of how he has used the money although the accountabilities are fake so there is no system for example like here we have the NGO forum we have CBO forums but they are all inactive, they don't do anything anyway, each CBO does what he wants, no-one does what, no one follows it up. The government should be doing that but doesn't do that.*
>
> (U2_CBO-F_8)

With an absence of formal rules/regulations, any inquiries about the complexities of aid (Williams 2019) tended instead to be guided by the belief system and informal rules, and responsibility emerged as highly subjective (Schlaile's 2018). Whether a donor was responsible for being the effective altruist (Pummer 2016; Schmidt 2017), or minimising anticipated unintended consequences (Zwart 2015), or if a boundary should exist to divide responsibility with a recipient were all blurred issues:

> *You know we [business donor] hold up our end and they [CBO recipient] hold up their end to us, what they do after that is nothing, you know, they could spend it well, they could spend it badly, you know, it's their money and that's that.*
>
> (U1_TB_3)

From a local perspective, despite awareness of problematic consequences of certain actors occupying certain roles and issues of responsibility, local communities nevertheless desired monetary gifts, and the consumption of donations. Hence, the system constrained whistleblowing and produced norms of silence (Bok 1981; Munro 2016; Von Bergen and Bressler 2017). For example:

> *They [potential responsibility whistleblower] don't have any forum, they don't have a way of following the donations because they are illiterate, they cannot write to anyone, and they cannot appeal to any local council for any help because they are kept in ignorance, they don't know what is happening, yeah so that's why they cannot challenge it.*
>
> (U2_CBO-F_8)

Furthermore, criticism of giving–receiving pathways or aid chains by many local stakeholders did not actually address issues of social, economic, or moral responsibility, with 'social silence' remaining prevalent:

> *You know we can see the electric wire, we use the electric but if I touch the wire directly by my hand I maybe lost my life, but I use the light, but I didn't touch the wire by itself you know, so the money is very important for local community, we are afraid and the channel is difficult, but.*
>
> (E2_NGO_61)

Understanding whether or how to address known irresponsibility, including individual or higher-level system issues, without being responsible for stopping the donations all together (Von Bergen and Bressler 2017), was a systemic challenge. Bounded rationality (Simon 1979) and wilful blindness (Heffernan 2012) have also proved useful in understanding the emergent issue of *responsibility* in the travel philanthropy system.

Both role occupants and role responsibility in this system were unstable (Rozuel 2011). For example, if a traveller happens to be an 'unenlightened altruist' (D'Souza and Adams 2014) then who is responsible for managing their decisions and behaviour in a local community should *they* decide to give a gift? Moreover, *should* anyone manage them? As one respondent said:

> *It's their [guide's] job [to manage tourist gift-giving] but not everybody is with a tour guide.*
>
> (E2_TB_62)

Lack of experience or knowledge may cause bounded responsibility, but "being on the hook to do more" (Sinclair 2018), such as managing a donation by a former tourist client, or meeting expectations of communities, also prompts it. Thus, one purposefully avoids involvement or responsibility:

> *If we get more involved then we always have to be more involved.*
>
> (U1_TB_3)

Once entering a role, responsibilities can emerge involuntarily and remain informally bound. For example, a local guide in Uganda emphasised involuntary responsibility towards an expectant donation recipient, expectations, and roles caused by a client promising a donation to a village resident during an informal guided tour. The pledged recipient then perceived the guide as the intermediary, who suffered when the promise was unfulfilled by the client:

> *I have taken the client in the field on a hike and I have got a burden because that parent [whom the tourist promised a donation to] will keep on coming, [name of guide], how far how far... I am talking directly to the client, and that one is not talking directly to the client, so that means parents we have met will think I have eaten money, or I am a liar such a stuff.*
>
> (U2_INTER_9)

The actions of one actor affected another in respect to responsibility, but a culture of compartmentalisation also existed, caused by a low level of collaboration (Croissant 2014; Rouzel 2011). Donors and intermediaries could draw the line of responsibility at giving rather than extending to incorporate the consequences of the transaction. If no one takes responsibility for a donation by not following it up in the right way, Louise (2009) argues, this affects the context of decision responsibility. In Ethiopia, intention, not impact bounds the role of *donor*, and there is no backward or forward blame for giving, even if irresponsibly:

> *It's just from being good, from good heart, I have no objection about being good to help they think it was for right purpose but it's the people who are using it for wrong purpose it's not their fault.*
>
> (E2_NGO_57)

On the other hand, others in Ethiopia discussing the issue believed the donor was responsible for monitoring their donation and if they did not follow it up, they were morally responsible for the effects of their donation:

> *When you support like kids or students you [donor] should have to follow them whether they attend or not.*
>
> (E1_TE_52)

> *The fault is the tourists.*
>
> (E2_TB_56)

In Uganda, a tourism hybrid argued for the responsibility of recipients to create 'good' impact, or a particular CBO in this case, not the donor:

> *It is frustrating knowing that it could potentially go in the right place but... they [CBO/community] are doing it to themselves.*
>
> (U1_TB_3)

In Zimbabwe, whilst perspectives varied and tourism businesses tended to manage their guests more closely, when it came to direct gifts by tourists some placed the onus on the donor:

> *Most of these projects still die when they are not supervised, but we also try and say we are not in the business of supervising these people, so if you [tourist] are donating this, you are donating this also at your own risk, it might be a good outcome, it might be a bad outcome.*
>
> (Z2_2_TB_44)

Our study supports the importance of Fassin's (2012) arguments on the reciprocity of responsibility, because focusing responsibility on one role or intervention is problematic in such an interconnected system of decision-makers. Rather as Fassin argues, there needs to be a shift in thinking towards *stakeholder* responsibility. Our study also draws on Schlaile et al. (2018) who say that we must shift our focus to shared responsibilities of all stakeholders.

10.4 Questioning impact quality: ex-ante and ex-post evaluation robustness

Levels of boundedness and privateness affect system-wide dynamics. Figure 10.9 illustrates the bigger picture of aid interventions in a travel philanthropy context and shows interconnections with strength of planning, monitoring, and evaluation (ex-ante and ex-post evaluation), as well as quality of impact from the exchange/intervention.

The quality of impact, if considered at all, was almost always judged on a single output/outcome – i.e., how much money was given, number of sponsored children, number of water tanks installed, or a school report, which reinforced reductionist concerns that infiltrate evaluation in international development as raised by Mutale et al. (2015). 'Projects' tended to focus on the means of addressing social problems, such as paying school fees, rather than questioning the ends (Gates 2016). Individual responsibility rather than systemic effects remained the focus for understanding impact (Jackson 2019), rather than collective impact (Patton 2019). For example, individuals commonly had multiple donors, which makes individual monitoring difficult:

> *It's not balanced, by the way a lot of sponsorship is here, [but] one student [gets] a lot of sponsorship, a lot of money is coming [to that individual], [they are] banking the sponsorship, [then] the students [do] not [get] good results.*
>
> (E1_TE_49)

When questioning impact (Figure 10.6), issues tended to focus on wasted resources, missed opportunities or negative impacts, rather than greater efficacy in a sustainable development context. From a local perspective, themes ranged from no impact to problematic issues. In each case study, many perceived interventions to be trivial or wasted. In Zimbabwe for example, a leading on-tourism NGO operating in the same area as tourism corporate interventions said that it was like "Throwing money into the dustbin…" (Z2_2_NGO_45). In Uganda, a tourism business, which was also a very significant donor to a CBO, confirmed that beyond good intentions impact was dubious claiming 'No impact at all' (U1_TB_3).

Travel philanthropy as a potential mechanism for sustainable development fell short in the studied locations. The aggregate value of resources generated by this system could be significant, but its current fragmented nature indicated that the impact hype of travel philanthropy as a development tool (Honey 2011) remained overemphasised.

In terms of the gift, interventions are often weighted towards causes close to the heart of donors or intermediaries, or ones that they are passionate about (Schmidt 2017). What will 'sell' in the business of poverty moves decisions, rather than ethics and efficacy informed by robust evaluation. For example, this was emphasised in the strategy of a tourism hybrid organisation in Zimbabwe in selecting beneficiaries of donations:

> *It's usually primary school students [that are the beneficiaries] because they are pretty little sweet whatever, we don't find it so much with high schools, you know they are not as endearing at that age [to donors/tourists].*
>
> (Z2_1_NGO_37)

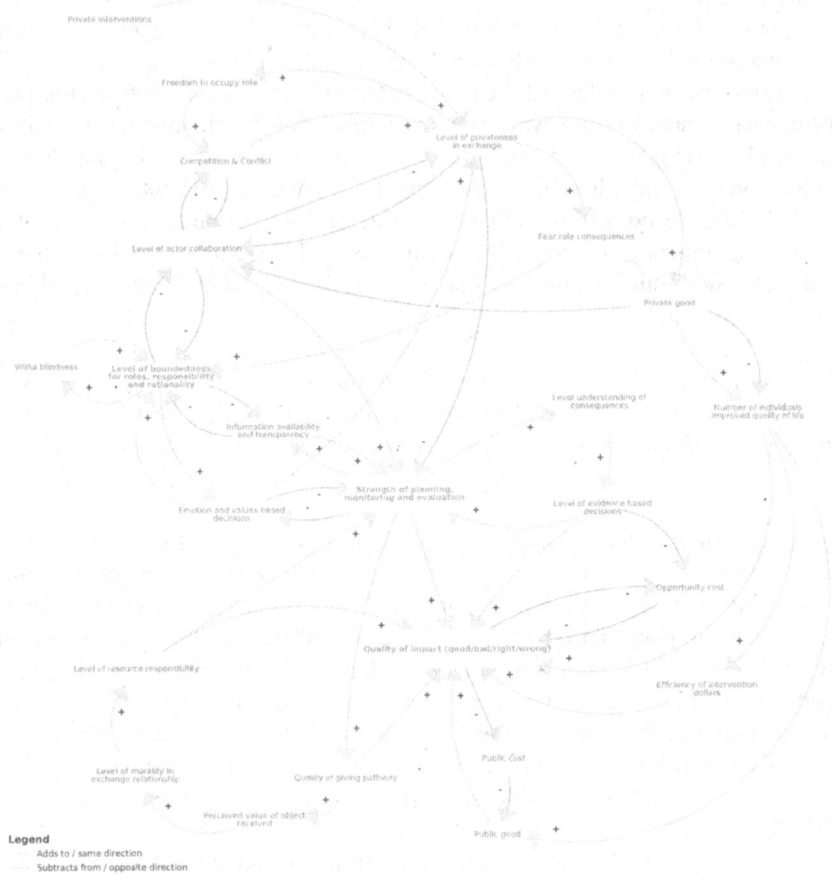

Figure 10.6 Mechanisms and events influencing knowledge and 'impact'.

Explanation for Figure 10.6

Reminder – the + arrows represent the same/positive relationships between elements – i.e., an increase in one element causes an increase in the interconnected element. The - lines show opposite/negative relationships – i.e., a decrease in one element causes an increase in another element.

High levels of boundedness and privateness tended to indicate low levels of M&E, for instance, which had a knock-on effect to quality of impact.

The diagram illustrates if the quality of impact reduces, public costs increase, and the public good decreases. Additionally, we must consider opportunity costs.

The quality of the giving pathway, steered partly by M&E activity, has implications on the perceived value of the resource, which interconnects with responsible management of a donated resource.

Overall, this part of the travel philanthropy system map emphasises the key elements that shape impact, such as ethics, efficiency, and efficacy of aid interventions.

Beyond the gift or object donated, or the cause, an important finding was the influence of the giving–receiving pathway on impact. The *quality of the giving pathway* is difficult to measure, as the definition of *quality* is contextual. Current discourse tends to focus more upon what was given and to whom, rather than *how*, influenced by weak ex-ante evaluation (Arensman et al. 2017; Richards 2019). Interrelated mechanisms underlying impact included the effect of the *quality of the giving pathway* on the perceived value of the object received. If the giving pathway is flawed, a well-intended impact may not be realised, or, worse, fuel negative behaviour:

> *He chews khat, drinks, relax, enjoy, but this faringi [tourist] continue to send the money.*
>
> (E1_B-COM_54)

> *They were sending money and the guy was using money for other things until they picked it up much later.*
>
> (Z1_TB_26)

> *All Ugandan's, you give it to someone to give to someone and it will not be 100%. There are very few who can do it.*
>
> (U2_TE_6)

> *They just want things that will happen that day, they want that kind of ill-gotten wealth and they don't think about tomorrow.*
>
> (U2_INTER_12)

> *No asking, no following, no control, just receiving money, who cares this money is coming from Europe, I don't want to go to school I don't do nothing.*
>
> (E1_TE_52)

Moreover, travel philanthropy is a classic example of the distanced donor issue (Burger and Owen 2008):

> *They [donors do] not come here, not see results, so that is one main point, one main problem... just they ask, even they don't come here, they stay in their country, then they ask maybe the driver, maybe the guide, the tour operator, they ask how is [name] he keep studying or not they ask [them].*
>
> (E1_TE_52)

Whilst a major theme was a lack of follow-up beyond the act of giving, not all donors behaved this way. In Uganda, for instance, some donors reportedly demand greater accountability from recipients than in the past. One recipient explained that:

> *The donors of these days are too demanding in terms of how the money has been used. You have to be accountable by sending the accounts, even one of them want audits account and they also want pictures, and testimonies in the form of reports, receipts, yeah.*
>
> (U2_CBO-F_8)

Mixed responses were gathered from local recipients with regards to donor monitoring in relation to impact but predominantly it was said to be 'a good thing' (U2_CBO-F_8) and as above, that "you [the donor] should have to follow them" (E1_TE_52).

On the other hand, recipients do not behave like 'automatons', as Silk (2004:235) noted, because even if these 'distanced donors' requested information, recipients can manipulate information going back up the aid chain. Therefore, provision of donation allocation does not imply quality. Evidence suggested an escalating norm of 'faking' or lying about resource allocation and accountability in travel philanthropy. In Uganda and Ethiopia, for example:

> *It's very easy to make up reports and to even fake pictures, take pictures of someone's kids, someone's buildings, and you send especially if you know your donor is not going to come so report can also be faked... most people do that in the world of donations, especially the beneficiaries, the primary beneficiaries do forge, do fake reports all the time.*
>
> (U2_CBO-F_8)

> *So the tourist is always asking but sometimes they receive fake papers so they keep sending.*
>
> (E2_TE_63)

The dependence or legitimacy of photographs, which seems to have become sufficient evidence of the good in giving, and a dominant form of information requested from up the aid chain, was not a reliable accountability approach:

> *They can go to the farm [of someone else] and take pictures and say they have done this.*
>
> (U2_CBO-F_10)

Evidence of such 'fake' information being provided for accountability, sent digitally, as well involving staged 'projects' if a donor does actually visit was prevalent. Similar patterns appeared in all cases. This was the most common issue in Ethiopia, where the direct giving pathway was dominant.

Shared agency in the giving–receiving pathway was arguably low and driven by a weak obligation to keep each other accountable (Tomasello 2020). This included low expectations of reciprocity in terms of impact, accountability, etc., on behalf of donors towards recipients, and different ideologies of the gift, in support of Parry

(1986), Toffin (2015), Sihlè (2015), and Erie (2016), between donor–intermediary/ recipient:

> *Someone gives you a donation of putting like a ventilated improved pit latrine, for you, you end up installing water in your house.*
>
> (U2_CBO-F_10)

> *They eat little [use some of the donation personally] but let the parent on the ground [know], not the donor.*
>
> (U2_INTER_9)

Even when a donor follows the impact of a gift, the result supports Picciotto's (2019) argument that a project does not qualify as successful if based on donor assumptions and are otherwise irrelevant, or lack efficiency through misallocation of resources. As one local put it:

> *If something has strings attached it may not work.*
>
> (U2_B-COM_13)

As an exchange economy operating predominantly on trust with very few checks and balances (Dunning 2012), travel philanthropy must shift beyond measuring prosocial behaviour to include the effects that encourage unhealthy behaviour (Legros and Cislaghi 2020):

> *Corruption has kind of eroded away our moral fibres so we are no more transparent ... as long as we have something on paper I can present it to the donor ... but in reality it is not the case, I used the money maybe to build a house or to buy a bicycle.*
>
> (U2_CBO-F_8)

In Uganda, strategies for beneficiary selection for instance tended not to derive from evidence and reason but on personal benefits for intermediaries. Thus, those who could afford to pay a bribe or had the right contacts in their community had a greater chance of their children benefitting from sponsorship than perhaps families in more need of support. For example:

> *Whether they are able to pay the children's fee or not ... they end up bribing these guides, these go-betweens, they give let's say a goat or maybe 100k UGX and say please kindly put my child on the programme [travel philanthropy sponsorship], I will give you this.*
>
> (U2_INTER_12)

Furthermore, the role of small-scale corruption within travel philanthropy and wider society, or how different cultures perceive resource responsibility cannot be ignored. It is critical to realise that travel philanthropy can become a mechanism for corruption. Whilst not the root cause, the travel philanthropy system can enable this behaviour to occur:

> *Tourism never created bad behaviour, it was there, it was there long ago, I am telling you during the time of Amin things were there like that, but here they started when tourism came, but long ago they were there.*
>
> (U2_TE_6)

Taking a personal share of donations was common practice in Uganda and Ethiopia. Within *informal* pathways, local intermediaries, or recipients tend to divert at least 40%–50% of donations for their own personal use. Although these are estimations based on anecdotes, not quantitatively verified, they are indicative of a pathway problem too often ignored.

> *When let's say they come up with a project, let's say a goat rearing project, if someone makes a budget of maybe UGX 10 million he will be knowing he will use 6 million and may get the 4 million balance if he is the one in it and if he is the one driving it, so he is targeting some cash somehow.*
>
> (U2_INTER_12)

> *The bad thing is when the money is transferred by the driver or by the guide they don't give properly like for example a person sends maybe £100 per month and then the person who bring that money maybe bring £50.*
>
> (E1_TE_52)

Our research supports Burger and Owen (2008), who cautioned against a naïve view of NGOs and the information they reported. Whilst Burger and Owen's NGO 'secrecy' meant NGO actors hid corruption and ineffectiveness to avoid public scrutiny of failures, the difference with travel philanthropy is that the giving pathways do not tend to be via independent individuals promoting private 'goals'. Burger and Owen's point about local communities having no incentive to scrutinise NGOs or CBOs and being pleased to receive what they do rather than nothing at all was a common pattern identified in our research as well, particularly in Uganda:

> *Because Ugandans are very poor, so when they are donated, they cannot [complain] ... that's why I said when you donate like UGX 10,000 and when eventually the one who has been donated gets a UGX 1,000 he will never mind. Instead of 10,000 he has got 1,000, it will be ok. Because that one is for nothing.*
>
> (U2_TE_6)

Despite being known that transparency is interrelated with good governance and successful feedback systems, 'information asymmetries' caused a lack of 'checks and balances' amongst decision-makers in the system of travel philanthropy

(Burger and Owen 2008). Recipients in travel philanthropy may be more powerful than in many NGO giving–receiving pathways. Many of the causal factors identified by Burger and Owen (2008) in regards to donors, such as their distance from the intervention and costs required to visit to observe outcomes, occurred here as well.

Overall, travel philanthropy is not driven by contemporary trends in sustainable development, such as tackling root causes, thinking beyond linear approaches to solving problems (Barbrook-Johnson et al. 2020), awareness of addressing wicked problems (Rittel and Weber 1973), and evaluating through systems and complexity (Gates 2016). Relevance, efficacy, and efficiency of interventions tended not to inform decisions (Picciotto 2019). Characteristics of D'Souza and Adams's (2014) unenlightened altruist appeared across all system actors. The situational knowledge and objectivity he argued for in his enlightenment solution rarely occur. Nonetheless, myopia and subjectivity were not necessarily malevolent but were potentially unintended consequences of the intellectual vacuum within which the majority of actors in travel philanthropy make decisions and act.

The absence of a formal regulatory body caused actors to rely on self-evaluation, as in Berry's (2014) study. Our research in Zimbabwe indicated very surface level ex-ante evaluation amongst a few interventions (Z2_1_NGO_37, Z1_TB_28), but no comprehensive ex-post evaluation was found, and there was an overall lack of involvement of local voices of the broader recipient community in determining impact (Richards 2019; Arensman et al. 2017). Self-evaluation in Uganda, when found, focused on how much or what had been donated (U1_CBO-F_2, U1_TB_3). The recipient organisations and individuals in the travel philanthropy system were purposely off the radar from these demands (E2_NGO_61). The M&E in travel philanthropy is overall arguably far from the UNFPA (2020) criteria discussed in Chapter 3, and certainly not found to generate impartial information on activities, projects, programmes, and performance, etc. The critical hindrance to information produced within this system, also discussed in the section above, hinders feedback for informed decision-making processes.

The negligible information gathered by different actors suggested travel philanthropy evaluation was another example of evaluation as the 'next trick' in the name of accountability (Arensman et al. 2017:222), paying lip service to results, rather than a genuine desire for sustainable change (Mowles 2013). Actors tended to reflect the trend of 'proving' expenditure and assumed successes, rather than 'revealing' resulting difficulties for the purpose of learning and catalysing rethinking in a sustainable development context (Paton 2019; Kirsch 2014; Picciotto 2019). For example:

> *If we [CBO] do it [give progress reports to the tourism business donor] in the future we will find that we are ahead of even other lodges in community activities because we would have more reports than any other people around, yeah!*
> (U1_CBO-F_2)

It was also found that for those providing some form of feedback up or down the aid chain, sufficient evidence to measure the 'impact' of travel philanthropy tended to be just a photograph sent over email of the gift being given or based on a single story:

> *We distribute it ourselves, while taking pictures, because you know when someone gives a donation they want to see how the people have shared, what happened, as evidence and that encourages someone to send more next time.*
>
> (Z2_2_TB_41)

Only one or two non-tourism foreign actors acknowledged unethical practices:

> *All the kids are trotted out to perform a little song and dance and then the gifts are handed over.*
>
> (Z1_NGO_21)

Accountability of interventions proved no more than a technical fix (Cavill and Sohail 2007) to keep all stakeholders upwards and downwards at bay, rather than improving performance of development interventions (Sarker and Rahman 2015). There was very much a combination of reluctance, unwillingness, and inability to provide information, as Burger and Owen (2008) argue happens in the NGO sector, but also a lack of awareness about the responsibility or benefit of evaluation. Any accountability mechanisms tended to be external, usually between recipient/intermediary-donor, driving an interest in identifying successes rather than failures (Kirsch 2014).

Green (2016) argued for greater willingness to learn from accidents and failures, but our research revealed very few incentives for actors to change behaviours when success tends to be defined by the interests of those specifically involved in giving, intermediating, and receiving, rather than broader impact. It also did not appear that donors needed to lose the 'stigma' of unintended outcomes being mistaken for ineffective management (Arensman et al. 2017), because there tended to be no external evaluation beyond whether or not they gave something. Thus, it was apparent the system risked overlooked critical emergent outcomes as warned by Arensman et al. (2017) and many other authors.

Awareness and understanding of evaluation was generally very low, with out-of-date log-frames linear thinking sporadically used (Douthwaite and Hoffecker 2017) and stakeholders operating in Vogel's (2012) culture of superficial visual representations, not even attempting to open the black box of causality. Whilst travel philanthropy may contribute to the achievement of some of the UN Sustainable Development Goals (UNSDGs) (Spalding 2011; Ardoin et al. 2016; Bertini et al. 2014), the vast majority of interventions take place without knowing what works and what does not (York 2020).

Interventions through travel philanthropy also characteristically focus on *expected* outcomes, not actual results, often ignoring unanticipated consequences (Arensman et al. 2017; Vogel 2012; Douthwaite and Hoffecker 2017). Our study furthers the discourse on 'complexity-blind' development interventions (Douthwaite and Hoffecker 2017), evidencing siloed (often questionable) practices, and confirming a general lack of systems thinking to drive the agenda for sustainable development in low-income settings.

In the next chapter the overall conceptual model of the travel philanthropy system, identifying prominent system outputs based on local dynamics is presented.

11 Redefining impact

The bigger picture, system outputs and the paradox of travel philanthropy

This chapter reflects on the changes required to enable travel philanthropy to ethically claim its contribution to sustainable development in Africa. In order to do so, we must understand potential levers for change operating at different levels. Five key system outputs are presented, with particular attention to the impacts emerging from the travel philanthropy system. This chapter calls for further critical assessment of the intended and unintended consequences of giving–receiving exchanges in travel philanthropy, which is defined as a paradox adding to the discourse on the 'unintended paradox of aid'.

11.1 The bigger picture of the travel philanthropy system

Chapters 9 and 10 focused on abstracted parts of the 'travel philanthropy' system, identified as most relevant in explaining the system's behaviour. With the intention of building an understanding of the emergent system structure (the system map) and of the emergence of system parts – 'emergence' intended as ontological social phenomena (Mingers 2014), this chapter unpacks causality and attempts to open "the travel philanthropy system black box" (Figure 11.1) by presenting the bigger picture in a conceptual model of the system whole and its interconnections (Meadows 2008; Sherwood 2002; Lambe et al. 2020). It is worth highlighting that a change in one part of the system may affect another (Banson et al. 2015) or may have the potential to produce consequential outputs if activated (Zachariadis et al. 2013). For example, our research showed that the quality of giving pathways, found to be influenced by the strength of ex-ante and ex-post evaluation, has effects on expectations, privateness in exchanges, and levels of wilful blindness, thus generating feedback into the quality of the giving pathway, and so on.

Through the identification of common emergent mechanisms across and within our study's cases and locations, the view of travel philanthropy as a system whole transcended the individual intervention level (Douthwaite and Hoffecker 2017) to a more collective view of the travel philanthropy phenomenon (Hailu 2015; Patton 2019; South et al. 2019). What this means is that in one way or another, directly or indirectly, every component of travel philanthropy emerged as being interconnected. As indicated in the previous chapter, travel philanthropy exchanges

DOI: 10.4324/9781003497929-11

can produce a system with questionable outputs. So, if the travel philanthropy system was to be transformed into a more ethical and responsible one contributing to sustainable development in Africa (Sen 1999; Ndlovu-Gatsheni 2015), we would need to better understand the levers for change operating at different levels (South et al. 2019). In order to achieve this, the system had to be fully understood (Jackson 2019) and related themes associated with it.

Figure 11.1 provides a graphic representation of the system roughly divided into four key system themes identified by letters A, B, C and D. *Section A* represents parts more directly driving the local consumption of donations; Section B – decision-making within exchanges and pathways; Section C – quality of giving–receiving pathway; and Section D –impact and evaluation.

Despite contextual variations, the system structure was found to remain relatively consistent, with recognisable patterns across case studies (Douthwaite and Hoffecker 2017; Senge 1990), leading to the construction of one system map encompassing all studied contexts. Whilst the system map is not intended to be representative of other tourism destinations, it may well provide a framework transferable to other parts of the African continent as well as other geographical locations where the travel philanthropy phenomenon is present. Here we challenge the positive narrative of giving–receiving monetary and non-monetary gifts, and we call for further attention to the need to raise some important new questions that delve into specific system areas in need of further critical investigation.

The system output found in our study is an example of emergence and reflects Aristotle's whole beyond its parts, with the plurality being not just a complete aggregate (Aristotle 1998). However, these findings come with some caution that binary judgements about the good/bad or right/wrong of outputs are not straightforward. Such judgements firstly depend on a range of social, cultural, economic, and personal contextual factors, to name just a few, because what is 'right' for someone may be 'wrong' for another person, and secondly, the systems' complexity may be obscured by linear judgements. Nevertheless, our study also warns that wilful blindness to deleterious effects of giving–receiving donations in Africa, not facing the reality beyond good intentions, is a precarious and possibly dangerous issue that cannot be ignored, which is captured by one Ethiopian participant, who said:

> *The culture gets spoiled when another culture comes and intervenes you know, on your home, so the difference from place to place, the giving and the taking, you may call it begging, you may call it theft, the meaning is contextual from place to place.*
>
> (EN_TB_66)

11.2 Travel philanthropy system outputs

Inspired by critical realism (Bhaskar 1975; Mingers 2014; Vandenbroeck 2015), systems thinking (Senge 1990; Meadows 2008; Sherwood 2002; Mingers 2014), and integrating interdisciplinary theoretical and empirical findings from Chapters 2–10, *Figure 11.2* illustrates an overall conceptual model, including five

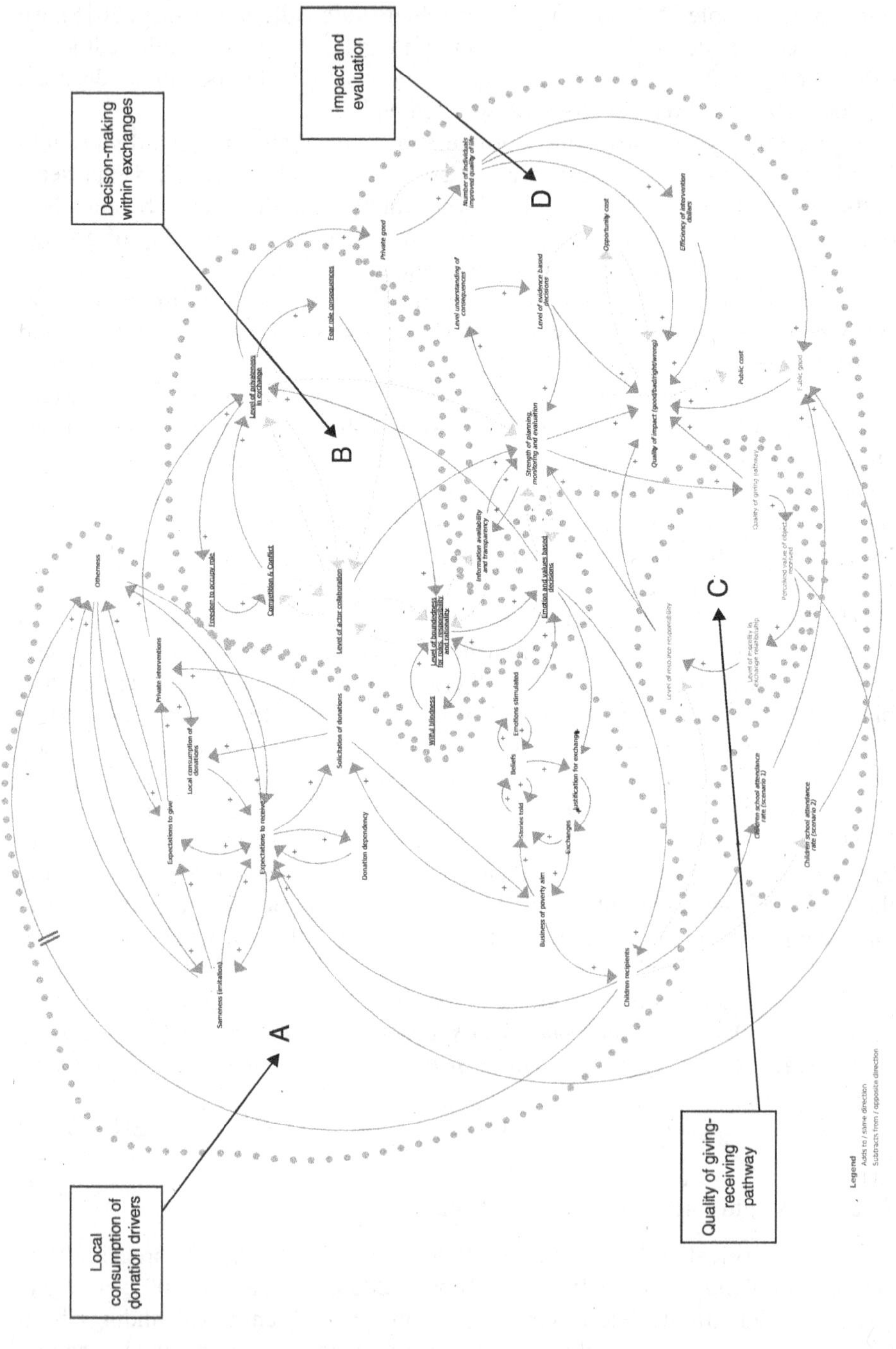

Figure 11.1 Travel philanthropy system – the bigger picture map.

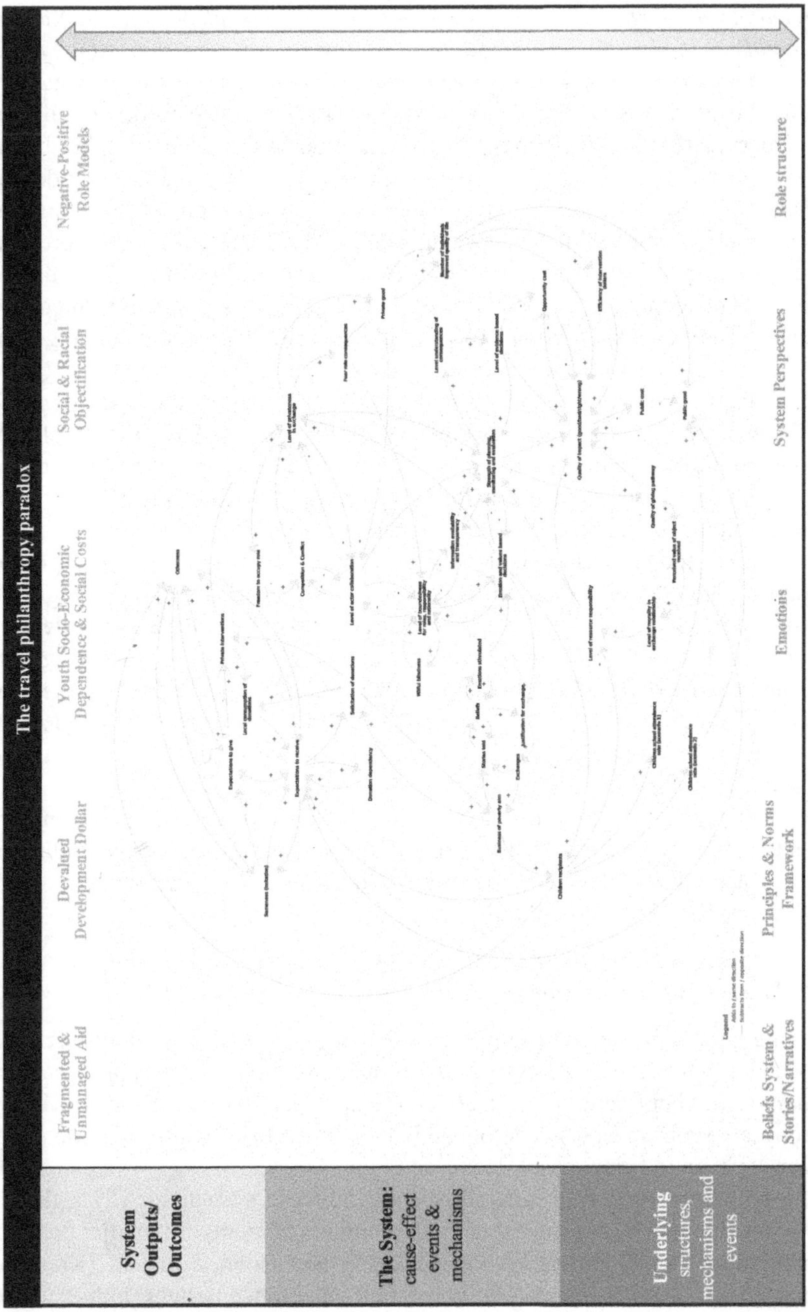

Figure 11.2 Causality and the travel philanthropy system – from the observable to the unobservable.

salient system outputs that emerged from our study – i.e., the *what* of the impacts emerging from the system. These are *fragmented and unmanaged aid landscape, devalued development dollar, youth socio-economic dependence and social costs, social and racial objectification*, and *negative–positive roles models*. At the bottom layer of the model are underlying mechanisms that were discussed in Chapter 8, such as the underlying belief system, stories and emotions, informal rules and norms, perspectives, as well as roles. To understand why the system (in the central layer) behaves as it does, it is imperative to also understand these underlying mechanisms. Further observable and unobservable mechanisms emerging as parts of the system structure, which were developed in Chapters 9 and 10, are represented in the system map in Figure 11.2, traversing the domains of the real, actual, and empirical. The interconnections indicate that a change in one system element has a tendency to have a system-wide knock-on effect, with the system behaving differently depending on contextual factors.

This chapter will now move on to findings that exist at the level of reality, which may appear more obvious as they are 'experienced' (Bhaskar and Callinicos 2003) – the most salient outputs of the travel philanthropy system. The outputs indicate new social phenomena that the system *produces* and that feed back into the system as inputs that in turn shape its behaviour.

There are undoubtedly countless social, economic, cultural, and political outcomes generated by travel philanthropy, so this chapter primarily highlights the *key* findings from our own research. As emphasised across the literature, *intended* and *unintended* consequences of the travel philanthropy giving–receiving exchanges within the sustainable development in Africa discourse demand critique (Fassin 2012; Picciotto 2019; Green 2016; D'Souza and Adams 2014; Silk 2004; Wessel 2018; Banson 2015; Mutale et al. 2015; Zwart 2015; Arensman et al. 2017).

Our in-depth investigation across the five specific Africa tourism destinations across three countries suggested that travel philanthropy is a paradox, adding further evidence on the good, the bad, and the ugly of a niche practice within Silk's (2004) 'unintended paradox of aid'.

11.2.1 Fragmented and unmanaged aid

Emerging from the travel philanthropy system, rooted in its differentiated perspectives, underlying belief system, norms, emotions, etc, a *fragmented and unmanaged aid* landscape was evident across all case studies – i.e., the travel philanthropy system produces this output and, through feedback, causes the travel philanthropy system to behave as it does (see Figure 11.3).

The findings revealed an alternative philanthropic system in Africa that has emerged within the growing trend of larger numbers of relatively smaller donations in international development (Atkinson 2009) and isolated impact (Kania and Kramer 2011). A profusion of relatively small value, inconsistent, individualised, and informal 'interventions', or transactions, characterised travel philanthropy donations and interventions:

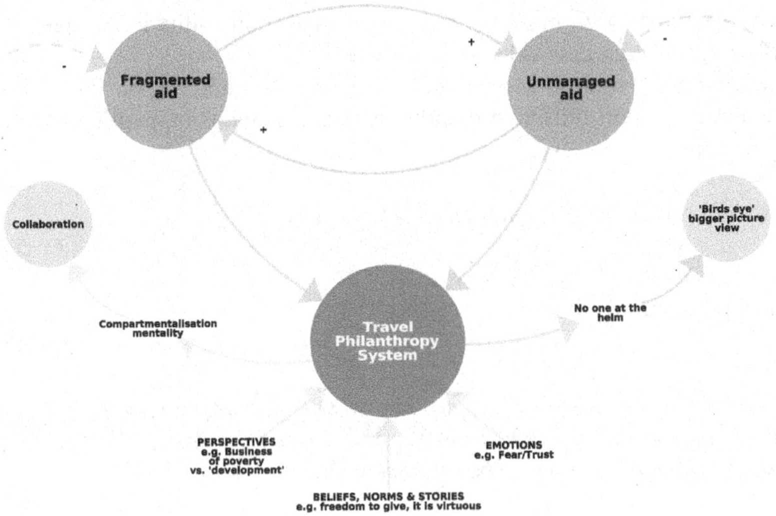

Figure 11.3 Emergent system property: fragmented and unmanaged aid landscape.

> *When we don't receive the clients who visit the community, no more to pay for the salaries, to pay the teachers, no food for the children at school, and then that leads to not getting here good teachers at school.*
>
> (U3_CBO-F_15)

> *When business is good you tend to do a lot more, when it's bad you tend to cut down on a lot of things, it also includes all those supplementary projects.*
>
> (Z1_TB_26)

Within this 'community' portfolio of interventions, compartmentalisation was found to be high and collaboration minimal (due to boundedness, privateness, etc), leading to a 'patchwork quilt' of interventions of varying size and quality, which generally lacked sustainability, as also argued by Edwards and Hulme (1998:10). With the absence of some form of central oversight and formal structures rather driven by its informal nature, the *bigger picture* of interventions remained out of sight to stakeholders, producing an *unmanaged aid* landscape. Therefore, the fragmented patchwork of interventions contributed to the landscape of unmanaged aid, and a reinforcing feedback loop meant the reverse. That is, 'unmanaged aid' drove 'fragmentation':

> *At the moment we don't have something that can really sustain the project [another school reliant on donations].*
>
> (U1_CBO-F_1)

The broad findings were that the financial *value* generated by the travel philanthropy system was relatively slight and underwhelming. Interventions were largely unrecorded, mostly informal and off the radar of accountability for all stakeholders, and their actual aggregate value is unknown.

Collaboration, such as shared goals, or clear roles and responsibilities, across this community 'portfolio' was found to be lacking, with competition and conflict being present:

> *It's still very individual, each one makes a plan.*
>
> (Z1_TB_26)

> *We don't work together.*
>
> (EN_TB_68)

> *[Donors are] like prostitutes... not kind of stable, they are running from charity to charity, or from organisation to organisation, or from individual to individual, which never used to be the case here.*
>
> (U2_CBO-F_8)

> *Collectively we all get involved... they [communities] are benefitting twice as opposed to benefitting once.*
>
> (Z2_2_TB_43)

Giving more rather than well seemed to be a theme. To expect the *entire* portfolio of interventions and transactions across a 'community' to be coordinated is unrealistic, as this has yet to be achieved in a broader development or corporate responsibility context too (Fassin 2012; Wettstein 2012; South et al. 2019; Patton 2019). Reducing fragmentation may increase potential efficacy or mitigate harmful effects; thus, greater levels of collaboration are one leverage point. But elements of the belief system, principles, and norms framework would need to change for an environment more conducive to collaboration:

> *We have managed to make them [two tourism businesses] understand that it is very very important for them to work together and ever since we managed to convince them to do that we have seen greater progress.*
>
> (Z2_1_B-INS_31)

> *Why can't you [two formal indirect pathways] put your resources together and do that project as a combined force.*
>
> (Z2_1_B-INS_31)

Beyond the numbers and missed opportunities of isolated development and/or business efforts, there are also a range of qualitative socio-economic and cultural issues to consider (Jamieson et al. 2004). These emerge partly from fragmented and uncontrolled aid.

11.2.2 Devalued development dollar

Both the *perceived* and *actual* value of donations had an impact on the system output, the production of a *devalued development* dollar (see Figure 11.4). As explained in Section 10.4, when the *perceived* value of a donation by recipients is low (e.g., the dollar is worth less than a dollar earned), similarly to what Von Bergen and Bressler (2017) refer to as unearned capital that comes as a windfall, this affects the level of care, morality, and attitudes towards being responsible for scarce resources by intermediaries/recipients.

In our assessment, the giving pathway played a significant role in this respect. The issue of the *perceived value* appeared to be more significant in Uganda and Ethiopia, seemingly affected by the giving–receiving pathway characteristics, as well as emergent cultures and role models, amongst other mechanisms discussed in Chapters 9 and 10. The overall issue identified was that 'free' money devalues its original or potential value:

> *They just take them as things they have not sweated for, which doesn't matter whatever you spend the money doing.*
>
> (U2_INTER_12)

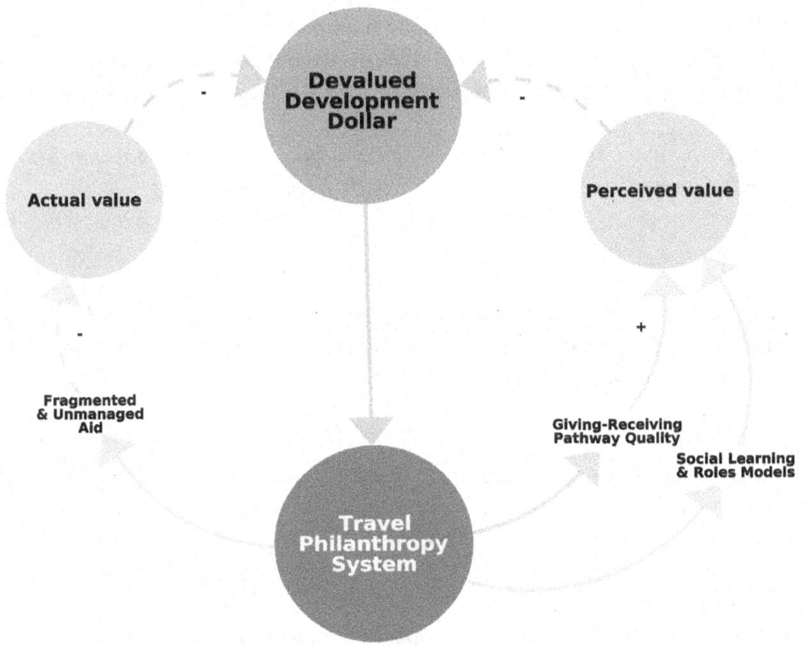

Figure 11.4 Emergent system property: devalued development dollar.

> *When people are given things, they take them to be just free things, they have not contributed anything about those things they are getting, so they take them for granted.*
>
> (U2_INTER_12)

> *Who cares this money [intended for my education] is coming from Europe, I don't want to go to school, I don't do nothing.*
>
> (E1_TE_52)

> *If you get money as a teenager a lot of money which you haven't really worked for then why would you care... after working [for your money], you would have been more cautious and more careful.*
>
> (E2_B-COM_64)

The *actual value*, driven partly by *fragmented and unmanaged aid* landscapes discussed above, refers to the relevance, efficiency, and efficacy of donations/ interventions, as argued by Picciotto (2019), for improving impact judgements of scarce resources. While in Section 10.4 impact was briefly discussed, here the point argued is how this contributes to the system output of a *devalued development dollar*. A dollar may have generated greater value to the achievement of some UN Sustainable Development Goals (UNSDGs) if allocated differently. Largely driven by emotion-based decisions, within the constraints of bounded rationality for all stakeholders (Simon 1956), and with system perspectives indicating purposes other than 'development' goals, *actual value* as well as *perceived value* contributed to this finding:

> *Most clients want something that is sustainable, something that they know will go on for a couple of years, without exactly being that taxing.*
>
> (Z2_2_CBO-F_40)

> *When people make a donation, sometimes they don't select or what, they just donate, so you will find that some of the books we have in the library are way above the scope of a primary school kid.*
>
> (Z2_1_B-INS_31)

> *Because once people start realising that they can get [donations], they expect to, and then they expect to be able to do nothing and receive money, and then of course they do as we now know, nothing.*
>
> (U1_TB_3)

Whether each dollar given/received helps the greatest number of beneficiaries possible, or the most relevant ones, or contributes to *structural change* to solve development problems sustainably (Jones 2017; Rorty 1993) is also questionable in travel philanthropy, partly due to being a 'private good'. While our study contributes new empirically grounded knowledge about implications of the ways development gifts are offered and received (Rudhrum et al. 2016; Gardner 2015; Sihlè 2015; Kragh 2016), further research is needed to ascertain if donations could be better allocated and/or achieve more, also given that what became evident through our study is that *free* (whether money or other tangible item) is rarely 'good' and the *pathway* has a significant effect on perceived and actual donation value.

11.2.3 Youth socio-economic dependence and social costs

Youth and children tend to be the main recipients of travel philanthropy donations (as also established in Chapters 8–10), whether directly or indirectly, passively or actively. The dominant cause tended to be affective decisions with regards to donations by donors and intermediaries, driven, for instance, by the business of poverty. As a consequence of exchanges that involved children, our study identified another key system output as *youth socio-economic dependence and social costs*. Once again produced by the system, this output also feeds back into it, reinforcing certain behaviours in travel philanthropy, such as telling poverty stories to tourists over attending school, having stronger expectations to receive donations, as well as impacting the broader society linked to the portfolio of interventions (see Figure 11.5).

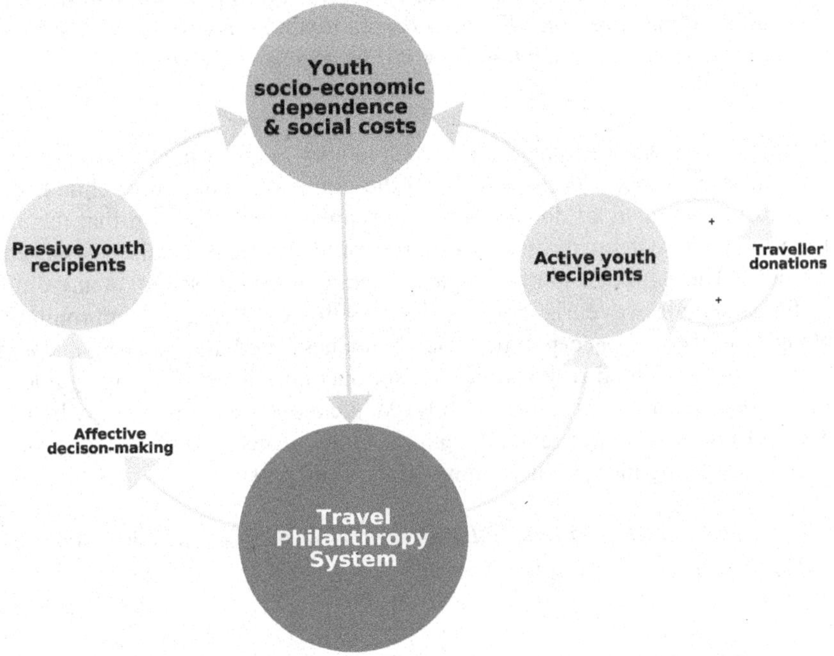

Figure 11.5 Emergent system property: youth socio-economic dependence and social costs.

Children are undoubtedly the most targeted beneficiaries in the travel philanthropy system by donors and intermediaries, thus emerging as *passive* recipients, and have also become the most *active* demographic of beneficiaries in some contexts, earning their appeal as a development 'solution' to potentially enlightened or unenlightened altruists (D'Souza and Adams 2014; Simon 1956). Whether intentions by donors are good or not (Weaver and Jin 2016; Ardoin et al. 2016), a range of *social costs* appeared to have explicitly emerged from this trend.

Whilst *youth socio-economic dependence and social costs* system output was detected in all case study contexts, in the Ethiopian one, this emerged as being of particular relevance because children were found to be the most active and negative consequences of their role in the system were comparatively blatant:

> *When the visitor knows the situation of the children, they automatically drop and at that moment the life of the children totally destructed.*
>
> (E2_NGO_61)

> *[The individual is] still looking money... because he's not getting his khat [local drug] or drink because he was addicted because the sponsor had stopped.*
>
> (E2_TB_56)

> *[When the donors stop sending money the individuals might] come into the houses, break the house, steal or kill... if they have money they have different needs, so if they are not fulfilling all these different needs and if they have not the money to do that, then they have to find another solution.*
>
> (E2_NGO_57)

Visiting schools has become a key contemporary tourism practice in Africa and emerged as an accepted norm, which, for the simple reasons that the child becomes the *passive recipients* of donations and the 'poverty product' and that there is an absence of consent by the parents of the child, irrefutably raises some ethical questions. The right of a tourist to enter a public or private school at any time and the fundamentally voyeuristic gaze (Sharp 1999) at children and impromptu gift-giving emerged to be acceptable, ethically unquestioned, and actively encouraged by the most acclaimed responsible tourism operators in Africa, local guides, and community-based organisations (CBOs). While these school visits may be fruitful in terms of gifts, there are generally ignored consequences associated with the most obvious one being the loss of learning on the child part:

> *They [children] have to go and meet visitors at the gate, that is time lost... the neighbouring school did not have that disturbance.*
>
> (Z2_1_B-INS_31)

Promoting the victim or the suffering child is common practice in the wider international philanthropy and development world and is not the subject of a new debate (Woods 2015). In travel philanthropy, however, when aid is so closely tied to commercial entities and exposure is no longer through TV advertisements but rather through face-to-face encounters, new questions arise in terms of safety of the children, the potentially exploitative nature of the visit, and the potentially self-driven intentions of the intermediary and the visitors. In simple terms, why should an outsider (more often than not an international tourist) be allowed into a local school without any pedagogical reasons? Would this be allowed in a European, American, or Australian school? The answer is simply 'no'. So why should that be allowed in an African context?

The research raised further controversy regarding Africans as the *subjects* of development (Lahiri-Dutt 2017) and affective decisions in charitable giving influenced by presenting single victims (Dickert et al. 2016), often at the hand of local intermediaries:

> *I think it's important we keep students in areas that are accessible, you know, so that's why we look at schools in that sort of area, because they will want to come and see them.*
>
> (Z2_1_NGO_37)

> *Sometimes you may see somebody dancing and you know he entices your eyes and you say oh I can support that child.*
>
> (U1_CBO-F_2)

While minor changes to practices have been identified by some tourism operators from learning, most encounters continue unquestioned:

> *He [a tourist] was one of our first sponsors and it was this little girl who came and put her hand on his hand in that whole ... now there's no opportunity for them to do things like that.*
>
> (Z2_1_NGO_37)

Well-meaning handouts by tourists to children are almost entirely resented by *local* adults, yet such gift-giving practices continue, often encouraged by tourism operators, with arguments against it pointing towards such practices as the cause of an emergent begging culture. Its longer-term implications to individuals, as well as inequality and conflict amongst recipients and physical dangers, were also raised as issues of concern:

> *This trend is becoming very common among the village kids so the moment a car is coming to their village they are running to the car so there is a possibility of having an accident definitely.*
>
> (EN_TB_65)

Social costs can also be considered in terms of *opportunity costs*, and allocating resources to educational scholarships in Africa as the most effective and relevant development intervention also remains unquestioned (Rodney 1972; Ndlovu-Gatsheni 2015). It seemed perplexing that this intervention was such a ubiquitous trend across all case studies, yet actual socio-economic impacts to society remained unknown, with few denouncing that:

> *Education has created unemployment.*
>
> (U2_B-COM_13)

> *It targets white collar jobs, so since they [colonialists/missionaries] wanted white collar jobs there was no need of getting into practicals, so because of that [the curriculum] has become theoretical [and questionably relevant].*
>
> (U2_CBO-F_8)

Yet the scarce resources are increasingly given blind as a kind of blanket solution and panacea for Africa's woes, without evidence. This is a critical area for future research if travel philanthropy as a development solution is to be considered ethically sound, truly altruistic, and effectively contributing to the UNSDGs.

In some contexts, children were found to be the most *active* recipients of donations, partly through processes of othering themselves as discussed in Chapter 9 (Sharma 2019), producing a reinforcing loop – the more youth recipients were active, the more social issues were identified. Whilst positive outcomes have the potential to emerge from donations, success stories amongst these active youth can also get lost in a profusion of social costs. This was evidently the case, especially in Ethiopia:

> *Through this kind of support there are a lot of people who are getting the support, but it increases the dependency syndrome rather than sustainable change on the life of the children, few children or few guides may have bought a car, may build a hotel but when you compare the informal money to individuals, to the existing life situation of the community, the impact is really is inclined to the negative one... I saw this situation.*
>
> (E2_NGO_61)

Giving by tourists in Ethiopia was found to be causing a culture of othering, laziness, alcoholism, and diminishing traditional cultural values amongst some actors, and significant donation dependency, led by expectations of easy money around the corner from the next tourist that enters the system. Gifts intended to provide educational opportunities for children, whether right or wrong as an intervention, were often having the reverse effect, adding new evidences to substantiate the 'paradox of aid' (Silk 2004):

> *Very few [that receive sponsorship] go to school.*
>
> (E1_TE_52)

> *If these youngsters didn't get any money they would go back to school, they would get an education, because there's no point going out on the streets if you are not getting any money.*
>
> (E2_TB_62)

Giving had increased the number of children begging on the street, and in turn it had led to the increased level of direct giving by tourists. As children grow their wealth, their power in their community becomes greater, and it is an addictive cycle. Children that do not attend school, or that would be *more likely* to attend if the donations did not exist, is not the only potential social cost. New social costs appeared to have emerged from the system, such as the inability to return to traditional life once 'left', difficulties leaving a social circle of soliciting, mental health problems and addiction, and crime, to name a few:

> *He [sponsored individual] can have many friends you know that is power, money has the power you know, so the town will be led by illiterates in the long run.*
>
> (E2_NGO_61)

> *You have to be there to feel the scale of the problem.*
>
> (E2_B-COM_64)

Whilst intentions of donors are unlikely to be malicious, rather the effect of bounded rationality, such issues fuelled by tourist donations are caused by a complexity of factors, such as locals' bounded rationality and so on. A generation that has arguably grown up on this system, and wilful blindness is one of the factors enabling this phenomenon to continue:

> *Once addicted it is difficult bring them back.*
>
> (E2_NGO_57)

> *Once the damage is there to change it, it takes generations.*
>
> (EN_TB_66)

This culture of active recipients also reinforces otherness and stereotyping (Powell and Menendian 2020; Ideland and Malmberg 2014), ingraining beliefs at a very young age (Wilson 2016), affecting feelings of inferiority, expectations, entitlement, etc. The effects of a single football or pen may cause temporary wrangles amongst recipients when the supply does not match demand, but a small gift, whether asked for or not, can plant the seed of new beliefs and behaviours and emerge as mechanisms for longer-term consequences.

11.2.4 Social and racial objectification

Social and racial objectification as a system output is effectively one of the key reasons identified for the system to exist. Many decisions are based on perceived or actual otherness, producing real ontological phenomena, defined as objectification. This feedback loop (see Figure 11.6) raises significant ethical questions about the basis of the entire travel philanthropy phenomenon and opens a new area for further critical debate.

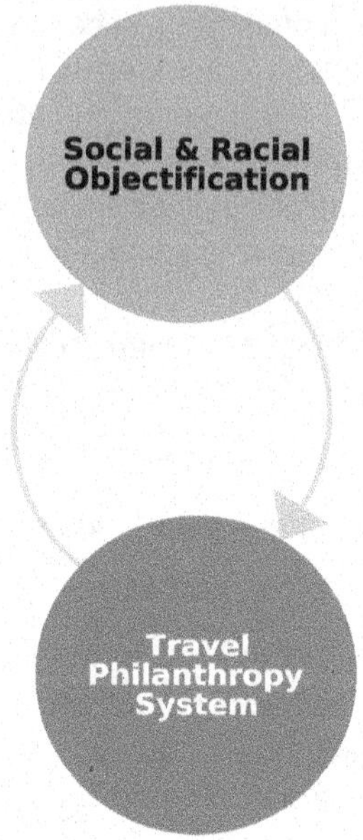

Figure 11.6 Emergent system property: social and racial objectification.

11.2.5 Negative–positive role models

Role models (Bandura 1977) were identified as a very influential mechanism and event in the system but also a key system output, recognised as a feedback loop back into the system too (see Figure 11.7). In the wider social system in which travel philanthropy is embedded, negative–positive role models were also found to exist, so even though they may be outside the travel philanthropy system, they also influence the behaviour of the travel philanthropy system and vice versa – e.g., how people behave in terms of trust and corruption. These role models may act as an input to the travel philanthropy system as well as an output as a feedback mechanism (Baden 2014).

The *negative–positive role model* identified as the most common system output refers to imitation of the Other (Arjoon et al. 2018) as negative social behaviours rather than pro-social (Koumakhov 2009; Thieme 2013). Role models were potential powerful leverage points in the system, and while imitation was powerful in all study settings, from children to businesses, because of spatial clustering of actors interacting within the travel philanthropy system, just one individual was found able to create the butterfly effect (interconnected system amplifying seemingly minor actions by one actor to become significant impacts through imitation), positive or negative.

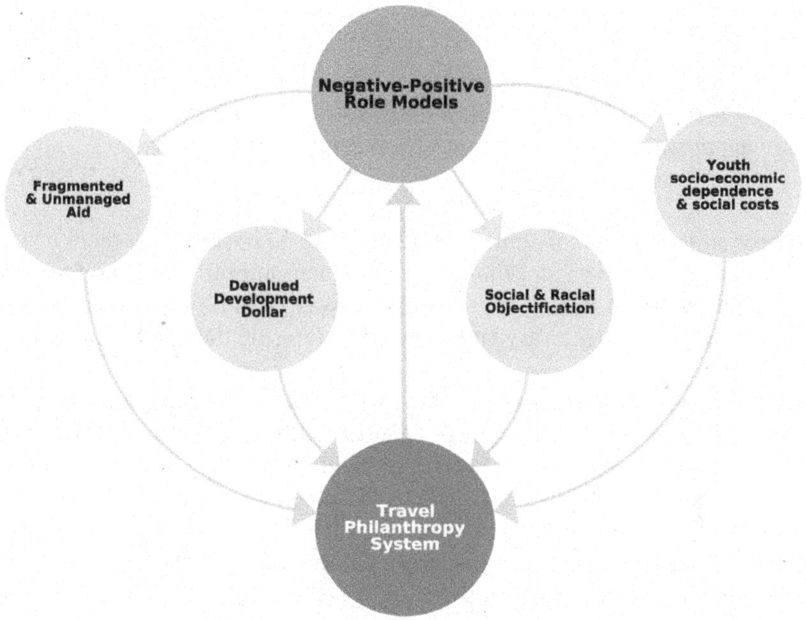

Figure 11.7 Emergent system property: negative–positive role models.

11.3 The paradox of travel philanthropy

In our study, the identification of travel philanthropy as a paradox yielded a range of dilemmas. 'Free' money, 'gifts', 'help', whatever the name and nature of resources donated through the travel philanthropy system, have emerged as an accessible 'solution' for both individuals and societies challenged economically and for ambitious individuals. Giving small gifts or doing one's best with good intentions in Africa may appear harmless, but evidence indicates that this can produce non-linear effects and contribute to individual and collective harm (Von Bergen and Bressler 2017; Hailu 2015; Patton 2019). If a system is found to cause social issues that would not otherwise emerge, one must question if the good of giving outweighs the bad, or if the bad outweighs the good. Reflecting on the Ethiopian findings above, particularly explicit social issues appeared to have created a serious dilemma:

> *One positive story must outweigh ninety-nine negative stories... nobody wants to youngsters not to get the opportunity.*
>
> (E2_TB_62)

They will be victims with a lot of things, some will be success, maybe 30–40%, is also difficult.

(E2_TB_56)

There have been many many students who have gained from tourism, who have gone to good school and been able to help their people and themselves by receiving money from foreigners, and some have lost themselves.

(E2_NGO_57)

Socio-economic impacts for some individuals may be positive, but there was also a significant fear amongst stakeholders about emerging effects. Therefore, a key question that emerged during our assessment was: should the system be left to behave this way? Is it necessary and more ethical to stop the good in order to stop the bad?

In Zimbabwe, social issues amongst the youth may have emerged as being comparatively less significant than in Ethiopia, but ethical dilemmas of travel philanthropy remain. It appeared that travel philanthropy offers an example of what Nickel and Eikenberry (2009) call marketised philanthropy and supports Pollock's (2016) argument that corporate social responsibility (CSR) and philanthropic practices in tourism need to stop justifying being wilfully blind (Heffernan 2012) to structural and systemic flaws of the current system.

It's always a double-edged sword, you might have the situation where tourism is impacting negatively on the environment in certain ways and you're the organisation [non-tourism NGO] that's trying to deal with that negative impact, so you're trying to derive an income from a tourism-based organisation through philanthropy but dealing with an impact that they created so it is a cycle.

(Z1_NGO_27)

Evidences also showed that travel philanthropy can also negatively impact the tourism sector of a country due to a vicious cycle of expectations–donations–solicitation, reinforcing not only a negative image of Africa (Ayittey 2015), but eventually the tourists experience, as evidenced through some business of poverty stories:

Our purpose is to change the image [of Ethiopia] and to say we did this, and that we achieved this and that, not we are having this problem, this problem, this problem ... they [local NGOs] don't have the right training on how to promote their projects.

(EN_TB_65)

The tourists hate it [solicitation by kids] and they are all commenting.

(E2_TB_62)

Therefore, if this perception of Africa generates resources and further cements a reputation of woes and lack of integrity, this may have knock-on effects elsewhere, such as discouraging sustainable investments outside tourism or deterring future visitations, with some substantial socio-economic repercussions for the country.

For instance, in Uganda, travel philanthropy increasingly had engendered inter-community conflict as they scrambled for the 'free' resources donated by tourists, enabling corruption and at times turbulent local politics. Therefore, if the system fuels this type of social behaviour, is it ok to continue? Theory over reality in terms of positive impact and surface over deeper understandings became evident during our empirical assessment, with participants denouncing widespread practices:

> *They [other locals] kind of tap the donors, so they use propaganda, they use lies to bring someone down who is in fact using the money well, so that they can take the advantage so, that one is so common here.*
>
> (U2_CBO-F_8)

> *In theory it's [donating to the CBO] such a great idea... you know, but at the end of the day it's extremely successful and it makes our business.*
>
> (U1_TB_3)

Inequality was identified as a core system element in *Figure 8.1*, and for the youth and other demographics, although unintended, travel philanthropy was in some circumstances fuelling inequality rather than solving it, having other non-linear effects:

> *Some people [have] started becoming richer than others, then others studying more than others, then statuses changing, social status, economic status and that is when this gossiping and propaganda started coming in.*
>
> (U2_CBO-F_8)

Ad hoc stories to attract travel philanthropists are many, but the blurred definition of what constitutes a negative impact is also a complex matter in itself, steered by what is *good* and *bad*, *right* and *wrong*, and most importantly, to whom (Burgold and Rolfes 2013; Smith and Sypher 2010; Ayittey 2015; Berry 2014). Currently, this remains a fine or fuzzy line between the binary categories of *right* and *wrong* in travel philanthropy, but for the reasons outlined in Chapters 9 and 10, such as bounded rationality (Simon 1956), most actors in the system remain wallowing in wilful blindness (Heffernan 2012).

Whilst it is an established fact that travel philanthropy has its advantages and disadvantages, and it is evident that there are known unknowns (Davies and McGoey 2012; Croissant 2014), at times it is unclear how to increase positive impact, but in the meantime, it is business as usual, as the belief system remains unchanged, actors keep telling each other the same stories, and the narrative of the virtuous nature of giving–receiving any gift through any pathway in Africa endures. Reflecting on our research, questions emerged about what the tipping point might be when system outputs (e.g., social costs amongst youth, or the opportunity costs and other implications of a fragmented and unmanaged aid landscape) attract attention more formally or broadly amongst the global and local actors or communities of these exchange economies. How many individuals gaining/losing directly or indirectly from travel philanthropy (at destination level or cumulatively across many places in Africa) will it take to shift actors' behaviours from business as usual to a different perspective that influences new behaviours? Who, if anyone, should be at the helm driving this? – The affected communities in Africa? Individual citizens, global or local? The travel and tourism industry suppliers and intermediaries? Government? An international body? Someone else? – Will the underlying belief system eventually change in that some exchanges become socially unacceptable, perceptions of accountability and responsibility shift, and we look back to old practices when we were more naive or wilfully blind? Whilst complex systems like this cannot be controlled, they can potentially be harnessed, tweaked, or constrained (Mueller 2020), so this understanding of the system contributes a new foundation for progress in this respect.

12 Whose business is poverty?

This chapter summarises what we found behind the scenes of travel philanthropy, what understanding the bigger picture as a collective impact generating system means, and how this advances knowledge beneficial to academia, policy making, and practice. It became increasingly evident in our study that to limit one's knowledge of the travel philanthropy system to 'donor' perspectives and behaviour is incomplete. We found donated resources are very often actively consumed – i.e., recipients or intermediaries are not always passive but active actors, and the 'gift' consumed is not always 'charitable' or 'altruistic' but can have multiple meanings – with different system stakeholders often applying for-profit strategic business principles and tactics, which has significant implications on the dynamics of travel philanthropy as a system and its outcomes. The business of poverty as a concept emerged as part of this complex and dynamic set of processes, involving different stakeholder groups depending on the context. We challenge existing myths about the 'donor-beneficiary' relationship and the top-down aid chain with our evidence showing that aid recipients have considerable agency within the travel philanthropy system when compared to evidence across international development literature and amongst practitioner 'development speak'. Travel philanthropy recipients are proactively engaged in driving the system's existence and shaping its structure. In travel philanthropy, individual exchanges are smaller in value compared to large-scale development aid but joining the long tail of philanthropy in Africa, together this creates an unmanaged, fragmented aid landscape with non-linear socio-economic effects that are packed with ethical and efficacy dilemmas.

By reflecting on other African 'geographies of compassion' as well, this chapter concludes that if one does not thoroughly understand the gift in these contexts and the reasons behind decision-making that are shaping giving–receiving pathways (from multiple perspectives), any knowledge about system dynamics or meanings of impact will be incomplete. We found that by understanding the interrelationships within travel philanthropy and using an interdisciplinary theoretical framework that looks beyond international development and gift-giving thinking, we can begin to improve our understanding. Based on our findings, including the critical socio-economic issues linked to the business of poverty, this chapter calls upon

DOI: 10.4324/9781003497929-12

those advocating for travel philanthropy as a sustainable development tool or those evaluating 'impact' to see the system, and to advance the frame of reference for this form of philanthropy beyond resource 'efficacy' in a development context, to the reality of the events and mechanisms inside the system that drive its outputs.

12.1 Travel philanthropy in Africa: an overview of key findings

Our study provides a much-needed critical investigation of the travel philanthropy phenomenon across African contexts and contributed a new conceptual model inspired by systems thinking and critical realism, with policy and practice implications for sustainable development (see Figure 11.2). What our research unpacked is that the exchanges within travel philanthropy are driven by multiple perspectives and complex decision-making processes embedded within heterogeneous moral economies and cognitive capacities, confirming that travel philanthropy is not simply based on a linear pathway of gift-giving by enlightened Global North altruists to grateful passive Global South beneficiaries. In fact, we advanced that it can be more accurately represented as an open system that has emerged informally without a central authority, and an evolving exchange economy of tangible and intangible objects. While the business of poverty unfolded as a particularly novel insight to the extant literature, new critical insights highlight how impacts of the travel philanthropy system demand an ethical and efficacy framework capable of capturing its greater complexity.

Having identified the evident knowledge gaps in Chapters 1 and 2, our research advanced new understandings of the concept of travel philanthropy by contributing a bigger picture view of its system in Africa. Although one could argue that evidence from three countries and five specific cases study settings may not be sufficient to represent what happens across the much wider African context, what we present here is to be used as a stepping stone into investigating an evolving phenomenon which is growing together with the rise of tourism in the African continent. Evidence from our research indicated that when the essence of the system for each case study was refined, the *same system* actually emerged in terms of structure. This type of middle-range theory was achieved by transcending isolated interventions and considering collective matters of causality. In this book's *Foreword*, Judy Kepher-Gona also provides a set of reflections on her own experience of travel philanthropy in Kenya where similar scenarios, dynamics, complexities and implications are present. Similarly, in contexts like Botswana, Burundi, Egypt, Eswatini, Ghana, Malawi, Namibia, Nigeria, Rwanda, South Africa, Tanzania, the Gambia, Tunisia, and Zambia, comparable travel philanthropy practices and consequences have been observed over the past 20 years by the authors of this book, backing the urgent need to delve more deeply into the topics identified inside the travel philanthropy system.

In our preliminary research phase, perspectives of local people were identified as a critical gap in knowledge; hence, the empirically grounded view of the system emerged from an analysis of how multiple stakeholders experience events that occur, which enabled a broader perspective on *what* was occurring in the system, *how* the system functions, and *why*. Through an analysis of existing literature, categories were identified in *Table 4.1* (Chapter 4), indicating parts of the overall travel philanthropy concept – input, vehicle, medium, intervention, output, and impact. To consider travel philanthropy as one or a selection of these parts, which is the approach authors have taken in the past, is incomplete so we must shift knowledge from the splintered reductionist thinking identified at the outset of this research. This led to understanding the behaviour of the whole which meant discovering emergent properties from these parts, rather than the sum of them.

Travel philanthropy must also be understood within the sustainable development debate and more adequately reflect contemporary theoretical advancements informing development interventions and development evaluation in Africa more generally, as per Chapters 2 and 3. The findings advanced knowledge beyond the role of tourism companies as a standalone stakeholder doing good in a community, towards their role amongst other actors.

Our study therefore looked behind the scenes of travel philanthropy, examining causal processes within the system, and what this means. By developing qualitative insights regarding causality through a systems approach inspired by critical realism, our research has advanced knowledge that is beneficial to academia, policy, and practice in the specific studied contexts and beyond. Transcending a surface level view of the system, our research highlights the importance of unpacking the observable, unobservable, and absent mechanisms that influence the structure and dynamics of the system, identifying what is or could be defining the impact of travel philanthropy on a local and national level and eventually continental and/or global scale. Tangible objects exchanged may be relatively small in value compared to mega philanthropy or large-scale development aid, but socio-economic effects can be non-linear, with a profusion of small exchanges contributing to the long tail of philanthropy in Africa, together creating an unmanaged, fragmented aid landscape packed with ethical and efficacy dilemmas. New questions have been raised through this research regarding giving and receiving pathways, outlining key emergent outputs of the travel philanthropy system, some of which are concerning and even deleterious. If travel philanthropy is to claim an ethical and professional niche of development aid or philanthropy in Africa (Honey 2011), the research needed to build upon Douthwaite and Hoffecker (2017), Banson (2015), Ramalingam (2013), Mutale et al. (2015), and Green's (2016) work that called upon stakeholders' better understanding of the complex systems in which they intervene. In addition to this, capturing system-wide causes and effects, or change is also needed. The mechanisms identified in our study represent potential leverage points for changing how the system behaves.

Our study led to a new conceptual model presented in Chapter 11 *Figure 11.2*, which incorporates the travel philanthropy system map (from Chapters 9 to 10), but also broadly identifies layers of causality spanning Bhaskar's (1975) ontological domains of the empirical, actual, and real. The boundaries of this model were defined by key themes that emerged from the data inductively, and an analysis of where patterns arose. The model illustrates deep underlying mechanisms as causes of the travel philanthropy system's existence, such as beliefs and norms (Chapter 8), and the emergent system structure of cause–effect interconnections indicating refined nodes such as sameness, wilful blindness and M&E, with feedback and typical patterns that create change (Chapters 9 and 10). The most salient outputs produced by system dynamics cannot be changed without understanding what underlies. These included the unmanaged fragmented aid landscape, devalued development dollar, socio-economic issues amongst youth, racial objectification, and role models (Chapter 11).

12.2 Travel philanthropy in Africa: new theoretical thinking

In addition to the conceptual contribution discussed above, the key areas of new knowledge generated from advanced understandings of the travel philanthropy system include the actual structure of the collective impact generating system, the business of poverty and active aid recipients, mechanised narratives and the issue of wilful blindness, the paradox of the gift, and the absence of monitoring and evaluation.

12.2.1 Collective impact generating system

This research effectively concludes with the bigger picture of travel philanthropy as an impact generating system. The conclusions shift the understanding of impact beyond evaluating individual 'project' interventions – e.g., paying children to attend school, installing a water pump or transferring business capital to the youth – to a representation of travel philanthropy that transcends isolated interventions and looks at collective effects, emergent issues, and impacts of the system whole. This conceptual model therefore contributes a more realistic understanding about the enabling/constraining environment that *shapes* the impact of travel philanthropy in tourism destinations, rather than defining the impact of individual travel philanthropy donations per se. The framework contributes knowledge about the root causes of impact as part of an overall system.

In the early stages of this research, it was expected that a unique system for each case study would be part of the conclusions, to aid the comparative analysis across countries. However, one of the most significant contributions is that, when the essence of the system for each case study was refined, the *same system* actually emerged in terms of structure. Initially, the stakeholders, system perspectives, and categories of elements found necessary for the exchange economy to *exist* as a social phenomenon emerged as comparable, including beliefs, principles and norms, emotions, 'the gift' or resource, stories, roles, technology, and inequality or perceived need. The emergent system discussed throughout Chapters 9 and 10, such as boundedness and otherness, is rooted in this deeper underlying structure, so without certain beliefs or emotions, for instance, the system would simply not exist. The system map illustrates the higher level emergent phenomena indicating how a change in one node tends to affect another. The similarities across cases provide insights into the structural conditions of the system and key causal elements linked to impact.

Whilst the systems were similar, there were contextual differences found influencing the *behaviour* of the system, such as role occupants and various factors affecting levels of resource responsibility and intervention quality. However, these differences occurred within the same system *structure*. Pathways could be of varying quality depending on individuals in the exchange, their capabilities, etc., which is not defined by stakeholder groups but a complex set of factors. Explanations of such processes found within the system have been inspired by linking theories that reach across the social sciences, which when interconnected help open the black box of causality and understand cause–effect of human behaviour in the travel philanthropy context.

By learning the system, our research aligned with current international development and evaluation debates in Africa arguing a weakness in understanding aggregate broader system level outcomes, impacts, and ethics in the development sector (Green 2016; Hailu 2015; Patton 2019). Our study advances the understanding of travel philanthropy beyond a demonstration of effectiveness or ineffectiveness, towards challenging the underlying logic of the complex system that it is (Gates 2016). Refining the essence of the travel philanthropy system by finding common mechanisms at work and developing this abstract analytic framework supports Pawson (2013) and Douthwaite and Hoffecker (2017) in contributing middle-range theories that transcend individual programme levels to a system potentially applicable to wider a set of interventions. Mechanisms identified within the model can help inform travel philanthropy, but also the broader philanthropy and development landscape. Discovering the travel philanthropy system as a concept applicable to all three countries in this study, paired with as previously mentioned empirical research previously conducted in the Gambia (Novelli et al 2015) and anecdotal evidences observed in other destinations across the continent, indicates potential relevance to other tourism destinations in Africa. As DFID (2019) called for evidence that can be transferred to other contexts in international development, this study contributes new knowledge in this merit.

12.2.2 Business of poverty and active aid recipients

Travel philanthropy is also concluded as an exchange economy shaped by *all actors*, but in particular arguing that the often-presumed *passive* beneficiaries of donations in African communities are typically not passive recipients of aid (Silk 2004; Korf 2007). Donations are very often *actively* consumed, not necessarily as charity, but using strategic business principles, for profit. Thus 'taking' can be a cause of giving, which opens new ethical queries. Our study challenged Silk's (2004) simplistic North–South encounter of Western donors choosing to give and what to give, because although this continues to be evident in some instances, gift recipients in travel philanthropy are actively steering donors. Therefore, we came to evidence that giving is also the adjunct of taking as Retsika (2016) proposed. To limit one's knowledge of the system to donor behaviour is incomplete. Aid recipients have considerable agency within the travel philanthropy system compared to evidence across international development literature, with travel philanthropy recipients proactively engaged in driving the system's existence and shaping its structure.

The business of poverty is part of this, but across all stakeholder groups. If one is to operate within the travel philanthropy system or change its dynamics to produce UN sustainable development goals' (SDG) impacts, the theory of unenlightened altruism (D'Souza and Adams 2014) must expand to include unenlightened recipients and intermediaries, and behaviours of all actors must be addressed. Decision-making across all stakeholder groups is not purely rational, rather rationality is bounded (Simon 2000), with travel philanthropy particularly influenced by emotions, beliefs, and other factors. If one does not thoroughly understand the gift and decisions shaping giving-receiving pathways from multiple perspectives, any knowledge about system dynamics or meaning of impact will be incomplete.

The implications of the travel philanthropy system require a frame of reference beyond resource efficacy in a development context because a range of critical socio-economic issues linked to the business of poverty emerged, such as local conflict, fear of personal consequences, racial objectification or troubling cultural changes that are interconnected and cannot be separated. One particular new insight is that whilst the 'Gaze' towards the West is far less dealt with in the literature (Jedlowski and Thomas 2017), our research shows it to be multi-directional, generating a significant source of influence on the travel philanthropy system. This means othering, for instance, does not only include the objectification of Africans by the West, but also the objectification of the West, the tourist, the rich or the white by Africans, as well as objectification of Africans by Africans in the business of poverty. Decision-making rooted in otherness is a significant example of learned behaviours driving this system, and amongst all actors. That the system is largely driven by social and racial objectification, promoting alterity, is an important system impact that may risk being omitted from efficacy metrics. 'Poverty' has become a significant resource in travel philanthropy raising moral questions that demand further research. Though, our study does not attempt to support or refute the ethics of this, but rather it is aimed to highlight the system.

12.2.3 Mechanised narratives and wilful blindness

Our study also concludes that most actors are perpetuating mechanised and unquestioned narratives, effortlessly recounted and consumed. Examples of Cornwall's (2010) 'development speak' emerged too, with very similar story themes entrenched across case studies, exchanged between all stakeholders to legitimatise the ethics and efficacy of this gift economy, whether for genuine altruistic reasons, the feel-good factor or profit, or ignorance. Everyone appeared to recount similar buzzwords so most stakeholders operated in the theatre of political participation, not just the beneficiaries of aid (Scott-Villiers (2011)), but the private sector seemed to have joined in too. These well-rehearsed narratives overpower important questions and sustain norms, driving bounded rationality (Simon 2000). Looking beyond the 'travel philanthropy speak' of *needs*, *wants,* and *help*, a disorganised landscape of giving-receiving of relatively small aid interventions was found, generally either superficial contributions to Africa's woes or a cause of new ones. The good in giving becomes masked in these stories and reinforces beliefs. We therefore conclude that it is critical to uproot the assumptions ingrained in the development speak of this system and move away from stakeholders exchanging stories they want to hear, reduce levels of wilful blindness towards more frank conversations that question the good in giving and the real meaning of the outputs produced by the system.

12.4 The paradox of the gift in Africa – 'free' tends not to activate positive system dynamics

Overall, the travel philanthropy system and its impacts across these case studies create a paradox, perhaps intended to be 'good' but raising ethical dilemmas, which substantiates sparse evidence found at the outset of the research but adds to the work of Burns and Barrie (2005) and Novelli (2015/6), for example. Whilst resources generated by the system could be beneficial, claims of travel philanthropy contributing a net benefit to sustainable development with these resources are debatable, especially because the research produces knowledge of some escalating negative impacts, such as social issues amongst the youth. It is also impossible to judge the good in sponsoring children without knowledge regarding comparative root values of this cause to other options, and its broader societal impacts. What truly constitutes 'doing good' demands further research, as the majority of 'gifts' continue to be channelled to identified 'needs', which tend to be more often than not based on unproven donor/local assumptions. For instance, the scenario encountered is that targeted beneficiaries may be simply asked what they would like to be 'given' or 'gift' being left in the hands of weak community-based organisations (CBOs)/individuals that do not extend benefits beyond elite capture, or being allocated directly to troubled minors in a community. While here the 'good in giving' (the intention) is unquestioned, what is problematic is how 'to give well' or whether 'to give at all'.

Travel philanthropy emerged as being not based on an effective altruism model; it remains highly subjective and an unenlightened development approach. The antecedents of an exchange plus what occurs in the wake of the gift – i.e., the resource pathway – have a greater influence on the impact of a donated resource, even more than the actual gift, or *what* is donated. It became clear from our research that 'free' very rarely means good. The resource pathway in travel philanthropy defines that value, such as monitoring characteristics, and many travel philanthropy gifts are regarded as free. This creates serious impact implications, including the 'devalued development dollar'.

Beyond donors' altruistic intentions, the giving-receiving pathways, when viewed from other perspectives such as local guides, recipients or the broader community, indicated that the travel philanthropy aid chain, its impacts and meanings were somewhat different to the extant literature. A key conclusion is that the quality of the resource pathway influences the meaning of a resource, and this is a systemic matter, generating system outputs identified in Chapter 11. Our research also emphasises that a resource pathway is not a simple chain of events. Whilst bottom-up approaches to development interventions may be important to tackle failings of top-down blueprint projects (Stirrat and Henkel 1997), it is also too simplistic and idealistic to assume locally driven interventions are the panacea for African development, because this is just another linear but only reversed perspective. With moral economies conflicting, otherness, low levels of M&E and bounded rationality playing out amongst other mechanisms, bottom-up or top-down interventions may both lead to negative impacts. Therefore the entire pathway impacting the meaning and value of a resource must be understood.

12.5 Absence of monitoring and evaluation

With the non-governmental organisation (NGO) sector in Africa already in a crisis of accountability (Burger and Owens 2008), our study warns of even higher levels of freedom and fragmentation in this niche of philanthropy or development aid, due partly to low entry barriers for all roles, the private and direct nature of exchanges and vague accountability across the entire system. Levels of ex-ante and ex-post evaluation have been found to be extremely low in travel philanthropy amongst individual interventions and very few external or internal pressures exist to even encourage evaluation whitewashing, common amongst traditional aid organisations. Beyond a few unenlightened numbers – such as total children sponsored in school, or 'what a community wants' – informed M&E in travel philanthropy effectively does not take place at all. Reports concerning 'evaluation', whether written, verbal or visual (i.e., photographic evidence), tend to be stuck on proving one's involvement in giving *per se*, to promote more giving, rather than steering or improving impacts. This lack of knowledge, understanding and informed management is enabling giving-receiving to continue regardless of results.

Our research contributes new insights into the critical importance of a bigger picture view going forwards, and with travel philanthropy as a living system, the importance of a developmental evaluation system in place on a destination level, where use is a design feature, including being accessible within the local community (South et al. 2019; Patton 2019; UNFP 2020). One issue behind the absence of this bigger picture perspective – and wilful blindness manifesting as it does – is the fact that no one is at the helm of travel philanthropy as a collective. This is partly because it is rooted in a lack of formal structures. In addition, significant confusion exists amongst all actors about accountability, procedures, and responsibility, which indicates pressing policy and practice implications.

12.6 Travel philanthropy: policy and practice implications

Having highlighted some of the key issues associated with the travel philanthropy system, our research may contribute to foster better policy and practice. The challenge remains how this informal, unmanaged, and fragmented aid landscape can be managed ethically, effectively, and collectively. One apparent mechanism to improve impact is to increase the levels of monitoring and evaluation from a complexity perspective to shape more beneficial system dynamics, driving transparency and informed decision-making. There remains a critical lack of knowledge transfer within the system driven partly by the significant lack of collaboration and privateness. Driving collaboration could be a key system intervention area. Imitation and the power of role models are also concluded as a key mechanism, thus shifting from negative–positive role models to positive–positive role models will be key. How 'positive' is defined undoubtedly remains unanswered (see opportunities for future research). Both giving and receiving are desired, and likely to continue, so designing appropriate and trusted pathways is critical, but who or what should be at the helm is a complex matter. Whether this patchwork of resources is aggregated into collective destination-based development efforts or not, system-wide and non-linear socio-economic effects must be incorporated into any design for successful management.

From a tourism sector perspective, it is clear that travel philanthropy is not an evolution of inclusive business models in Africa, but a separate phenomenon not solving that issue. Many tourism companies feel the right or need to intervene by giving 'free' gifts and are placing travel philanthropy higher on the agenda to restructuring structural flaws of businesses, which prevent more sustainable economic or social opportunities through value chains, as cautioned by Pollock (2016) and Ashley et al. (2007). Many continue to gain legitimacy as sustainable ethical businesses models through this philanthropic rhetoric, channelling tourists' donations, without understanding or wanting to understand sustainable development. With positive impacts currently questionable and negative impacts apparent, it is suggested that interrogating knowledge beyond what a business contributes, to questioning resource pathways such as involving visiting children in schools and broader impacts is critical. On the other hand, many tourism businesses (especially in Ethiopia and Uganda) are not engaging formerly in corporate philanthropy at all, partly because of fear, as well as knowledge and capacity limitations. If more ethical giving models were to be achieved, the role of tourism businesses or their potential as private actors in the development landscape requires careful attention.

Additionally, beyond a few individual locals benefitting from their business of poverty, most of the economic value from tourism remains in the accounts of international or elite nationals' owned tourism businesses. CBOs or the sale of stories based on problems have effectively become the local tourist product. Local people bordering protected areas in these case studies are not genuinely benefitting economically from the tourism value chain. Rather, they are 'altruistically' given a tiny part, which is unreliable and condescending, and may or may not generate returns for their time and efforts – i.e., a village visit tends to rely on the possible donations or handouts from tourists, thus remains at the mercy of an emotion-based decision. Where amounts that are more substantial are donated, such as some of the hybrid models, it is a charity model not tourism. Travel philanthropy as it stands is not a sustainable solution for protected area communities to be involved in and benefit from tourism.

It is also critical that tourists are factored-in as influential *direct* donors in these communities, as well as drivers of otherness; this is not an insignificant nor former issue as the literature suggests. Tourists tend to give 'free' gifts including to children, without monitoring or evaluation mechanisms, and combined with unfulfilled promises too, these actors are key in generating local-level economic, social, and psychological knock-on effects. In-destination and post-trip harassment of visitors needs to be managed, and visitor expectations and behaviour need to be better managed, something host communities need to learn. The collective effects of former tourists, such as handouts during their trip and when they go home, are also significantly part of the cause of socio-economic issues. All stakeholders must stop avoiding questioning the good in giving beyond intentions, which remain a critical weakness and subject of potential contention. The business of poverty also involves Africans and foreigners reinforcing the detrimental long-standing narrative of its woes, which may further affect its global reputation and hamper more sustainable investments on the continent beyond charity. The opportunity to change beliefs through global tourism is being missed.

As a relatively new player within private philanthropy for international development and within the complex arena of sustainability and ethics, travel philanthropy is an open system, so will remain unpredictable. However, the recognisable systemic patterns presented in our research indicate how to better understand its interconnections, thus informing policy and practice about mechanisms that may be adjusted to achieve greater ethics and efficacy (Douthwaite and Hoffecker 2017; Mueller 2020). This living system cannot therefore be tightly controlled, but it is currently a 'man-made' system that is effectively allowed to remain in the wild, so steps need to be taken to tame it, to avoid irreversible damage and to generate sustainable benefits in Africa.

It would have been impossible to understand travel philanthropy as a system in such depth without transcending the empirical domain and understanding unobservable phenomena. Methodologically, a contribution made by our research was therefore its philosophical position of critical realism able to inform its approach to causality and generate qualitative insights beyond observable facts to where patterns arise. Combined with systems thinking, this contributed a new approach to understanding travel philanthropy. Reflecting on the original work of Bhaskar (1975), the mechanisms and structures became the object of our scientific enquiry, guiding the research process of stepping inside the system (Hummelbrunner 2011), learning its behaviour (Jackson 2019), and enriching the interdisciplinary lens on the subject of travel philanthropy, inspired particularly by theories discussed in Chapter 5 (Lowe et al. 2013). To understand such a system, one cannot just look at sustainable development literature but must look at social systems and socio-economic behaviour more broadly. Furthermore, our study provides some critical links to the why's behind human behaviour, as different people's decisions fundamentally shape the system and 'sustainable development'. Drawing upon theories such as imitation, bounded rationality, and wilful blindness helps to make sense of why the system behaves as it does and importantly these become key leverage points if one is to sustainably change the system.

12.7 Travel philanthropy research: limitations and future research opportunities

While some research limitations and some challenges are acknowledged in Chapter 5 such as the number of case studies possible, research participants, and boundary decisions, it was discovered that within the complexity of the travel philanthropy system, or arguably within any open system of the social world, by 'opening the black box', one tends to discover another black box within that, and then within that, and so on. Furthermore, whilst the interdisciplinary theoretical approach employed in our study provided a richer and deeper understanding of the travel philanthropy phenomenon, time and space limitations meant setting relatively tight boundaries in explanation, when considered in the context of *all* academic disciplines. To further understand the 'why's' behind the 'what' is to better inform positive change. To set boundaries for a curious scientist is to be disciplined.

Change, whether right or wrong, is also always happening in the world. Therefore, another emerging limitation is that the travel philanthropy system map will of course be constantly changing too, so combined with the fallibility of socially produced knowledge, the findings presented here can never be completely correct. Rather our study provides grounds for discussion and further investigation, and its relevance may vary through time and space.

Lastly, a key limitation to deepening our understanding of the implications of most travel philanthropy donations being allocated to children's education is in not having existing empirical evidences to draw upon and better investigate the real-world impacts of 'formal education' in Africa, which led to the discovery of a further fundamental gap in the understanding of intervention impacts. On the other hand, the reflection of this challenge also inspired a new perspective on the conceptualisation of 'impact' which may have been overlooked if we were not pushed in this direction, when focussing on the impact of travel philanthropy interventions per se, as anticipated at the outset. While the identified research limitations may have caused temporary challenges, it was found that they eventually tended to pave the way for new research opportunities proposed here.

a. Deepening knowledge of causal mechanisms – Time and budget limitations restricted the range of destinations across Africa that could be incorporated in our study, the depth with which each emergent mechanism could be explored empirically, as well as the extent to which further theoretical lenses explaining causal interconnections could be drawn upon. Our study suggests a travel philanthropy framework as an impact generating system, but each element or node identified in the system would benefit from further research to enrich the knowledge, whether that is deepening our understanding of otherness, imitation, boundedness, evaluation constraints, and so on, or investigating the belief system and informal rules underlying. Having said this, travel philanthropy as an evolving phenomenon would benefit from further critical research to develop this model, as well as investigate its transferability to other locations across destinations in Africa and beyond. Additional countries, tourism destinations or other contexts could be studied to see if similar patterns arise in different contexts, how the system behaves in particular circumstances, and what varying system dynamics mean for impact. It is potentially the case that the same overall conceptual model would emerge, but certainly a more in-depth understanding of contextual factors would be invaluable.

b. Quantitative and specific impact research – There still remains a lack of quantitative data available on this subject and although assessing travel philanthropy was never at the centre of our research, with foundations now set through the new conceptual model, the collection of statistical data would be complementary to subsequent theoretical developments. This may include the size of travel philanthropy resources and impact metrics aggregated across a destination, quantifying money being channelled to children's education sponsorship education, via guides and CBOs, or tourism businesses, for example. A comprehensive econometric study aggregating impacts across this fragmented aid landscape, to measure travel philanthropy beyond observations and qualitative insights would be timely, particularly as Environmental, Social and Governance (ESG) reporting processes becomes increasingly important across sectors. As outlined in Chapter 2, impacts of philanthropy must be considered within the context of what does/should sit under standard business practice, such as value chains to avoid masking necessary changes to businesses, which is more sustainable, so understanding how to approach this is critical. Importantly, our study also cautions against only effectiveness metrics and the blindness to unintended consequences. Research about how to measure the impact of observable and unobservable mechanisms for this specific system is also needed, quantitatively and qualitatively – as well as research for a better understanding of how to make the impacts manageable.

Our study identified the critical importance of enriching and maintaining a bigger picture. Research into how best to develop and implement a developmental evaluation system for learning and accountability in travel philanthropy, which not only proves but improves and develops theories of change, and how information could be best understood and *used* by all stakeholders including policymakers, should be a high priority.

c. Ethics and efficacy of giving-receiving pathways – As pathways of giving-receiving are so pivotal, it is recommended that whilst sustainable development root 'solutions' remains a critical area for research as part of the pathway, further research is required to define what an effective or ethical giving-receiving pathway means in different contexts and in greater depth. It would be invaluable to better understand how the pathway defines the meaning of a resource and its implications for impact in different contexts. Children were also not included as research participants for ethical reasons (see Chapter 5), and as key direct and indirect actors in the system, the role of children in the pathway and their perspectives remain a key area for future research. Further research is also recommended to understand the issue of individual and collective ethics such as moral responsibility for collectively produced harm as well (Wettstein 2012); there remains a knowledge gap informing who exactly should and could be doing what in this respect, and how.

d. Human behaviour in travel philanthropy – Further empirical research specifically applying behavioural economics methods, and other insights from psychology and moral philosophy, would contribute depth to the topic by better understanding human behaviour that drives the system, and what the system drives. Perhaps building on the above, issues of anti-epistemology (Davies and McGoey 2012; Croissant's 2014) or the ontology of ignorance, including enriched knowledge regarding wilful blindness (Heffernan 2012), are recommended because this is strongly interconnected with responsibility, accountability, and impact. Further knowledge on this would be imperative to ethical and sustainable development debates.

To conclude, our study critically addressed travel philanthropy as an under researched phenomenon so to understand the meaning, significance, impact, as well as processes influencing pathways to giving-receiving engagements; the research navigated the challenges of a relatively blank theoretical page. Limited to present its rich qualitative findings within the bounds of our book, the conceptual model contributes new foundations and inspires further investigations on the subject, rather than providing all the answers. Our study is therefore not an end point but rather provides fertile new grounds that pave the way for future research – each element of the travel philanthropy bigger picture system, its underlying mechanisms, and impacts are now open for deeper investigation, as well as examining different contexts, and it is hoped that this model is developed as knowledge is enriched and the social world continues to evolve.

References

Abernethy, D. (2011). Engaging travelers in travelers' philanthropy: Unintended consequences of the traveler's best intentions. In: M. Honey, ed. *Travelers' philanthropy handbook*. Washington, DC: Center for Responsible Travel. pp. 179–190.

Acs, Z. and Phillips, R. (2002). Entrepreneurship and philanthropy in American capitalism. *Small Business Economics*. **19**(3), 189–204.

Adamides, E., Papachristos, G. and Pomonis, N. (2012). Critical realism in supply chain research: Understanding the dynamics of a seasonal goods supply chain. *International Journal of Physical Distribution & Logistics Management*. **42**(10), 906–930.

Adams, K.M. (2004). The genesis of touristic imagery: Politics and poetics in the creation of a remote Indonesian island destination. *Tourist Studies*. **4**(2), 115–135.

Adu-Ampong, E.A. and Kimbu, A.N. (2020). *Sustainable tourism policy and planning in Africa*. London: Routledge.

Alcoff, L. (2007). Epistemologies of ignorance: Three types. In: S. Sullivan and N. Tuana, ed. *Race and epistemologies of ignorance*. Albany: State University of New York Press. pp. 39–57.

Allik, J. and Realo, A. (2004). Individualism-collectivism and social capital. *Journal of Cross-Cultural Psychology*. **35**(1), 29–49.

Alt, M., Gallier, C., Schlüter, A., Nelson, K. and Anggraini, E. (2018). Giving to versus taking from in- and out-group members. *Games*. **9**(3), 57.

Antoni-Komar, I., Lautermann, C. and Pfriem, R. (2012). Unternehmen und Konsumenten in Verantwortungsgemeinschaft jenseits des Konsumismus: Nachhaltigkeit erfordert, das Zusammenwirken zwischen Produzieren und Konsumieren neu zu denken [Companies and consumers in a community of responsibility beyond consumerism: Sustainability requires rethinking the interaction between production and consumption]. *Zeitschrift fu¨r Wirtschafts- und Unternehmensethik* [*Journal for Business, Economics & Ethics*]. **13**(3), 297–316.

Apesteguia, J., Huck, S. and Oechssler, J. (2004). Imitation—Theory and experimental evidence. *Journal of Economic Theory*. **136**, 217–235.

Arcand, J.-L. and Wagner, N. (2015). Does community-driven development improve inclusiveness in peasant organizations? – Evidence from Senegal. *World Development*. **78**, 105–124.

Archer, M. (1982). Morphogenesis versus structuration: On combining structure and action. *The British Journal of Sociology*. **33**(4), 455–483.

Archer, M. (1995). *Realist social theory: The morphogenetic approach.* Cambridge: Cambridge University Press.

Archer, M., Decoteau, C., Gorski, P., Little, D., Porpora, D., Rutzou, T., Smith, C., Steinmetz, G. and Vandenberghe, F. (2016). What is critical realism? Available at: www.asatheory.org/current-newsletter-online/what-is-critical-realism

Ardoin, N., Wheaton, M., Hunt, A., Schuh, J. and Durham, W. (2016). Post-trip philanthropic intentions of nature-based tourists in Galapagos. *Journal of Ecotourism.* **15**(1), 21–35.

Arendt, H. (1958/1998). *The human condition.* Chicago & London: University of Chicago Press.

Arendt, H. (1994). *Essays on understanding.* New York: Harcourt, Brace.

Arensman, B., van Waegeningh, C. and van Wessel, M. (2017). Twinning "practices of change" with "theory of change": Room for emergence in advocacy evaluation. *American Journal of Evaluation.* **39**(2), 221–336.

Argrou, V. (1999). Sameness and the ethnological will to meaning. *Current Anthropology.* **40**(Supplement), 29–41.

Aristotle. (1998). *The metaphysics.* Translated from the Greek by H. Lawson-Tancred. London: Penguin Books.

Arjoon, S., Turriago-Hoyos, A. and Thoene, U. (2018). Virtuousness and the common good as a conceptual framework for harmonizing the goals of the individual, organizations, and the economy. *Journal of Business Ethics.* **147**, 143–163.

Arndt, C. and Jones, S. (2015). Assessing foreign aid's long-run contribution to growth and development. *World Development.* **69**, 6–18.

Ashley, C. and Haysom, G. (2006). From philanthropy to a different way of doing business: Strategies and challenges in integrating pro-poor approaches into tourism business. *Development Southern Africa.* **23**(2), 265–280.

Ashley, C., De Brine, P., Lehr, A. and Wilde, H. (2007). The role of the tourism sector in expanding economic opportunity. Available at: www.odi.org/publications/1117-role-tourism-sector-expanding-economic-opportunity

Ashley, C., Poultney, C., Haysom, G., McNab, D. and Harris, A. (2005). Tips and tools for South African tourism companies on local procurement, products and partnerships: Investing in local linkages – A win-win. Available at: www.odi.org/sites/odi.org.uk/files/odi-assets/publications-opinion-files/2592.pdf

Atkinson, A. (2009). Giving overseas and public policy. *Journal of Public Economics.* **93**, 647–653.

Aumann, R. (1997). Rationality and bounded rationality. *Games and Economic Behaviour.* **21**, 2–14.

Ayittey, G. (2015). Post-MDGs and Africa's development conundrum. *Journal of International Development.* **27**, 345–361.

Babutsidze, Z. and Chai, A. (2018). Look at me saving the planet! The imitation of visible green behavior and its impact on the climate value-action gap. *Ecological Economics.* **146**, 290–303.

Baden, D. (2014). Look on the bright side: A comparison of positive and negative role models in business ethics education. *Academy of Management Learning & Education.* **13**(2), 154–170.

Baert, P. (1998). *Social theory in the twentieth century.* New York: New York University Press.

Bailie, J., Laycock, A., Peiris, D., Bainbridge, R., Matthews, V., Cunningham, F., Conte, K., Abimbola, K., Passey, M. and Bailie, R. (2020). Using developmental evaluation to enhance continuous reflection, learning and adaptation of an innovation platform in Australian Indigenous primary healthcare. *Health Research Policy and Systems.* **18**, 45.

Bandura, A. (1977). *Social learning theory*. Englewood Cliffs, NJ: Prentice Hall.

Banerjee, A. (1992). A simple model of herd behavior. *The Quarterly Journal of Economics*. **107**, 797–817.

Banks, G., Scheyvens, R., McLennan, S. and Bebbington, A. (2016). Conceptualising corporate community development. *Third World Quarterly*. **37**(2), 245–263.

Banson, K., Nguyen, N., Bosch, O. and Nguyen, T. (2015). A systems thinking approach to address the complexity of agribusiness for sustainable development in Africa: A case study in Ghana. *Systems Research and Behavioral Science*. **32**, 672–688.

Baptista, J.A. (2017). *The good holiday: Development, tourism and politics of Benevolence in Mozambique*. New York: Berghahn Books.

Barbrook-Johnson, P., Proctor, A., Giorgi, S. and Phillipson, J. (2020). How do policy evaluators understand complexity? *Evaluation*. **26**(3), 315–332.

Barkan, J. (2013). Plutocrats at work: How big philanthropy undermines democracy. *Social Research*. **80**(2), 635.

Barnes, M. and Eagles, P. (2004). Examining the relationship between ecotourists and philanthropic behavior. *Tourism Recreation Research*. **29**(3), 35–38.

Baron, D. (2001). Private politics, corporate social responsibility and integrated strategy. *Journal of Economics and Management Strategy*. **10**, 7–45.

Baron, J. (2018). Social norms for citizenship. *Social Research: An International Quarterly*. **85**(1), 229–253.

Baskin, E., Wakslak, C., Trope, Y. and Novemsky, N. (2014). Why feasibility matters more to gift receivers than to givers: A construal-level approach to gift giving. *Journal of Consumer Research*. **41**(1), 169–182.

Bates, S. (2006). Making time for change: On temporal conceptualizations within (critical realist) approaches to the relationship between structure and agency. *Sociology*. **40**(1), 143–161.

Bateson, G. (1971). The cybernetics of "self": A theory of alcoholism. *Psychiatry*. **34**(1), 1–18.

BBC. (2020). Comic relief will stop sending celebrities to Africa. Available at: www.bbc.co.uk/news/entertainment-arts-54716750 (Accessed 28/10/2020).

Bebbington, A. (2005). Donor–NGO relations and representations of livelihood in non-governmental aid chains. *World Development*. **33**(6), 937–950.

Beckert, J. (2013). Imagined futures: Fictional expectations in the economy. *Theory and Society*. **42**(3), 219–240.

Beckert, J. (2016). *Imagined futures: Fictional expectations and capitalist dynamics*. Cambridge, MA: Harvard University Press.

Bekkers, R. and Wiepking, P. (2011). A literature review of empirical studies of philanthropy: Eight mechanisms that drive charitable giving. *Nonprofit and Voluntary Sector Quarterly*. **40**(5), 924–973.

Belk, R. (1979). Gift giving behavior. *Research in Marketing*. **2**, 95–126.

Ben-Ze'ev, A. (2000). *The subtlety of the emotion*. Cambridge, MA: MIT Press.

Bengtsson, B. and Hertting, N. (2014). Generalization by mechanism: Thin rationality and ideal-type analysis in case study research. *Philosophy of the Social Sciences*. **44**(6), 707–732.

Bennett, E. and McWhorter, R. (2016). Opening the black box and searching for smoking guns: Process causality in qualitative research. *European Journal of Training and Development*. **40**(8/9), 691–718.

Benton, T. and Craib, I. (2001). *Philosophy of social science*. London: Palgrave.

Berry, N. (2014). Did we do good? NGOs, conflicts of interest and the evaluation of short-term medical missions in Sololá, Guatemala. *Social Science & Medicine*. **120**, 344–351.

Bertini, I., Malone, C. and Revell, T. (2014). *Guide to sustainable philanthropy* (2nd ed.). Blue & Green Communications. Available at: https://blueandgreentomorrow.com/environment/the-guide-to-sustainable-philanthropy-2014/

Bhaskar, R. (1975). *A realist theory of science*. Leeds: Leeds Books.

Bhaskar, R. (1986). *Scientific realism and human emancipation*. London: Verso.

Bhaskar, R. (1989). *Reclaiming reality: A critical introduction to contemporary philosophy*. London: Verso.

Bhaskar, R. (2008). *A realist theory of science, with a new introduction*. Oxon: Routledge.

Bhaskar, R. and Callinicos, A. (2003). Marxism and critical realism. *Journal of Critical Realism*. **1**(2), 89–114.

Biggs, G. (2008). *Ecotourism as a conservation strategy for funders*. New York: Environmental Grantmakers Association.

Bikhchandani, S., Hirshleifer, D. and Welch, I. (1998). Learning from the behavior of others: Conformity, fads, and informational cascades. *Journal of Economic Perspectives*. **12**, 151–170.

Bill, A. (2011). Communications and marketing of travelers' philanthropy programs. In: M. Honey, ed. *Travelers' philanthropy handbook*. Washington, DC: Center for Responsible Travel. pp. 138–140.

Bimonte, S. and Punzo, L.S. (2016). Tourist development and host – Guest interaction: An economic exchange theory. *Annals of Tourism Research*. **58**, 128–139.

Binder, C. (2019). Redistribution and the individualism – Collectivism dimension of culture. *Social Indicators Research*. **142**, 1175–1192.

Blanco, E., Lopez, M.C. and Coleman, E.A. (2012). Voting for environmental donations: Experimental evidence from Majorca, Spain. *Ecological Economics*. **75**, 52–60.

Bloom, P. (2016). *Against empathy: The case for rational compassion*. London: Vintage Publishing.

Blowfield, M. and Dolan, C. (2014). Business as a development agent: Evidence of possibility and improbability. *Third World Quarterly*. **35**(1), 22–42.

Boella, G. and Torre, L. (2007). A game-theoretic approach to normative multi-agent systems. In: *Proceedings of the Dagstuhl Seminar: Normative multi-agent systems*. Dagstuhl, Germany.

Bok, S. (1981). Blowing the whistle. In: J. Fleishman, L. Liebman and M. Moore, eds. *Public duties: The moral obligations of government officials*. Cambridge, MA: Harvard University Press. pp. 204–220.

Boland, T. and Griffin, R. (2016). Making sacrifices: How ungenerous gifts constitute jobseekers as scapegoats, Distinktion. *Journal of Social Theory*. **17**(2), 174–191.

Bornstein, E. (2009). The impulse of philanthropy. *Cultural Anthropology*. **24**(4), 622–651.

Bourdieu, P. (1977). *Outline of a theory of practice*. Translated by R. Nice. Cambridge: Cambridge University Press.

Bouville, M. (2008). Whistle-blowing and morality. *Journal of Business Ethics*. **81**, 579–585.

Bowman, K., Chettleborough, H.J., Rowlands, J. and Whitehead, J. (2015). *Systems thinking: An introduction for Oxfam programme staff*. Available at: https://gsdrc.org/document-library/systems-thinking-an-introduction-for-oxfam-programme-staff/ (Accessed October 2015).

Brandt, M., Sibley, C. and Osborne, D. (2019). What is central to political belief system networks? *Personality and Social Psychology Bulletin*. **45**(9), 1352–1364.

Brenya, E. and Adu-Gyam, S. (2015). External forces on Africa's democracy and development. *Journal of Transdisciplinary Studies*. **8**(3), 28–55.

Brest, P. (2015). Strategic philanthropy and its discontents. *Stanford Social Innovations Review*. Available at: https://ssir.org/up_for_debate/article/strategic_philanthropy_and_its_discontents

Brightsmith, D., Stronza, A. and Holle, K. (2008). Ecotourism, conservation biology, and volunteer tourism: A mutually beneficial triumvirate. *Biological Conservation*. **141**, 2832–2842.

Brod, M., Tesler, L. and Christensenet, T. (2009). Qualitative research and content validity: Developing best practices based on science and experience. *Quality of Life Research*. **18**, 1263–1278.

Brown, A. (2018). Adam Smith's psychological and political insight: The theory of moral sentiments. *Adam Smith Social Research: An International Quarterly*. **85**(3), 541–555.

Brundtland, G. (1987). *Report of the world commission on environment and development: Our common future*. United Nations General Assembly document A/42/427.

Brunnermeier, M. and Parker, J. (2005). Optimal expectations. *The American Economic Review*. **95**(4), 1092–1118.

Buchtel, E., Ng, L. and Norenzayan, A. (2018). A sense of obligation: Cultural differences in the experience of obligation. *Personality and Social Psychology Bulletin*. **44**(11), 1545–1566.

Burger, R. and Owens, T. (2008). Promoting transparency in the NGO sector: Examining the availability and reliability of self-reported data, CREDIT Research Paper, No. 08/11, The University of Nottingham, Centre for Research in Economic Development and International Trade (CREDIT), Nottingham.

Burgold, J. and Rolfes, M. (2013). Of voyeuristic safari tours and responsible tourism with educational value: Observing moral communication in slum and township tourism in Cape Town and Mumbai. *Die Erde – Journal of the Geographical Society of Berlin*. **144**(2), 161–174.

Burke, L. and Logsdon, J.M. (1996). How corporate social responsibility pays off. *Long Range Planning*. **29**(4), 495–502.

Burns, P. and Barrie, S. (2005). Race, space and 'our own piece of Africa': Doing good in Luphisi Village? *Journal of Sustainable Tourism*. **13**(5), 468–485.

Burns, T. and DeVillé, P. (2017). Socio-economics: The approach of social systems theory in a forty year perspective. *Economics and Sociology*. **10**(2), 11–20.

Butcher, J. (2003). *The moralisation of tourism: Sun, sand and saving the world*. London: Routledge.

CAF. (2019). CAF UK giving 2019, an overview of charitable giving in the UK. Charities Aid Foundation. Available at: www.cafonline.org (Accessed May 2019).

Caldwell, C. and Millen, A. (2009). Social learning mechanisms and cumulative cultural evolution: Is imitation necessary? *Psychological Science*. **20**(12), 1478–1483.

Campergue, C. (2015). Gifts and the selfless work ethic in Tibetan Buddhist centres in France. *Religion Compass*. **9**(11), 443–461.

Cantis, S., Parroco, A., Ferrante, M. and Vaccina, F. (2015). Unobserved tourism. *Annals of Tourism Research*. **50**, 1–18.

Carlisle, C. (2011). Living in the light of religious ideals. *Royal Institute of Philosophy Supplement*. **68**, 245–255.

Carlson, D. (2012). Why you should consider cancelling your short-term mission trips. Available at: www.thegospelcoalition.org/article/why-you-should-consider-cancelling-your-short-term-mission-trips

Carroll, A. (1991). The pyramid of corporate social responsibility: Toward the moral management of organizational stakeholders. *Business Horizons*. **34**(4), 39–48.
Castoriadis, C. (1987). *The imaginary institution of society*. Cambridge: Polity.
Cavanaugh, L., Gino, F. and Fitzsimons, G. (2015). When doing good is bad in gift giving: Mis-predicting appreciation of socially responsible gifts. *Organizational Behavior and Human Decision Processes*. **131**, 178–189.
Cavill, S. and Sohail, M. (2007). Increasing strategic accountability: A framework for international NGOs. *Development in Practice*. **17**(2), 231–248.
Chakraborty, K., Saha, B. and Jammulamadaka, N. (2017). Where silence speaks-insights from Third World NGOs. *Critical Perspectives on International Business*. **13**(1), 38–53.
Chambers, R. (2012). *Provocations for development*. Bourton on Dunsmore: Practical Action Publishing.
Checkland, P. (1981). *Systems thinking, systems practice*. New York: Wiley.
Checkland, P. (2000). Soft systems methodology: A thirty year retrospective. *Systems Research and Behavioral Science*. **17**, S11–S58.
Chen, B. (2012). Moral and ethical foundations for sustainability: A multidisplinary approach. *Journal of Global Citizenship*. **2**(2), 1–20.
Chen, H. (1990). Issues in constructing program theory. *New Directions for Program Evaluation*. 1990, 7–18.
Chen, X. and Li, S. (2005). Cross-national differences in cooperative decision-making in mixed-motive business contexts: The mediating effect of vertical and horizontal individualism. *Journal of International Business Studies*. **36**, 622–636.
Choi, J. and Wang, H. (2007). The promise of a managerial values approach to corporate philanthropy. *Journal of Business Ethics*. **75**(4), 345–359.
Churchman, C. (1968). *The systems approach*. New York: Dell.
Cialdini, R. (2007). Descriptive social norms as underappreciated sources of social control. *Psychometrika*. **72**, 263–268.
Cialdini, R. and Trost, M. (1998). Social influence: Social norms, conformity and compliance. In: D. Gilbert et al., eds. *The handbook of social psychology*. New York: McGraw-Hill. pp. 151–192.
Cleveland, T. (2021). *A history of tourism in Africa: Exoticization, exploitation, and enrichment*. Athens: Ohio University Press.
Cochrane, L. and Thornton, A. (2016). Charity rankings: Delivering development or dehumanising aid? *Journal of International Development*. **28**, 57–73.
Coghlan, A. and Filo, K. (2013). Using constant comparison method and qualitative data to understand participants' experiences at the nexus of tourism, sport and charity events. *Tourism Management*. **35**, 122–131.
Cohen, N. and Arieli, T. (2011). Field research in conflict environments: Methodological challenges and snowball sampling. *Journal of Peace Research*. **48**(4), 423–425.
Cold-Ravnkilde, S., Engberg-Pedersen, L. and Fejerskov, A. (2018). Global norms and heterogeneous development organizations: Introduction to special issue on new actors, old donors and gender equality norms in international development cooperation. *Progress in Development Studies*. **18**(2), 77–94.
Coleman, J. (1990). *Foundations of social theory*. Cambridge: Harvard University Press.
Collet, F. (2009). Does habitus matter? A comparative review of Bourdieu's habitus and Simon's bounded rationality with some implications for economic sociology. *Sociological Theory*. **27**(4), 419–434.
Collier, A. (1994). *Critical realism: An introduction to Roy Bhaskar's philosophy*. London: Verso.

Collins, M. (2017). Nation, state and agency: Evolving historiographies of African decolonization. In: A.W.M. Smith and C. Jeppesen, eds. *Britain, France and the decolonization of Africa: Future imperfect?* London: UCL Press. pp. 17–42.

Contu, A. and Girei, E. (2014). NGOs management and the value of 'partnerships' for equality in international development: What's in a name. *Human Relations*. **67**(2), 205–223.

Cooley, C. (1902). *Human nature and the social order*. New York: Scribner's.

Cordery, C. and Baskerville. R. (2011). Charity transgressions, trust and accountability. *Voluntas: International Journal of Voluntary and Nonprofit Organizations*. **22**(2), 197–213.

Cornwall, A. and Brock, K. (2005). What do buzzwords do for development policy? A critical look at 'participation', 'empowerment' and 'poverty reduction'. *Third World Quarterly*. **26**(7), 1043–1060.

Cornwall, C. (2010). Introductory overview – Buzzwords and fuzzword, deconstructing development discourse. In: C. Cornwall and D. Eade, eds. *Deconstructing development discourse: Buzzwords and fuzzwords*. Rugby and Oxford: Practical Action Publishing and Oxfam GB. pp. 1–19.

Croissant, J. (2014). Agnotology: Ignorance and absence or towards a sociology of things that aren't there. *Social Epistemology*. **28**(1), 4–25.

Crouch, J. (2011). Engaging communities and local organizations in travelers' philanthropy. How to manage interaction with community projects. In: M. Honey, ed. *Travelers' philanthropy handbook*. Washington, DC: Center for Responsible Travel. pp. 162–166.

Crutchfield, R. (1955). Conformity and character. *American Psychologist*. **10**, 191–198.

D'Souza, J. and Adams, K. (2014). On unenlightened altruism. *Journal of Human Values*. **20**(2), 183–191.

Danermark, B., Ekstrom, M. and Karlsson, J. (2002). *Explaining society: Critical realism in social sciences*. London: Routledge.

Dare, T. (2016). Robust role-obligation: How do roles make a moral difference? *Journal Value Inquiry*. **50**, 703–719.

Davies, W. and McGoey, L. (2012). Rationalities of ignorance: On financial crisis and the ambivalence of neo-liberal epistemology. *Economy and Society*. **41**(1), 64–83.

Davis, B., Handa, S., Hypher, N., Rossi, N., Winters, P. and Yablonski, J. (eds.) (2016). *From evidence to action: The story of cash transfers and impact evaluation in Sub-Saharan Africa*. Rome, New York and Oxford: FAO, UNICEF and Oxford University Press. Available at: www.unicef.nl/files/From%20Evidence%20to%20Action%20Rapport.pdf

Dawson, J., Johnston, M.J., Stewart, E.J., Lemieux, C.J., Lemelin, R.H., Maher, P.T. and Grimwood, B.S.R. (2011). Ethical considerations of last chance tourism. *Journal of Ecotourism*. **10**(3), 250–265.

Deifelt, W. (2007). Intercultural ethics: Sameness and otherness revisited. *Dialog: A Journal of Theology*. **46**(2), 112–119.

Denzin, N. and Lincoln, Y. (eds.) (2018). *Handbook of qualitative research* (5th ed.). London: Sage.

Derrida, J. and Kamuf, P. (1992). Given time: The time of the king. *Critical Inquiry*. **18**(2), 161–187.

Deutsch, M. and Gerard, H. (1955). A study of normative and informational social influences upon individual judgment. *Journal of Abnormal and Social Psychology*. **51**(3), 629.

Dev Singh, K. (2015). Creating your own qualitative research approach: Selecting, integrating and operationalizing philosophy, methodology and methods. *Vision*. **19**(2), 132–146.

DFID. (2019). *Evaluation annual report 2018/19.*

Di Bella, J. and Grant, A. (2013). The private sector and development: Key concepts. The North-South Institute. Policy Brief.

Dickert, S., Kleber, J., Västfjäll, D. and Slovic, P. (2016). Mental imagery, impact, and affect: A mediation model for charitable giving. *PLoS ONE.* **11**(2), e0148274.

Dieke, P.U.C. (2000). *The political economy of tourism development in Africa.* New York: Cognizant Communication Corporation.

DiMaggio, P.J. and Powell, W.W. (1983). The iron cage revisited: Institutional isomorphism and collective rationality in organizational fields. *American Sociological Review.* **48**(2), 147–160.

Dolnicar, S. and McCabe, S. (2022). Solidarity tourism how can tourism help the Ukraine and other war-torn countries? *Annals of Tourism Research.* **94**. https://doi.org/10.1016/j.annals.2022.1033

Donmoyer, R. (2012). Attributing causality in qualitative research: Viable option or inappropriate aspiration? An introduction to a collection of papers. *Qualitative Inquiry.* **18**(8), 651–654.

Dornschneider, S. (2019). High-stakes decision-making within complex social environments: A computational model of belief systems in the Arab Spring. *Cognitive Science.* **43**, e12762.

Douthwaite, B. and Hoffecker, E. (2017). Towards a complexity-aware theory of change for participatory research programs working within agricultural innovation systems. *Agricultural Systems.* **155**, 88–102.

Dubbink, G., Graafland, J. and Liederkerke, L. (2008). CSR: Transparency and the role of intermediate organisations. Munich Personal RePEc Archive – Tilburg University, Netherlands – MPRA Paper No. 17892–2008.

Dundar, E. (2016). Safe distance, dark tourism, and ambivalent spatial relation in Gillian Plowman's Yours Abundantly, from Zimbabwe, and Christine Evans's Trojan Barbie. *Journal of Contemporary Drama in English.* **4**(2), 201–219.

Dunning, D. (2012). Trust as a social and emotional act: Noneconomic considerations in trust behavior. *Journal of Economic Psychology.* **33**, 686–694.

Durand, R. and Calori, R. (2006). Sameness, otherness? Enriching organizational change theories with philosophical considerations on the same and the other. *The Academy of Management Review.* **31**(1), 93–114.

Durkheim, E. (1897). *Le Suicide: Étude de Sociologie.* Paris: Félix Alcan.

Earl, P. (1983). *The economic imagination: Towards a behavioral theory of choice.* Brighton: Wheatsheaf.

Easton, G. (2010). Critical realism in case study research. *Industrial Marketing Management.* **39**, 118–128.

Edwards, M. and Hulme, D. (1998). Too close for comfort? The impact of official aid on nongovernmental organizations. *World Development.* **24**, 961–973.

Ehnert, I. (2014). Paradox as a lens for theorizing sustainable HRM. In: I. Ehnert, W. Harry, and K.J. Zink (eds.) *Sustainability and human resource management.* Berlin: Springer. pp. 247–271.

Ehrnstrom-Fuentes, M. (2016). Delinking legitimacies: A pluriversal perspective on political CSR. *Journal of Management Studies.* **53**(3), 433–462.

El Kassar, N. (2018). What ignorance really is. Examining the foundations of epistemology of ignorance. *Social Epistemology.* **32**(5), 300–310.

Elder-Vass, D. (2010). *The causal power of social structures: Emergence, structure and agency.* Cambridge: Cambridge University Press.

Destinations, E. (2012). What is travel philanthropy? Available at: www.huffingtonpost.com/elevate-destinations/what-is-travel-philanthro_b_1405994.html
Elster, J. (1991). Envy in social life. In: R. Zeckhauser, ed. *Strategy and choice*. Cambridge, MA: MIT Press. pp. 49–82.
Elster, J. (1999). *Alchemies of the mind: Rationality and the emotions*. Cambridge: Cambridge University Press.
Emerson, R. (2020). Power dynamics in international development evaluations: A case study of the Girls Education Challenge programme. *African Evaluation Journal*. **8**(1), a459.
Erie, M.S. (2016) Sharia, charity, and minjian autonomy in Muslim China: Gift giving in a plural world. *American Ethnologist*. **43**(2), 311–324.
Escobar, A. (1992). Reflections on 'development': Grassroots approaches and alternative politics in the Third World. *Futures*. **24**(5), 411–436.
Escobar, A. (1995). *Encountering development: The making and unmaking of the Third World*. Princeton, NJ: Princeton University Press.
ESRC. (2021). *What is social science?* Available at: https://esrc.ukri.org/about-us/what-is-social-science/
Etang, A., Fielding, D. and Knowles, S. (2012). Giving to Africa and perceptions of poverty. *Journal of Economic Psychology*. **33**, 819–832.
Evans, P. (2004). Development as institutional change: The pitfalls of monocropping and the potentials of deliberation. *Studies in Comparative International Development*. **38**(4), 30–52.
Exquisite Safaris. (2010). Understanding travelers philanthropy – Defining philanthropy's role in travel. Available at: https://travelmatters.wordpress.com/
Fassin, Y. (2012). Stakeholder management, reciprocity and stakeholder responsibility. *Journal of Business Ethics*. **109**, 83–96.
Fastenrath, F., Orban, A. and Trampusch, C. (2018). From economic gains to social losses: How stories shape expectations in the case of German municipal finance. *Köln Z Soziol*. **70**, 89–116.
Fejerskov, A. (2015). From unconventional to ordinary? The Bill and Melinda Gates Foundation and the homogenizing effects of international development cooperation. *Journal of International Development*. **2**, 1–15.
Ferguson, J. (1994). *The anti-politics machine: Development, depoliticisation, and bureaucratic power in Lesotho*. Minneapolis: University of Minnesota Press.
Fernandez, K. and Hager, M. (2014). Public and private dimensions of grantmaking Foundations. *Public Administration Quarterly*. **38**, 405–439.
Fernandez. M. (2010). 6 Questions to ask before planning a philanthropic trip. Available at: https://philanthropyindaba.wordpress.com/2009/12/03/6-questions-to-ask-before-going-on-a-philanthropic-trip-part-1/
Fletcher, A. (2016). Applying critical realism in qualitative research: Methodology meets method. *International Journal of Social Research Methodology*. **20**(2), 181–194.
Floate, H., Durham, J. and Marks, G. (2019). Moving on from logical frameworks to find the 'missing middle' in international development programmes. *Journal of Development Effectiveness*. **11**(1), 89–103.
Flood, R. (2010). The relationship of 'systems thinking' to action research. *Systemic Practice and Action Research*. **23**, 269–284.
Fodor, J. and Pylyshyn, Z. (1981). How direct is visual perception?: Some reflections on Gibson's 'Ecological Approach'. *Cognition*. **9**(2), 139–196.
Forrester, J. (1994). System dynamics, systems thinking, and soft OR. *System Dynamics Review*. **10**(2–3), 245–256.

Foucault, M. (1977). *Discipline and punish*. New York: Pantheon Books.

Foucault, M. (1980). *Power-knowledge: Selected interviews and other writings*. New York: Pantheon Books.

Foundation Center. (2015). U.S. foundation funding for Africa. Available at: http://foundationcenter.org/gainknowledge/research/pdf/funding_for_africa2015.pdf

Francis, R., Armstrong, A. and Foxley, I. (2015). Whistleblowing: A three part view. *Journal of Financial Crime*. **22**(2), 208–218.

Francis, S., Francis, M. and Akinola, A. (2016). The edge of the periphery: Situating the ≠Khomani San of the Southern Kalahari in the political economy of Southern Africa. *African Identities*. **14**(4), 1–14.

Freeman, E. (2015). Stakeholder theory. In: C.L. Cooper, ed. Wiley *e*ncyclopedia of *m*anagement, vol. 2. Chichester: Wiley.

Friberg, K. (2015). Accounts along the aid chain: Administering a moral economy. *Journal of Global Ethics*. **11**(2), 246–256.

Friedman, M. (1953). The methodology of positive economics. In: M. Friedman, ed. *Essays in positive economics*. Chicago, IL: University of Chicago Press. pp. 3–43.

Gao, W., Qiu, L., Chiu, C. and Yang, Y. (2015). Diffusion of opinions in a complex culture system: Implications for emergence of descriptive norms. *Journal of Cross-Cultural Psychology*. **46**(10), 1252–1259.

Garcia, G. (1997). Why we choose to be stupid. Ed. D. diss., Montana State University, Bozeman.

Gardner, K. (2015). Chevron's gift of CSR: Moral economies of connection and disconnection in a transnational Bangladeshi village. *Economy and Society*. **44**(4), 495–518.

Gates, E. (2016). Making sense of the emerging conversation in evaluation about systems thinking and complexity science. *Evaluation and Program Planning*. **59**, 2–73.

GDC. (2019). *Fewer Germans are donating to charity—But why?* Available at: www.dw.com/en/fewer-germans-are-donating-to-charity-but-why/a-52627077

Getz, D. (1987). Tourism planning and research: Traditions, models and futures. Paper presented at Australian Travel Research Workshop, Bunbury, Western Australia, November 5–6.

Gibbs, E., Jones, C., Atkinson, J., Attfield, I., Bronwin, J., Hinton, R., Potter, A. and Savage, L. (2020). Scaling and 'systems thinking' in education: Reflections from UK aid professionals. *Compare: A Journal of Comparative and International Education*. **51**(1), 137–156.

Giddens, A. (1993). *Sociology* (2nd ed.). Cambridge: Polity Press.

Girard, R. (1977). *Violence and the sacred*. Baltimore, MD: Johns Hopkins University Press.

Giroux, H. (1992). Post-colonial ruptures and democratic possibilities: Multiculturalism as antiracist pedagogy. *Cultural Critique*. **21**, 5–39.

Giving USA. (2020). Giving USA Foundation. Indiana University Lilly Family School of Philanthropy at IUPUI.

Gladwell, M. (2000). *Tipping point: How little things can make a big difference*. New York: Little, Brown and Company.

Gligor-Cimpoieru, D., Munteanu, V., Nitu-Antonie, R., Schneider, A. and Preda, P. (2017). Perceptions of future employees toward CSR environmental practices in tourism. *Sustainability*. **9**(9), 1631–1614.

Global Giving. (2014). *Nonprofit partner handbook*. Available at: http://tools.blog.globalgiving.org

Goddard, A. (2020). Accountability and accounting in the NGO field comprising the UK and Africa – A Bordieusian analysis. *Critical Perspectives on Accounting*. Awaiting publication details: https://doi.org/10.1016/j.cpa.2020.102200

Goel, A. and Thakor, A. (2005). Green with envy: Implications for corporate investment distortions. *Journal of Business*. **78**(6), 2255–2288.

Goldberg, A. (1986). Classification, evolution and phylogeny of the families of Dicotyledons. *Smithsonian Contributions to Botany*. **58**, 1–314.

Goldfarb, N. (2011). Josiah Royce's philosophy of loyalty as philanthropy. *Nonprofit and Voluntary Sector Quarterly*. **40**(4), 720–739.

Gomberg, P. (1986). Self and others in Bentham and Sidgwick. *History of Philosophy Quarterly*. 3(4), 437–448.

Goodwin, H. (2011). *Taking responsibility for tourism*. Oxford: Goodfellow Publishers.

Goodwin, H., McCombes, L. and Eckardt, C. (2009). Advances in travel philanthropy: Raising money through the travel and tourism industry for charitable purposes. World Travel Market ICRT Occasional Paper.

Gouldner, A. (1960). The norm of reciprocity: A preliminary statement. *American Sociological Review*. **25**(2), 161–178.

Green, D. (2016). *How change happens*. Oxford: Oxford University Press.

Green, E., Deschamps, J. and Paez, D. (2005). Variation of individualism and collectivism within and between 20 countries: A typological analysis. *Journal of Cross-Cultural Psychology*. **36**(3), 321–339.

Griffiths, M. (2015). A compelling and flawed story of power-body relations. A review of: Volunteer tourism: Popular humanitarianism in neoliberal times, by Mary Mostafanezhad. *Tourism Geographies*. **17**(4), 627–629.

Grossman, Z. and van der Weele, J. (2017). Self-image and willful ignorance in social decisions. *Journal of the European Economic Association*. **15**(1), 173–217.

Guba, E. and Lincoln, Y. (2005). Paradigmatic controversies, contradictions, and emerging confluences. In: N.K. Denzin and Y.S. Lincoln, eds. *The Sage handbook of qualitative research* (3rd ed.). Thousand Oaks, CA: Sage. pp. 191–215.

Hailu, D. (2015). An ethnography of social assistance to Ethiopian children by non-governmental organizations: Types, adequacy and predictability. *Social Policy & Administration*. **51**(5), 817–836.

Hall, S. (1992). The west and the rest: Discourse and power. In: S. Hall and B. Geiben, eds. *Formations of modernity*. Cambridge: Polity Press. pp. 185–225.

Hall, S. (1997). *Representations: Cultural representations and signifying practices*. London: Sage.

Halliwell, S. (2002). *The aesthetics of mimesis: Ancient texts and modern problems*. Princeton, NJ: Princeton University Press.

Ham, S. (2011). The ask – Or is it the offer? In: M. Honey, ed. *Travelers' philanthropy handbook*. Washington, DC: Center for Responsible Travel. pp. 141–149.

Hambrick, D. and Mason, P. (1984). Upper echelons: The organization as a reflection of its top managers. *Academy of Management Review*. **9**, 193–206.

Hann, C. (2017). Review article: Can the gift be sponsored? *Journal of the Royal Anthropological Institute*. **22**, 984–989.

Hanoch, Y. (2002). 'Neither an angel nor an ant': Emotion as an aid to bounded rationality. *Journal of Economic Psychology*. **23**, 1–25.

Harbaugh, W. (1998). What do donations buy? A model of philanthropy based on prestige and warm glow. *Journal of Public Economics*. **67**, 269–284.

Harrison, G. (2013). Campaign Africa: Exploring the representation of Africa and its role in British identity. *The British Journal of Politics and International Relations*. **15**, 528–547.

Harrow, J. and Jung, T. (2016). Philanthropy and community development: The vital signs of community foundation? *Community Development Journal*. **51**(1), 132–152.

Hart, S., Lane, D. and Chinn, S. (2018). The elusive power of the individual victim: Failure to find a difference in the effectiveness of charitable appeals focused on one compared to many victims. *PLoS ONE*. **13**(7), e0199535.

Hayek, F. (1967). The results of human action but not of human design. In: *Studies in philosophy: Politics and economics*. London: Routledge and Kegan Paul. pp. 96–105.

Heffernan, M. (2012). *Wilful blindness: Why we ignore the obvious at our peril*. London: Simon & Schuster.

Heiner, R. (1983). The origin of predictable behavior. *American Economic Review*. **73**, 560–595.

Held, V. (1970). Can a random collection of individuals be morally responsible? *Journal of Philosophy*. **67**(14), 471–481.

Henderson, D. (2002). Norms, normative principles, and explanation. *Philosophy of the Social Sciences*. **32**(3), 329–364.

Heyer, G. (2015). The making of a tragedy: Perversion in the perception of truth. *Journal of Analytical Psychology*. **60**(5), 642–656.

Higgins-Desbiolles, F. (2022). The question of solidarity in tourism. *Journal of Policy Research in Tourism, Leisure and Events*. 1–19. https://doi.org/10.1080/19407963.2022.2107657

Hollander, C. and Wu, A. (2011). The current state of normative agent-based systems. *Journal of Artificial Societies and Social Simulation*. **14**(2), 6.

Homer-Dixona, T., Mocka, S., Quilleyd, S., Mocka, S., Quilleyd, Q., Maynard, J., Schröder, T., Mildenbergerc, M. and Thagarde, P. (2013). A complex systems approach to the study of ideology: Cognitive-affective structures and the dynamics of belief systems. *Journal of Social and Political Psychology*. **1**(1), 337–363.

Honey, M. (2011). *Travelers' philanthropy handbook*. Washington, DC: Center for Responsible Travel.

Howard, J. and Burns, D. (2015). Volunteering for development within the new ecosystem of international development. *IDS Bulletin*. **46**(5), 5–16 .

Hu, Y., Wang, J. and Chen, S. (2018). Managerial humanistic attention and CSR: Do firm characteristics matter? *Sustainability*. **10**(11), 4029.

Hughes, E. and Scheyvens, R. (2016). Corporate social responsibility in tourism post-2015: A Development First approach. *Tourism Geographies*. **18**(5), 469–482.

Hume, D. (1978). *A treatise on human nature (1740)*. Oxford: Clarendon Press.

Hummelbrunner, R. (2011). Systems thinking and evaluation. *Evaluation*. **17**(4), 395–403.

Hüncke, A. and Koot, S. (2012). The presentation of Bushmen in cultural tourism: Tourists' images of Bushmen and the tourism provider's presentation of (Hai//om) Bushmen at Treesleeper Camp, Namibia. *Critical Arts*. **26**(5), 671–689.

Hyun, N., Park, Y. and Park, S. (2016). Narcissism and gift giving: Not every gift is for others. *Personality and Individual Differences*. **96**, 47–51.

Ideland, M. and Malmberg, C. (2014). 'Our common world' belongs to 'Us': Constructions of otherness in education for sustainable development. *Critical Studies in Education*. **55**(3), 369–386.

Inskeep, E. (1991). *Tourism planning: An integrated and sustainable development approach*. New York: Van Nostrand Reinhold.

Ison, R. (2008). Systems thinking and practice for action research. In: P. Reason and H. Bradbury, eds. *The SAGE handbook of action research* (2nd ed.). London: Sage Publications. pp. 139–158.

Jackson, M. (2019). *Critical systems thinking and the management of complexity: Responsible leadership for a complex world.* Oxford: Wiley.

Jackson, M. and Sambo, L. (2019). Health systems research and critical systems thinking: The case for partnership. *Systems Research and Behavioral Science.* **37**, 3–22.

Jahn, J. and Bruhl, R. (2018). How Friedman's view on individual freedom relates to stakeholder theory and social contract theory. *Journal of Business Ethics.* **153**, 41–52.

Jamal, T. and Getz, D. (1995). Collaboration theory and community tourism planning. *Annals of Tourism Research.* **22**(1), 186–204.

Jamali, D. and Karam, C. (2016). Corporate social responsibility in developing countries as an emerging field of study. *International Journal of Management Reviews.* **20**(1), 32–61.

Jamieson, W., Goodwin, H. and Edmunds, C. (2004). Contribution of tourism to poverty alleviation. Pro-poor tourism and the challenge of measuring impacts – For Transport Policy and Tourism Section Transport and Tourism Division UN ESCAP.

Jaworski, A. and Thurlow, C. (2010). Language and the globalizing habitus of tourism: A sociolinguistics of fleeting relationships. In: N. Coupland, ed. *The handbook of language and globalization.* Oxford: Blackwell. pp. 256–286.

Jedlowski, A. and Thomas, M. (2017). Representing 'otherness' in African popular media: Chinese characters in Ethiopian video-films. *Journal of African Cultural Studies.* **29**(1), 1–18.

Johnson, A., Papi-Thornton, D. and Stauch, J. (2019). Student guide to mapping a system. Skoll Centre for Social Entrepreneurship, Saïd Business School, University of Oxford.

Jones, A. (2017). Band Aid revisited: Humanitarianism, consumption and philanthropy in the 1980s. *Contemporary British History.* **31**(2), 1–21.

Jose, T. (2013). *Saving the world, one trip at a time: Different responsible travels.* Available at: www.wanderink.com/archives/saving-the-world-one-trip-at-a-time-different-responsible-travels/

Jungkunz, V. and White, J. (2013). Ignorance, innocence, and democratic responsibility: Seeing race, hearing racism. *The Journal of Politics.* **75**(2), 436–450.

Kahneman, D. and Tversky, A. (1979). Prospect theory: An analysis of decision under risk. *Econometrica.* **47**(2), 263–292.

Kalu, K. (2017). State-society relations, institutional transformation and economic development in Sub-Saharan Africa. *Development Policy Review.* **35**(51), 234–245.

Kamstra, J. and Schulpen, L. (2014). Worlds apart but much alike: Donor funding and the homogenization of NGOs in Ghana and Indonesia. *Studies in Comparative International Development.* **50**(3), 331–357.

Kania, J. and Kramer, M. (2011). Collective impact. *Stanford Social Innovation Review.* Available at: https://ssir.org/articles/entry/collective_impact

Kant, I. (2011). *Metaphysics of morals.* Translated from the German by M. Gregor and J. Timmermann. Cambridge: Cambridge University Press.

Katsikopoulos, K. (2014). Bounded rationality: The two cultures. *Journal of Economic Methodology.* **21**(4), 361–374.

Kaulingfreks, F. and Kaulingfreks, R. (2013). Open-access communism. *Business Ethics: A European Review.* **22**(4), 417–429

Kelman, H. (1958). Compliance, identification, and internalization: Three processes of attitude change. *Journal of Conflict Resolution.* **2**, 51–60.

Kepher-Gona, J. and Atieno, L. (2022). *Sustainable tourism dialogues in Africa*. Berlin; Boston: De Gruyter.

King, A. (2009). Overcoming structure and agency: Talcott Parsons, Ludwig Wittgenstein and the theory of social action. *Journal of Classical Sociology*. **9**(2), 260–288.

Kirk, M. (2012). Beyond charity: Helping NGOs lead a transformative new public discourse on global poverty and social justice. *Ethics & International Affairs*. **26**(2), 245–263.

Kirsch, D. (2014). How do the approaches to accountability compare for charities working in international development? *Healthcare Policy*. **10**(Special Issue), 145–149.

Klein, S. (2014). Sameness and the self: Philosophical and psychological considerations. *Frontiers in Psychology*. **5**(29), 29.

Koot, S. and Fletcher, R. (2021). Donors on tour: Philanthrotourism in Africa. *Annals of Tourism Research*. **89**, 103216. https://doi.org/10.1016/j.annals.2021.103216

Korf, B. (2007). Antinomies of generosity: Moral geographies and post-tsunami aid in Southeast Asia. *Geoforum*. **38**, 366–378.

Korzybski, A. (1931). A non-Aristotelian system and its necessity for rigour in mathematics and physics. Paper presented at the American Mathematical Society meeting at New Orleans, Louisiana meeting of the American Association for the Advancement of Science, December 28, 1931. Reprinted in Science and Sanity, 1933, pp. 747–761.

Kosansky, O. (2002). Tourism, charity, and profit: The movement of money in Moroccan Jewish Pilgrimage. *Journal of Cultural Anthropology*. **17**, 359–400.

Kotler, P. and Lee, N. (2005). *Corporate social responsibility: Doing the most good for your company and your cause*. Hoboken, NJ: Wiley.

Koumakhov, R. (2009). Conventions in Herbert Simon's theory of bounded rationality. *Journal of Economic Psychology*. **30**, 293–306.

Kourula, A. and Halme, M. (2008). Types of corporate responsibility and engagement with NGOs: An exploration of business and societal outcomes. *Corporate Governance*. **8**(4), 557–570.

Kovacs, A. (2017). A phenomenological attempt at understanding otherness. *Bulletin of the Transilvania University of Braşov Series VII: Social Sciences and Law*. **10**(59)(2), 79–86.

Kragh, S. (2016). Tribe and village in African organizations and business. *Personnel Review*. **45**(1), 51–66.

Kraut, R. (1973). Effects of social labeling on giving to charity. *Journal of Experimental Social Psychology*. **9**(6), 551–562.

Krijnen, C. (2016). Values – Limits of economic rationality and imperialism of culture. *Economic and Political Studies*. **4**(2), 101–121.

Krippendorf, J. (1987). *The holiday makers: Understanding the impact of leisure and travel*. Oxford: Butterworth-Heinemann.

Kumu. (2020). *Systems mapping software*. Access at: www.kumu.io

Kurki, M. (2007). Critical realism and causal analysis in international relations. *Journal of International Studies*. **35**(2), 361–378.

Kutz, C. (2000). *Complicity: Ethics and law for a collective age*. Cambridge: Cambridge University Press.

Kyriacou, A. (2015). Individualism-collectivism, governance and economic development. MPRA Paper No. 65151.

Lahiri-Dutt, K. (2017). Thinking "differently" about a feminist critical geography of development. *Geographical Research*. **55**(3), 326–331.

Lambe, F., Ran, Y., Jurisoo, M., Holmlid, S., Muhoza, C., Johnson, O. and Osborne, M. (2020). Embracing complexity: A transdisciplinary conceptual framework for understanding behavior change in the context of development-focused interventions. *World Development*. **126**, 104703.

Lambert, D. and Lester, A. (2004). Geographies of colonial philanthropy. *Progress in Human Geography*. **28**(3), 320–341.

Langer, E. (1989). *Mindfulness*. Reading, MA: Merloyd Lawrence.

Lantos, G.P. (2001). The boundaries of strategic corporate social responsibility. *Journal of Consumer Marketing*. **18**(7), 595–632.

Lavena, C. (2016). Whistle-blowing: Individual and organizational determinants of the decision to report wrongdoing in the federal government. *American Review of Public Administration*. **46**(1), 113–136.

Lawson, T. (1997). *Economics and reality*. Oxon: Routledge.

Lawson, T. (2003). *Reorienting economics*. London: Routledge.

Layder, D. (1990). *The realist image in social science*. New York: St. Martin's Press.

Legros, S. and Cislaghi, B. (2020). Mapping the social-norms literature: An overview of reviews. *Perspectives on Psychological Science*. **15**(1), 62–80.

Lennox, R. and Jurdi-Hage, R. (2016). Beyond the empirical and the discursive: The methodological implications of critical realism for street harassment research. *Women's Studies International Forum*. **60**, 28–38.

Lets Go Travel Kenya. (2011). Travellers philanthropy: Giving back to our communities. Available at: http://letsgotravelkenya.blogspot.ug/2011/09/travellers-philanthropy-giving-back-to.html.

Levi-Strauss, C. (1969). *The elementary structures of kinship*. Boston, MA: Beacon Press.

Levi-Strauss, C. (1987). *Introduction to the work of Marcel Mauss*. Translated by F. Baker. London: Routledge and Kegan Paul.

Levi, I. (1997). *The covenant of reason*. Cambridge: Cambridge University Press.

Levinas, E. (1999). *Alterity & transcendence*. Translated by M.B. Smith. New York: Columbia University Press.

Lewis, P. (2000). Realism, causality and the problem of social structure. *Journal for the Theory of Social Behaviour*. **30**(3), 249.

Lieberman, M. and Shigeru, A. (2006). Why do firms imitate each other? *Academy of Management Review*. **31**, 366–385.

Liebowitz, S. and Margolis, S. (1995). Path dependence, lock-in, and history. *Journal of Law, Economics, & Organization*. **11**(1), 205–226.

Liket, K. and Simaens, A. (2015). Battling the devolution in the research on corporate philanthropy. *Journal of Business Ethics*. **126**(2), 285–308.

Lincoln, Y.S. and Guba, E.G. (2005). Paradigmatic controversies, contradictions, and emerging confluences. In N.K. Denzin and Y.S. Lincoln, eds. *The Sage handbook of qualitative research*, 3rd edn. Thousand Oaks, CA: Sage. pp. 191–215.

Lindkvist, L. (2011). Working with donors in travel-based philanthropy: Lessons from basecamp. In: M. Honey, ed. *Travelers' philanthropy handbook*. Washington, DC: Center for Responsible Travel. pp. 123–137.

Lisle, D. (2006). *The global politics of contemporary travel writing*. New York: Cambridge University Press.

Liu, F., Ma, J. and Li, R. (2019). Which role model is more effective in entrepreneurship education? An investigation of storytelling on individual's entrepreneurial intention. *Frontiers in Psychology*. **10**, 837.

Locke, J. (1846). *An essay concerning human understanding*. Book II. Philadelphia: Kay & Troutman.

Locke, J. (1690) *An essay concerning human understanding*. London: Thomas Bassett.

Lockwood, P., Marshall, T. and Sadler, P. (2005). Promoting success or preventing failure: Cultural differences in motivation by positive and negative role models. *Personality and Social Psychology Bulletin*. **31**(3), 379–392.

Loewenstein, G. and Schkade, D. (1999). Wouldn't it be nice? Predicting future feelings. In: D. Kahneman, E. Diener, E. and N. Schwarz, eds. *Well-being: The foundation of hedonic psychology*. New York: Russell Sage Foundation. pp. 85–105.

Louise, J. (2009). I won't do it! Self-prediction, moral obligation and moral deliberation. *Philosophical Studies: An International Journal for Philosophy in the Analytic Tradition*. **146**(3), 327–348.

Lowe, P., Phillipson, J. and Wilkinson, K. (2013). Why social scientists should engage with natural scientists. *Contemporary Social Science*. **8**(3), 207–222.

Lucea, R. (2010). How we see them versus how they see themselves. A cognitive perspective of firm – NGO relationships. *Business & Society*. **49**(1), 116–139.

Lynch, K. (2016). Willful ignorance and self-deception. *Philosophical Studies: An International Journal for Philosophy in the Analytic Tradition*. **173**(2), 505–523.

Lyon, A., Hunter-Jones, P. and Warnaby, G. (2017). Are we any closer to sustainable development? Listening to active stakeholder discourses of tourism development in the Waterberg Biosphere Reserve, South Africa. *Tourism Management*. **61**, 234–247.

Lyons, K.D. and Wearing, S. (2008). Reflections on the ambiguous intersections between volunteering and tourism. *Leisure Sciences*. **34**(1), 88–93.

Maathai, W. (2011). Foreward. In: M. Honey, ed. *Travelers' philanthropy handbook*. Washington, DC: Center for Responsible Travel. pp. 1–2.

Mabsout. R. (2015). Abduction and economics: The contributions of Charles Peirce and Herbert Simon. *Journal of Economic Methodology*. **22**(4), 491–516.

MacIntosh, J. (1994). Belief-in revisited: A reply to Williams. *Religious Studies*. **30**(4), 487–503.

Maclean, M., Harvey, C., Gordon, J. and Shaw, E. (2015). Identity, storytelling and the philanthropic journey. *Human Relations*. **68**(10), 1623–1652.

Macy, P. (2011). Unwanted philanthropy. In: M. Honey, ed. *Travelers' philanthropy handbook*. Washington, DC: Center for Responsible Travel. pp. 177–178.

Mahmoud, M., Ahmad, M., Yusoff, M. and Mustapha, A. (2014). A review of norms and normative multiagent systems. *Scientific World Journal*. **2014**, Article ID 684587.

Mahoney, J. (2001). Review: Beyond correlational analysis: Recent innovations in theory and method. *Sociological Forum*. **16**(3), 575–593.

Malinowski, B. (1922). *Argonauts of the Western Pacific: An account of native enterprise and adventure in the archipelagoes of Melanesian New Guinea*. London: Routledge.

Mallard, G. (2014). *Fallout: Nuclear diplomacy in an age of global fracture*. Chicago, IL: University of Chicago Press.

Mallard, G. and McGoey, L. (2018). Strategic ignorance and global governance: An ecumenical approach to epistemologies of global power. *The British Journal of Sociology*. **69**(4), 884–909.

Maltby, J. and Rutterford, J. (2016). Investing in charities in the nineteenth century: The financialization of philanthropy. *Accounting History*. **21**(2–3), 263–280.

Maness, M. and Cirillob, C. (2016). An indirect latent informational conformity social influence choice model: Formulation and case study. *Transportation Research Part B: Methodological*. **93**(PA), 75–101.

Mann, L. (1969). *Social psychology*. New York: Wiley.

Manski, C. (2004). Measuring expectations. *Econometrica*. **72**(5), 1329–1376.

Manz, C., Cameron, K., Manz, K. and Marx, R. (eds.) (2008). *The virtuous organization: Insights from some of the world's leading management thinkers*. Hackensack, NJ: World Scientific.

Manzini, P. and Mariotti, M. (2014). Welfare economics and bounded rationality: The case for model-based approaches. *Journal of Economic Methodology*. **21**(4), 343–360.

March, J. (1978). Bounded rationality, ambiguity, and the engineering of choice. *The Bell Journal of Economics*. **9**, 587–608.

Marginson, S. (2011). Higher education and public good. *Higher Education Quarterly*. **65**(4), 411–433.

Mathie, A., Cameron, J. and Gibson, K. (2017). Asset-based and citizen-led development: Using a diffracted power lens to analyze the possibilities and challenges. *Progress in Development Studies*. **17**(1), 54–66.

Matthiesen, N. and Klitmøller, J. (2019). Encountering the stranger: Hannah Arendt and the shortcomings of empathy as a moral compass. *Theory & Psychology*. **29**(2), 182–199.

Matunhu, J. (2011). A critique of modernisation and dependency theories in Africa: Critical assessment. *African Journal of History and Culture*. **3**(5), 65–72.

Maturana, H. and Varela, F. (1980). *Autopoiesis and cognition: The realization of the living*. Dordrecht: D. Reidel.

Mauss, M. (1925) Essai sur le don. Forme et raison de l'échange dans les sociétés archaïques. *L'Année Sociologique*. **1**(1923–1924), 30–186.

Mauss, M. (2016). *The gift*. Selected, annotated, and translated from the French by J. I. Guyer. London: Hau.

Mawdsley, E. (2014). A new development era? The private sector moves to the centre. The Norwegian Peacebuilding Resource Centre (NOREF).

Maxwell, J. (2012a). The importance of qualitative research for causal explanation in education. *Qualitative Inquiry*. **18**(8), 655–661.

Maxwell, J. (2012b). *A realist approach for qualitative research*. Thousand Oaks, CA: Sage.

Maxwell, J. and Dornan, J. (1997). *Becoming a person of influence: How to positively impact the lives of others*. HarperCollins Leadership.

May, L. (1992). *Sharing responsibility*. Chicago, IL: University of Chicago Press.

McDonald, K. and Scaife, W. (2011). Print media portrayals of giving: Exploring national "cultures of philanthropy". *International Journal of Nonprofit and Voluntary Sector Marketing*. **16**(4), 311–324.

McElwee, B. (2010). The rights and wrongs of consequentialism. *Philosophical Studies: An International Journal for Philosophy in the Analytic Tradition*. **151**(3), 393–412.

McEwan, C., Mawdsley, E., Banks, G. and Scheyvens, R. (2017). Enrolling the private sector in community development: Magic bullet or sleight of hand? *Development and Change*. **48**(1), 28–53.

McGoey, L. (2007). On the will to ignorance in bureaucracy. *Economy and Society*. **36**(2), 212–235.

McGoey, L. (2012). Strategic unknowns: Towards a sociology of ignorance. *Economy and Society*. **41**(1), 1–16.

McGoey, L. and Thiel, D. (2018). Charismatic violence and the sanctification of the super-rich. *Economy and Society*. **47**(1), 111–134.

McIntyre, M., Murphy, S. and Sirsly, C. (2015). Do firms seek social license to operate when stakeholders are poor? Evidence from Africa. *Corporate Governance*. **15**(3), 306–314.

McLachlan, S. and Binns, T. (2014). Tourism, development and corporate social responsibility in Livingstone, Zambia. *Local Economy*. **29**(1–2), 98–112.

McLeod, S.A. (2016, Jan 14). What is conformity? Available at: www.simplypsychology.org/conformity.html

McNamara, N. and Morse, S. (2004). Voices from the aid 'chain': The personal dynamics of care. *Social & Cultural Geography*. **5**(2), 253–270.

McNamara, T. (2017). Do the Chinese bring Chitukuko? Rural Malawian understandings of Chinese development. *Journal of International Development*. **29**(8), 1149–1165.

Mead, G. (1934). *Mind, self, and society from the standpoint of a social behaviourist*. Chicago, IL: University of Chicago Press.

Meadows, D. (2008). *Thinking in systems: A primer*. Edited by D. Wright. White River Junction, VT: Chelsea Green Publishing.

Meadows, D., Randers, J. and Meadows, D. (1972). *The limits to growth*. London: Potomac Associates.

Menzel, D. (1999). The morally mute manager: Fact or fiction? *Public Personnel Management*. **28**(4), 515–527.

Merriam, S. (1998). *Qualitative research and case study applications in education*. San Francisco, CA: Jossey-Bass.

Merton, R. (1948). The self-fulfilling prophecy. *The Antioch Review*. **8**(2), 193–210.

Meyer, J. and Rowan, B. (1977). Institutionalized organizations: Formal structure as myth and ceremony. *American Journal of Sociology*. **83**(2), 340–363.

Michaelson, C. (2008). Moral luck and business ethics. *Journal of Business Ethics*. **83**, 773–787.

Michelson, E. (2013). *Adventure philanthropist: Great adventures volunteering abroad*. New York: emCom LLC.

Mickiewicz, T., Sauka, A. and Stephan, U. (2016). On the compatibility of benevolence and self-interest: Philanthropy and entrepreneurial orientation. *International Small Business Journal*. **34**(3), 303–328.

Mingers, J. (2011). The contribution of systemic thought to critical realism. *Journal of Critical Realism*. **10**(3), 303–330.

Mingers, J. (2014). *Systems thinking, critical realism and philosophy: A confluence of ideas*. Ontological explorations. London: Routledge.

Mitchell, J. (2012). Value chain approaches to assessing the impact of tourism on low-income households in developing countries. *Journal of Sustainable Tourism*. **20**(3), 457–475.

Mitschow, M. (2000). Unfocused altruism: The impact of iconography on charitable activity. *Journal of Business Ethics*. **23**, 73–82.

Mlynek, A. (2017). Give and take. *Today's Parent*. **33**(12), 72–74.

Mohan, G. and Stokke, K. (2000). Participatory development and empowerment: The dangers of localism. *Third World Quarterly*. **21**(2), 247–268.

Moore, M., Parsons, B. and Jessup, P. (2019). How evaluators can use a complex systems lens to get "untrapped" from limiting beliefs and assumptions. *American Journal of Evaluation*. **40**(1), 75–91.

Morgan, T. (2018). This 14th-century African emperor remains the richest person in history. Available at: www.history.com

Morgenroth, T., Peters, K. and Ryan, M. (2015). The motivational theory of role modeling: How role models influence role aspirants' goals. *Review of General Psychology*. **19**(4), 465–483.

Morrison, E. (2014). Employee voice and silence. *Annual Review of Organizational Psychology and Organizational Behavior*. **1**(1), 173–197.

Morse, S. and McNamara, N. (2012). Partnership and inter-dependence in aid chains: Social pathology or social health. *Journal of International Development*. **24**, 909–921.

Mostafanezhad, M. (2012). The geography of compassion in volunteer tourism. *Tourism Geographies*. **15**(2), 318–337.

Mostafanezhad, M. (2013). 'Getting in touch with your inner Angelina': Celebrity humanitarianism and the cultural politics of gendered generosity in volunteer tourism. *Third World Quarterly*. **34**(3), 485–499.

Motilal, S. (2015). Sustainable development goals and human moral obligations: The ends and means. *Journal of Global Ethics*. **11**(1), 24–31.

Mowles, C. (2013). Evaluation, complexity, uncertainty: Theories of change and some alternatives. In: T. Wallace, F. Porter and M. Ralph-Bowman, eds. *Aid, NGOs and the reality of women's lives: A perfect storm*. Rugby: Practical Action Publishing.

Mowles, C. (2014). Complex, but not quite complex enough: The turn to the complexity sciences in evaluation scholarship. *Evaluation*. **20**(2), 160–175.

Moyo, D. (2009). *Dead aid: Why aid is not working and how there is a better way for Africa*. New York: Farrar, Straus and Giroux.

Mueller, B. (2020). Why public policies fail: Policymaking under complexity. *EconomiA*. **21**, 311–323.

Munro, I. (2016). Whistleblowing and the politics of truth: Mobilizing 'truth games' in the WikiLeaks case. *Human Relations*. **70**(5), 519–543.

Murphy, P. (1988). Community driven tourism planning. *Journal of Tourism Management*. **9**(2), 96–104.

Mutale, W., Ayles, H., Bond, V., Chintu, N., Chilengi, R., Mwanamwenge, M.T., Taylor, A., Spicer, N. and Balabanova, D. (2015). Application of systems thinking: 12-month postintervention evaluation of a complex health system intervention in Zambia: The case of the BHOMA. *Journal of Evaluation in Clinical Practice*. **23**(2), 439–452.

Nankervis, A. (2009). Vulnerability analysis and sustainability in tourism: lessons from Phuket. In: C. Pforr and P. Hosie, eds. *Crisis management in the tourism industry*. Ashgate Publishing. pp. 93–106.

Ndlovu-Gatsheni, S. (2015). Decoloniality as the future of Africa. *History Compass*. **13**(10), 485–496.

Near, J. and Miceli, M. (1995). Effective whistle-blowing. *The Academy of Management Review*. **20**(3), 679–708.

Nickel, P. and Eikenberry, A. (2009). A critique of the discourse of marketized philanthropy. *American Behavioral Scientist*. **52**(7), 974–989.

Niehaus, P. (2014). *A theory of good intentions*. San Diego: University of California San Diego.

Novelli, M. (2015). *Tourism and development in Sub-Saharan Africa*. London: Routledge.

Novelli, M., Adu-Ampong, E.A. and Ribeiro, M.A. (eds.) (2021). *Handbook on tourism in Africa*. African Studies Series. Oxford: Routledge.

Novelli, M., Morgan, N., Mitchell G. and Ivanov, K. (2015). Travel philanthropy and sustainable development: The case of the Plymouth Banjul Challenge. *Journal of Sustainable Tourism*. **24**(6), 824–845.

Null, C. (2011). Warm glow, information, and inefficient charitable giving. *Journal of Public Economics*. **95**, 455–465.

Oakley, B. (2013). Concepts and implications of altruism bias and pathological altruism. *Proceedings of the National Academy of Sciences USA*. **110**(Suppl 2), 10408–10415.

Olson, V. (2004). The significance of sameness: An overview of standardization and imitation. *Medieval Art, Visual Resources*. **20**(2–3), 161–178.

Paliwal, P. (2005). Sustainable development and systems thinking: A case study of a heritage city. *The International Journal of Sustainable Development & World Ecology*. **12**(2), 213–220.

Parry, J. (1986). The gift, the Indian gift and the 'Indian Gift'. *Royal Anthropological Institute of Great Britain and Ireland*. **21**(3), 453–473.

Partzsch, L. (2017). Powerful individuals in a globalized world. *Global Policy*. **8**(1), 1–9.

Partzsch, L. and Fuchs, D. (2012). Philanthropy: Power within international relations. *Journal of Political Power*. **5**(3), 359–376.

Patokorpi, E. and Ahvenainen, M. (2009). Developing an abduction-based method for futures research. *Futures*. **41**, 126–139.

Patomaki, H. and Wight, C. (2000). After postpositivism? The promises of critical realism. *International Studies Quarterly*. **44**(2), 213–237.

Patton, M. (2011). *Developmental evaluation: Applying complexity concepts to enhance innovation and use*. New York: Guilford Press.

Patton, M. (2016). From evangelist to utilization-focused evaluator. In: D.D. Williams, ed. *Seven North American evaluation pioneers: New directions for evaluation*, vol. **150**. San Francisco, CA: Jossey-Bass. pp. 69–76.

Patton, M. (2019). Proposal for evaluation of Women Affirming Motherhood. In: C. Christie and M. Alkin, eds. *Theorists' models in action: A second look. new directions for evaluation*, vol. **163**. San Francisco, CA: Jossey-Bass. pp. 27–47.

Pawson, R. (2013). *The science of evaluation: A realist manifesto*. Los Angeles, CA and London: Sage.

Pawson, R. and Tilley, N. (1997). *Realistic evaluation*. London: Sage Publications.

Payton, R. (1988). *Philanthropy: Voluntary action for the public good*. New York: American Council on Education/Macmillan.

Pearson, T. (2020). *What is mimetic theory? A summary of things hidden since the foundation of the world by Rene Girard*. Available at: www.taylorpearson.me/bookreview/mimetic-theory-things-hidden-since-the-foundation-of-the-world/

Peric, M. and Djurkin, J. (2014). Systems thinking and alternative business model for responsible tourist destination. *Kybernetes*. **43**(3/4), 480–496.

Pettit, S. and Beresford, A. (2009). Critical success factors in the context of humanitarian aid supply chains. *International Journal of Physical Distribution & Logistics Management*. **39**(6), 450–468.

Phocuswright. (2015). *Good travels: The philanthropic profile of the American traveler*. Accessed at: www.phocuswright.com/Free-Travel-Research/Good-Travels-The-Philanthropic-Profile-of-the-American-Traveler

Picard, S. (1997). *Affective computing*. Cambridge, MA: MIT Press.

Picciotto, R. (2019). Towards a 'New Project Management' movement? An international development perspective. *International Journal of Project Management*. **38**(8), 474–485.

Pinelli, M. and Maiolini, R. (2016). Strategies for sustainable development: Organizational motivations, stakeholders' expectations and sustainability agendas. *Sustainable Development*. **25**(4), 288–298.

Pleasants, N. (1999). *Wittgenstein and the idea of a critical social theory: A critique of Giddens, Habermas, and Bhaskar*. London: Routledge.

Polanyi, K. (1944). *The great transformation: The political and economic origins of our time*. New York: Farrar & Rinehart.

Pollock, A. (2016). *Time has come to rethink travel philanthropy*. Posted online at Conscious Travel website May 2016: https://conscious.travel/2016/05/19/time-has-come-to-rethink-travel-philanthropy/

Porter, M. and Kramer, M. (2011). Creating shared value: How to reinvent capitalism—And unleash a wave of innovation and growth. *Harvard Business Review*. **89**(1/2), 62–67.

Powell, J. and Menendian, S. (2020). *The problem of othering: Towards inclusiveness and belonging*. Accessed at: https://otheringandbelonging.org/the-problem-of-othering/

Proctor, R. (2008). *Agnotology: The making and unmaking of ignorance*. Stanford, CA: Stanford University Press.

Pummer, T. (2016). Whether and where to give. *Philosophy & Public Affairs*. **44**(1), 77–95.

Putnam, H. (1990). *Realism with a human face*. Edited by J. Conant. Cambridge, MA: Harvard University Press.

Rabin, M. (2002). A perspective on psychology and economics. *European Economic Review*. **46**, 657–685.

Rajak, D. (2011). *In good company: An anatomy of corporate social responsibility*. Stanford, CA: Stanford University Press.

Rajak, D. and Dolan, C. (2016). Remaking Africa's informal economies: Youth, entrepreneurship and the promise of inclusion at the bottom of the pyramid. *The Journal of Development Studies*. **52**(4), 514–529.

Ramalingam, B. (2013). *Aid on the edge of chaos: Rethinking international cooperation in a complex world*. Oxford: Oxford University Press.

Rapport, N. (2015). Anthropology through Levinas. *Current Anthropology*. **56**(2), 256–276.

Ratner, B. (2020). Kenya fishermen say they are squeezed by Ethiopian mega-dam. *Reuters World News*. July 20, 2020.

Reginster, B. (2013). Three views of otherness. *Pastoral Psychology*. 62, 437–449.

Reinholz, D. and Andrews, T. (2020). Change theory and theory of change: What's the difference anyway? *International Journal of STEM Education*. **7**(2), 1–12. doi: 10.1186/s40594-020-0202-3

Retsikas, K. (2016). The other side of the gift: Soliciting in Java. Heidelberg Ethnology. Occasional Paper No. 4.

Richards, R. (2019). The value of theory of change in large-scale projects and complex interventions. K4D Helpdesk Report. Brighton: Institute of Development Studies.

Richerson, P. and Boyd, R. (2005). *Not by genes alone: How culture transformed human evolution*. Chicago, IL: University of Chicago Press.

Richey, L. and Ponte, S. (2011). *Brand aid: Shopping well to save the world*. Minneapolis: University of Minnesota Press.

Ricoeur, P. (1981). *Hermeneutics and the human sciences: Essays on language, action, and interpretation*. Cambridge: Cambridge University Press.

Rittel. H. and Webber, M. (1973). Dilemmas in a general theory of planning. *Policy Sciences*. **4**, 155–169.

Roberts, J.M. (2014). Critical realism, dialectics, and qualitative research methods. *Journal for the Theory of Social Behaviour*. **44**(1), 1–23. doi: 10.1111/jtsb.12056

Robichaud, P. (2017). Is ignorance of climate change culpable? *Science and Engineering Ethics*. **23**, 1409–1430.

Rodney, W. (1972). *How Europe underdeveloped Africa*. London: Bogle-L'Ouverture Publications.

Roehrich, J. (2014). Reputational risks and sustainable supply chain management. *International Journal of Operations & Production Management*. **34**(5), 695–719.

Rona-Tas, A. and Guseva, A. (2001). Uncertainty, risk, and trust: Russian and American credit card markets compared. *American Sociological Review*. **66**, 623–646.

Rorty, R. (1993). Human rights, rationality, and sentimentality. In: S. Shute and S. Hurley, eds. *On human rights: The Oxford Amnesty Lectures*. New York: Basic Books. pp. 111–134.

Rostow, W. (1960). *The stages of economic growth: A non-communist manifesto*. London: Cambridge University Press.

Routley, L. (2016). Teaching Africa, presenting, representing and the importance of who is in the classroom. *Politics*. **36**(4), 482–494.

Rozuel, C. (2011). The moral threat of compartmentalization: Self, roles and responsibility. *Journal of Business Ethics*. **102**(4), 685–697.

Rudrum, S., Brown, H. and Oliffe, J.L. (2016). Understanding the meaning and role of gifts given to Ugandan mothers in maternity care settings: "The help they give when they've seen how different you are." *Sociology of Health & Illness*. **38**(8), 1318–1335.

Rutzou, T. (2017). Finding Bhaskar in all the wrong places? Causation, process, and structure in Bhaskar and Deleuze. *Journal Theory Social Behaviour*. **47**, 402–417.

Sahlins, M. (1968). *Tribesmen*. Englewood Cliffs, NJ: Prentice Hall.

Sahlins, M. (1974). *Stone age economics*. London: Tavistock.

Sahlins, M. (2013). On the culture of material value and the cosmography of riches. *Journal of Ethnographic Theory*. **3**(2), 161–195.

Said, E. (1978). *Orientalism*. New York: Pantheon Books.

Salamon, L. and Anheier, H. (1992). In search of the non-profit sector. I: The question of definitions. *International Journal of Voluntary and Nonprofit Organizations*. **3**(2), 125–151.

Salazar, N. (2013). Imagineering otherness: Anthropological legacies in contemporary tourism. *Anthropological Quarterly*. **86**(3), 669–696.

Saldana, J. (2016). *The coding manual for qualitative researchers* (3rd ed.). London: Sage.

Sanghera, B. (2016). Charitable giving and reflexive individuals: How personal reflexivity mediates between structure and agency. *Social Science Information*. **56**(1), 28–48.

Sarch, A. (2018). Willful ignorance in law and morality. *Philosophy Compass*. **13**, e12490.

Sarker, A. and Rahman, M. (2015). The role of social accountability in poverty alleviation programs in developing countries: An analysis with reference to Bangladesh. *Public Organization Review*. **15**, 317–333.

Sartre, J. (1943). *Being and nothingness: An essay on phenomenological ontology. A phenomenological essay on ontology*. Translated by H.E. Barnes. New York: Philosophical Library.

Sayer, A. (1992). *Methods in social science: A realist approach* (2nd ed.). New York: Routledge.

Sayer, A. (2000). *Realism and social science*. London: Sage.

Scarth, A. and Novelli, M. (2019). Travel philanthropy and development. In: R.A. Sharpley and D.A. Harrison, eds. *Research agenda for tourism and development*. Cheltenham: Elgar. pp. 88–108.

Scarth, A. and Novelli, M. (2024). Travel Philanthropy: A multifaceted 'exchange economy'. *Annals of Tourism Research*. **106**, 103750.

Scheffer, S. (2001). *Boundaries and allegiances: Problems of justice and responsibility in liberal thought*. Oxford and New York: Oxford University Press.

Schiffer, E. and Hauck, J. (2010). Net-map: Collecting social network data and facilitating network learning through participatory influence network mapping. *Field Methods*. **22**(3), 231–249.

Schlag, K. (1996). Why imitate, and if so, how? A bounded rational approach to multi-armed bandits. *Journal of Economic Theory*. **78**(1), 130–156.

Schlag, K. (2012). Imitation and social learning. In: N.M. Seel, ed. *Encyclopedia of the sciences of learning* (2012 ed.). Boston, MA: Springer. pp. 1489–1493.

Schlaile, P., Bock, W. and Klein, K. (2018). From bounded morality to consumer social responsibility: A transdisciplinary approach to socially responsible consumption and its obstacles. *Journal of Business Ethics*. **149**, 561–588.

Schmidt, A. (2017). Being good by doing good: Goodness and the evaluation of persons. *Utilitas*. **29**(1), 3–26.

Schmidt, I. and Seele, P. (2012). Consumer responsibility in business ethics. A lifestyle centered approach. *Journal for Business, Economics & Ethics*. **13**(2), 169–191.

Schurman, R. (2018). Micro(soft) managing a 'green revolution' for Africa: The new donor culture and international agricultural development. *World Development*. **112**, 180–192.

Schwitzgebel, E. (2006). Belief. In: *The stanford encyclopedia of philosophy*. Available at: https://plato.stanford.edu/entries/belief/ (Accessed: 16 Feb 2020).

Scott-Villiers, P. (2011). We are not poor! Dominant and subaltern discourses of pastoralist development in the Horn of Africa. *Journal of International Development*. **23**, 771–781.

Scott, J. and Selwyn, T. (eds.) (2010). *Thinking through tourism*. London: Bloomsbury Academic.

Seddon, J. (2008). *Systems thinking in the public sector: The failure of the reform regime ... and a manifesto for a better way*. Axminster: Triarchy Press.

Seltzer M. (2004). *Travelers' philanthropy: Helping communities build economic assets & sustain environmental & cultural resources in an era of rapid globalization*. Washington, DC: Center for Responsible Travel.

Sen, A. (1999). *Development as freedom*. Oxford: Oxford University Press.

Sen, S. and Airiau, S. (2007). Emergence of norms through social learning. In: *Proceedings of the 20th International Joint Conference on Artificial Intelligence* (*IJCAI* '07), pp. 1507–1512, Hyderabad, India.

Senge, P. (1990). *The fifth discipline: The art and practice of the learning organization*. New York: Doubleday/Currency.

Seu, I., Flanagan, F. and Orgad, S. (2015). The Good Samaritan and the marketer: Public perceptions of humanitarian and international development NGOs. *International Journal of Nonprofit and Voluntary Sector Marketing*. **20**(3), 211–225.

Sgalitzer, H., Brownlee, M., Zajchowski, C., Bricker, K. and Powell, R. (2016). Modelling travellers' philanthropy: Tourists' motivations to donate at Sweetwater Chimpanzee Sanctuary. *Journal of Ecotourism*. **15**(1), 1–20.

Shabad, P. (2010). The suffering of passion: Metamorphoses and the embrace of the stranger. *Psychoanalytic Dialogues*. **20**(6), 710–729.

Shackle, G. (1961). *Decision, order and time in human affairs*. Cambridge: Cambridge University Press.

Shah, K., Arjoon, S. and Rambocas, M. (2017). Aligning corporate social responsibility with green economy development pathways in developing countries. *Sustainable Development*. **24**, 237–253.

Sharma, B. (2019). Discursive representations of difference and multilingualism in Himalaya with Michael Palin. *International Journal of Multilingualism*. **16**(3), 226–243.

Sharp, J. (1999). Writing over the map of Provence: The touristic therapy of a year in Provence. In: J. Duncan and D. Gregory, eds. *Writes of passage: Reading travel writing*. London: Routledge. pp. 200–218.

Sherwood, D. (2002). *Seeing the forest for the trees: A manager's guide to applying systems thinking*. London: Nicholas Brealey.

Shivji, I. (2006). The silences in the NGO discourse: The role and future of NGOs in Africa. *Africa Development*. **31**(4), 22–51.

Sihlè, N. (2015). Towards a comparative anthropology of the Buddhist gift (and other transfers). *Religion Compass*. **9**(11), 352–385.

Silk, J. (2004). Caring at a distance: Gift theory, aid chains and social movements. *Social & Cultural Geography*. **5**(2), 229–251.

Siltaoja, M. and Lähdesmäki, M. (2015). From rationality to emotionally embedded relations: Envy as a signal of power in stakeholder relations. *Journal of Business Ethics*. **128**(4), 837–850.

Simon, H. (1956). Rational choice, and the structure of the environment. *Psychological Review*. **63**, 129–138.

Simon, H.A. (1965) *Administrative behavior: A study of decision-making processes in administrative organizations* (2nd edn.). New York: Macmillan.

Simon, H. (1978). Rationality as process and as product of thought. *The American Economic Review*. **68**(2), 1–16.

Simon, H. (1979). Rational decision making in business organizations. *The American Economic Review*. **69**(4), 493–513.

Simon, H. (1982). *Models of bounded rationality*. Cambridge, MA: MIT Press.

Simon, H. (1983). *Reason in human affairs*. Stanford, CA: Stanford University Press.

Simon, H.A. (1987) 'Making management decisions: The role of intuition and emotion', *Academy of Management Executive*, **1**(1), 57–64.

Simon, H. (1991). Bounded rationality and organizational learning. *Organization Science*. 2(1, Special Issue), 125–134.

Simon, H. (1994). Decision-making and administrative organization. *Public Administration Review*. **4**(1), 16–30.

Simon, H. (2000). Bounded rationality in social science: Today and tomorrow. *Mind & Society*. **1**, 25–39.

Simpson, M. (2009). An integrated approach to assess the impacts of tourism on community development and sustainable livelihoods. *Community Development Journal*. **44**(2), 186–208.

Sims, C. (2003). Implications of rational inattention. *Journal of Monetary Economics*. **50**(3), 665–690.

Sinclair, T. (2018). Are we conditionally obligated to be effective altruists? *Philosophy & Public Affairs*. **46**(1), 36–59.

Sinervo, A. (2015). Brokering aid through tourism: The contradictory roles of volunteer coordinators in Cusco, Peru. *Tourist Studies*. **15**(2), 156–174.

Singelis, T., Triandis, H., Bhawuk, D. and Gelfand, M. (1995). Horizontal and vertical dimensions of individualism and collectivism: A theoretical and measurement refinement. *Cross-Cultural Research*. **29**, 240–275.

Singer, P. (1972). Famine, affluence, and morality. *Philosophy and Public Affairs*. **1**(1), 229–243.

Singh, P. and Chudasama, H. (2020). Evaluating poverty alleviation strategies in a developing country. *PLoS ONE*. **15**(1), e0227176.

Singh, T. (2002). Altruistic tourism: Another shade of sustainable tourism. *Tourism (Zagreb)*. **50**, 361–370.

Sinwell, L. (2008). Using Giddens's theory of 'structuration' and Freirean philosophy to understand participation in the Alexandra Renewal Project. *Development Southern Africa*. **25**(3), 245–258.

Skarbek, E. (2016). Aid, ethics, and the Samaritan's dilemma: Strategic courage in constitutional entrepreneurship. *Journal of Institutional Economics*. **12**(2), 371–393.

Sleek, S. (2016). The science of sameness. Available at: www.psychologicalscience.org/observer/the-science-of-sameness

Smaldino, P. (2014). The cultural evolution of emergent group-level traits. *Behavioural and Brain Sciences*. **37**, 243–295.

Smith, A. (1759). *The theory of moral sentiments*. London: A. Kincaid and J. Bell.

Smith, A. and Jeppesen, C. (2017). Introduction: Development, contingency and entanglement: decolonization in the conditional. In: A.W.M. Smith and C. Jeppesen, eds. *Britain, France and the decolonization of Africa: Future imperfect?* London: UCL Press. pp. 1–15.

Smith, C. (1994). The new corporate philanthropy. *Harvard Business Review*. **72**, 105–116.

Smith, J. and Sypher, B. (2010). Philanthropy in the workplace: How a financial institution communicates charitable giving values. *Southern Communication Journal*. **75**(4), 370–391.

Smithson, M. (1989). *Ignorance and uncertainty: Emerging paradigms*. New York: Springer-Verlag.

Song, H., Liu, J. and Chen, G. (2013). Tourism value chain governance: Review and prospects. *Journal of Travel Research*. **52**(1), 15–28.

South, J., Button, D., Quick, A., Bagnall, A., Trigwell, J., Woodward, J., Coan, S. and Southby, K. (2019). Complexity and community context: Learning from the evaluation design of a national community empowerment programme. *International Journal of Environmental and Public Health*. **17**(1), 91.

Spalding, M.J. (2011). What is successful philanthropy? In: M. Honey, ed. *Travelers' philanthropy handbook*. Washington, DC: Center for Responsible Travel. pp. 19–24.

Spencer, H. (1879). *The data of ethics*. London: Williams and Norgate.

Spicer, A. (2012). Shooting the shit: The role of bullshit in organizations. *Management (France)*. **16**(5), 653–666.

Spivak, G. (2012). *An aesthetic education in the era of globalization*. Cambridge, MA: Harvard University Press.

Staber, U. (2010). Imitation without interaction: How firms identify with clusters. *Organization Studies*. **31**(2), 153–174.

Stake, R. (1995). *The art of case study research*. Thousand Oaks, CA: Sage Publications.

Staszak, J. (2009). Other/otherness. In: R. Kitchin and N. Thrift, eds. *Publié dans International Encyclopedia of Human Geography*. Oxford: Elsevier. pp. 43–47.

Steiner, P. (2017). Religion and the sociological critique of political economy: Altruism and gift. *The European Journal of the History of Economic Thought*. **24**(4), 876–906.

Steinmetz, G. (2004). Odious comparisons: Incommensurability, the case study, and 'small N's' in sociology. *Sociological Theory*. **22**(3), 371–400.

Stenger, M. (2019). How expectations influence performance. InformedED. Available at: www.opencolleges.edu.au/informed

Sterman, J.D. (2006). Learning from evidence in a complex world. *American Journal of Public Health*. **96**(3), 505–514.

Stern, D. (1991). Heraclitus' and Wittgenstein's river images: Stepping twice into the same river. *The Monist*. **74**(4), 579–604.

Stirrat, R. and Henkel, H. (1997). The development gift: The problem of reciprocity in the NGO world. *The Annals of the American Academy of Political and Social Science*. **554**, 66–80.

Strachwitz, R. (2015). The role of foundations in public debates in Germany. *American Journal of Economics and Sociology*. **74**(4), 826–852.

Stutzer, A. (2013). Economic consequences of mispredicting utility. IZA Discussion Paper No. 7430.

Suchman, M. (1995). Managing legitimacy: Strategic and institutional approaches. *Academy of Management Review*. **20**, 571–610.

Sulek, M. (2010). On the modern meaning of philanthropy. *Nonprofit and Voluntary Sector Quarterly*. **39**(2), 193–212.

Sundstrom, A. (2019). Why do people pay bribes? A survey experiment with resource users. *Social Science Quarterly*. **100**(3), 725–735.

Sustainable Travel International. (2008). Best practices in travel philanthropy manual V1. – Tour operator sector. Accessed at: https://voluntourismgal.files.wordpress.com/2009/04/sti_bp.pdf

Swamy, R. (2017). Humanitarianism and unequal exchange. *Journal of World-Systems Research*. **23**(2), 353–371.

Tabellini, G. (2008). Institutions and culture. *Journal of the European Economic Association*. **6**, 255–294.

Talladay, J. (2011). Why travelers become philanthropists: Donor motivations. In: M. Honey, ed. *Travelers' philanthropy handbook*. Washington, DC: Center for Responsible Travel. pp. 193–196.

Talpalaru, M. (2014). Blake mycoskie, TOMS, and life narratives of conspicuous giving. *Biography*. **37**(1), 168–190.

Tammelleo, S. and Lombardi, L. (2014). Consumer social responsibility? *Business & Professional Ethics Journal*. **33**(1), 99–126.

Taylor-Gooby, P. (1998). Economic beliefs and social policy behaviour. In: P. Taylor-Gooby, ed. *Choice and public policy: The limits to welfare markets*. London: Macmillan. pp. 1–23.

Taylor, B. (2014). Who wants to give forever? Giving meaning to sustainability in development. *Journal of International Development*. **26**(8), 1181–1196.

Taylor, C. (2004). *Modern social imaginaries*. Durham: Duke University Press.

Taylor, J. (2018). Internal whistle-blowing in the public service: A matter of trust. *Public Administration Review*. **78**(5), 717–726.

Thieme, T. (2013). The "hustle" amongst youth entrepreneurs in Mathare's informal waste economy. *Journal of Eastern African Studies*. **7**(3), 389–412.

Thomas, F. (2014). Addressing the measurement of tourism in terms of poverty reduction: Tourism value chain analysis in Lao PDR and Mali. *International Journal of Tourism Research*. **16**, 368–376.

Thomas, G. (2010). Doing case study: Abduction not induction, phronesis not theory. *Qualitative Inquiry*. **16**(7), 575–582.

Thomas, T., Schermerhorn, J. and Dienhart, J. (2004). Strategic leadership of ethical behavior in business. *Academy of Management Executive*. **18**(2), 56–65.

Thomassen, B. (2015). Begging Rome: Norms at the margins, norms of the in-between. *Critique of Anthropology*. **35**(1), 94–113.

Thrift, N. (1985). Flies and germs: A geography of knowledge. In: D. Gregory and J. Urry, eds. *Social relations and spatial structures*. London: Macmillan. pp. 366–403.

Tirivanhu, P. (2020). A realist assessment of the facilitation process for improving social accountability by community based organizations. *Development Southern Africa*. **37**(6), 953–970.

Toffin, G. (2015). Donations and gift-giving among Newar Buddhists, Nepal. *Religion Compass*. **9**(11), 423–442.

Tomas, L., Girgenti, S. and Jackson, C. (2017). Pre-service teachers' attitudes toward education for sustainability and its relevance to their learning: Implications for pedagogical practice. *Environmental Education Research*. **23**(3), 324–347.

Tomasello, M. (2020). The moral psychology of obligation. *Behavioral and Brain Sciences*. **43**(56), 1–58.

Trampusch, C. and Palier, B. (2016). Between X and Y: How process tracing contributes to opening the black box of causality. *New Political Economy*. **21**(5), 437–454.

Triandis, H. (1995). *Individualism and collectivism*. Boulder, CO: Westview.

Tuomela, R. (1995). *The importance of us: A philosophical study of basic social norms*. Stanford, CA: Stanford University Press.

Turner, R., Miller, G. and Gilbert, D. (2001). The role of UK charities and the tourism industry. *Tourism Management*. **22**, 463–472.

Tyrrell, T., Paris, C. and Biaett, V. (2012). A quantified triple bottom line for tourism: Experimental results. *Journal of Travel Research*. **52**(3), 279–293.

Uche, L. (1994). Some reflections on the dependency theory. *Africa Media Review*. **8**(2), 41–55.

Ulrich, W. (1988). Churchman's process of unfolding and its significance for policy analysis and evaluation. *Systems Practice*. **1**(4), 415–428.

Ulrich, W. (2005). *A brief introduction to critical systems heuristics (CSH)*. Available at: http://projects.kmi.open.ac.uk/ecosensus/publications/ulrich_csh_intro.pdf

UN. (2010). International recommendations for tourism statistics 2008. New York: United Nations. Available at: https://unstats.un.org/unsd/publication/Seriesm/SeriesM_83rev1e.pdf#page=21

UN. (2020a). *COVID-19 and transforming tourism*. Policy brief, August 2020. Available at: www.un.org/sites/un2.un.org/files/sg_policy_brief_covid-19_tourism_august_2020.pdf

UN. (2020b). *The sustainable development goals report 2020*. Available at: https://unstats.un.org/sdgs/report/2020/The-Sustainable-Development-Goals-Report-2020.pdf

UNDP. (2015). Sustainable development goals. Available at: www.undp.org

UNFPA. (2020). *Assessing the quality of developmental evaluations at UNFPA*. United Nations Population Fund, Evaluation Office.

UNWTO. (2013). World Tourism Organisation: *Aid for trade and value chains in tourism*. UNWTO, Madrid.

UNWTO. (2014). AM reports, volume nine – *Global report on adventure tourism*. UNWTO, Madrid.

UNWTO. (2019). *Global and regional tourism performance*. Available at: www.unwto.org/global-and-regional-tourism-performance

UNWTO. (2024). World tourism barometer. *World Tourism Barometer*. **22**(Issue 3), September 2024.

Urry, J. (1992). The tourist gaze "revisited". *The American Behavioral Scientist*. **36**(2), 172–186.

Usó-Doménech, J. and Nescolarde-Selva, J. (2016). What are belief systems? *Foundations of Science*. **21**, 147–152.

Valdez, M. and Magio, K. (2016). Travellers' philanthropy: A new transformative trend in Mazatlán, México. In: World Tourism Organization, ed. *Affiliate members global reports, volume fourteen – The transformative power of tourism: A paradigm shift towards a more responsible traveller*. UNWTO, Madrid. pp. 97–101.

Valente, M. and Crane, A. (2010). Public responsibility and private enterprise in developing countries. *California Management Review*. **12**(3), 52–78.

van de Ven, N., Zeelenberg, M. and Pieters, R. (2009). Leveling up and down: The experiences of benign and malicious envy. *Emotion*. **9**(3), 419–429.

van der Riet, M. (2008). Participatory research and the philosophy of science. Beyond the moral imperative. *Qualitative Inquiry*. **14**(4), 546–565.

Vandenbroeck, P. (2015). *Systems thinking and four forms of complexity*. White Paper. Shift.

Vogel, I. (2012). Review of the use of 'Theory of Change' in international development. UK Department of International Development.

Von Bergen, C. and Bressler, M. (2017). Helping behaviors can negatively impact long-term well-being how "skin in the game" more effectively helps others. *Society and Business Review*. **13**(1), 112–123.

W.K. Kellogg Foundation. (2004). Using logic models to bring together planning, evaluation, and action. Logic model development guide. Michigan: W.K. Kellogg Foundation.

Wearing, S. and McGhee, N. (2013). Volunteer tourism: A review. *Tourism Management*. **38**, 120–130.

Weaver, D. and Jin, X. (2016). Compassion as a neglected motivator for sustainable tourism. *Journal of Sustainable Tourism*. **24**(5), 657–672.

Weber, M. (1919/1994). The profession and vocation of politics. In: P. Lassman and R. Speirs, eds. *Weber: Political writings*. Cambridge: Cambridge University Press. pp. 309–369.

Weber, M. (1922). *Wirtschaft und Gesellschaft* (Grundriß der Sozialo ¨konomik; Abt. 3). Tu ¨bingen: Mohr.

Webster, J., Exley, J., Copestake, J., Davies, R. and Hargreaves, J. (2018). Timely evaluation in international development. *Journal of Development Effectiveness*. **10**(4), 482–508.

Weiss, C. (1972). *Evaluation research: Methods of assessing program effectiveness*. Englewood Cliffs, NJ: Prentice-Hall.

Weiss, C. (1997). How can theory-based evaluation make greater headway? *Evaluation Review*. **21**(4), 501–524.

Weller, N. and Barnes, J. (2016). Pathway analysis and the search for causal mechanisms. *Sociological Methods & Research*. **45**(3), 424–457.

Wessel, M. (2018). Narrative assessment: A new approach to evaluation of advocacy for development. *Evaluation*. **24**(4), 400–418.

Western, D. (2011). Travelers' philanthropy and the Good Samaritan. In: M. Honey, ed. *Travelers' philanthropy handbook*. Washington, DC: Center for Responsible Travel. pp. 13–18.

Wettstein, F. (2012). Corporate responsibility in the collective age: Toward a conception of collaborative responsibility. *Business and Society Review*. **117**(2), 155–184.

Whiten, A., McGuigan, G., Marshall-Pescini, S. and Hopper, L. (2009). Emulation, imitation, over-imitation and the scope of culture for child and chimpanzee. *Philosophical Transactions of the Royal Society Biological Sciences*. **364**(1528), 2417–2428.

Wieland, J. (2017a). Willful ignorance. *Ethical Theory and Moral Practice*. **20**(1), 105–119.

Wieland, J. (2017b). Responsibility for strategic ignorance. *Synthese*. **194**, 4477–4497.

Williams, B. and Hummelbrunner, R. (2011). *Systems concepts in action: A practitioner's toolkit*. Stanford, CA: California University Press.

Williams, C. (2019). Developing critical realist comparative methods for researching religions, belief-systems, and education. *Revista Española de Educación Comparada*. **33**, 78–105.

Williams, G. (1961). *Criminal law: The general part* (2nd ed.). London: Stevens & Sons.

Wilson, C. (2016a). Managing expectations: Locke on the material mind and moral mediocrity. *Royal Institute of Philosophy Supplement*. 78, 127–146.

Wilson, T. (2016b). Buffett's son funds hydropower plant to help save Congo Gorillas. *Bloomberg*. Access at www.bloomberg.com/news/articles/2016-04-06/congo-park-builds-hydropower-plants-to-protect-mountain-gorillas

Wisor, S. (2012). How should INGOs allocate resources? *Ethics & Global Politics*. **5**(1), 27–48.

Wolff, J. (2011). Survey of tour operators in Arusha, Tanzania. In: M. Honey, ed. *Travelers' philanthropy handbook*. Washington, DC: Center for Responsible Travel. pp. 150–158.

Wood, C. (2020). *Why imitation is at the heart of being human*. Available at: https://greatergood.berkeley.edu/article/item/why_imitation_is_at_the_heart_of_being_human.

Woods, A. (2015). The limits of moral intuitions for human rights advocacy. *Law & Ethics of Human Rights*. **9**(1), 91–111.

World Bank. (1998). *Assessing aid: What works, what doesn't and why*. Washington, DC: World Bank.

WTTC. (2024). Travel & tourism economic impact 2024: Global trends.

WTTC. (2015). Connecting global climate action. www.wttc.org

WTTC. (2019). Travel & tourism regional performance 2019. *Economic Impact Reports*. www.wttc.org

Xueming, L. (2005). A contingent perspective on the advantages of stores' strategic philanthropy for influencing consumer behaviour. *Journal of Consumer Behaviour*. **4**(5), 390–401.

Yin, R. (1984). *Case study research: Design and methods*. Beverly Hills, CA: Sage.

Yin, R. (1989). *Case study research: Design and methods*. Newbury Park: Sage.

Yin, R. (2003). *Case study research: Design and methods*. Newbury Park: Sage.

York, P. and Bamberger, M. (2020). *Measuring results and impact in the age of big data: The nexus of evaluation, analytics, and digital technology*. New York: The Rockefeller Foundation.

Young, I. (2006). Responsibility and global justice: A social connection model. *Social Philosophy and Policy*. **23**(1), 102–130.

Young, I. (2011). *Responsibility for justice*. Oxford: Oxford University Press.

Yu, H. (2017). Philanthropy on the move: Mobile communication and neoliberal citizenship in China. *Communication and the Public*. **2**(1), 35–49.

Zachariadis, M., Scott, S. and Barrett, M. (2013). Methodological implications of critical realism for mixed-methods research. *Management Information Systems Quarterly*. **37**(3), 855–879.

Zane, D. (2016, May 1). "Barbie challenges the 'white saviour complex'". *BBC News*. Retrieved February 21, 2017: www.bbc.co.uk/news/world-africa-36132482

Zevallos, Z. (2020). What is otherness? Available at: https://othersociologist.com/otherness-resources/

Zhan, Y. (2016). Science as gift: NGO-supported informal science education for rural migrant children in contemporary China. *Mind, Culture, and Activity*. **23**(3), 247–258.

Zijlma, A. (2016). Travel giving tips for visitors to Africa: Africa travel do's and don'ts – What gifts to bring to Africa? Available at: http://goafrica.about.com/od/africatraveltips/a/travelgiving.htm

Zwart, F. (2015). Unintended but not unanticipated consequences. *Theory and Society*. **44**(3), 283–297.

Index